Intelligent and Sustainable Power and Energy Systems

"Intelligent and Sustainable Power and Energy Systems" delves into the critical advancements shaping the future of global energy. This compilation presents cutting-edge research and innovative solutions addressing the urgent need to transition towards environmentally responsible and technologically sophisticated energy infrastructures. Explore the integration of artificial intelligence, machine learning, and advanced control systems in optimising energy generation, distribution, and consumption. Discover novel approaches to renewable energy integration, smart grid technologies, and energy storage solutions, all geared towards enhancing efficiency and minimising environmental impact. From theoretical frameworks to practical implementations, this work offers a comprehensive overview of the latest developments, providing essential insights for researchers, engineers, and policymakers striving to build a resilient and sustainable energy future. This book is a vital resource for navigating the complex challenges and opportunities in the evolving landscape of power and energy systems.

Intelligent and Sustainable Power and Energy Systems

Editors

Dr. M. Premkumar

Dr. Pasumarthi Usha

Dr. Sujit Kumar

Dr. Manikanta Gopisetti

CRC Press
Taylor & Francis Group
Boca Raton London New York

CRC Press is an imprint of the
Taylor & Francis Group, an **informa** business

First edition published 2026
by CRC Press
4 Park Square, Milton Park, Abingdon, Oxon, OX14 4RN

and by CRC Press
2385 NW Executive Center Drive, Suite 320, Boca Raton FL 33431

CRC Press is an imprint of Informa UK Limited

British Library Cataloguing-in-Publication Data
A catalogue record for this book is available from the British Library

ISBN: 978-1-041-10313-4 (hbk)
ISBN: 978-1-041-10314-1 (pbk)
ISBN: 978-1-003-65446-9 (ebk)

DOI: 10.1201/9781003654469

Typeset in Times LT Std
by Aditiinfosystems

Intelligent and Sustainable Power and Energy Systems – Dr. M. Premkumar et al. (eds)
© 2026 Taylor & Francis Group, London, ISBN 978-1-041-10314-1

Contents

.

Intelligent and Sustainable Power and Energy Systems – Dr. M. Premkumar et al. (eds)
© 2026 Taylor & Francis Group, London, ISBN 978-1-041-10314-1

List of Figures

Intelligent and Sustainable Power and Energy Systems – Dr. M. Premkumar et al. (eds)
© 2026 Taylor & Francis Group, London, ISBN 978-1-041-10314-1

List of Tables

Foreword

The "2nd International Conference on Intelligent and Sustainable Power and Energy Systems (ICISPES2024)" is aimed to address the state-of-the-art multidisciplinary research needs and interdisciplinary aspects of intelligent and sustainable technologies in power and energy systems in the form of research papers from industry, faculty, research scholars, and PG/UG students, along with keynote lectures and several invited talks from reputed speakers. The ICISPES 2024 aims to provide a platform to present their work to share experiences and ideas in the areas of Renewable Energy, Electric Transportation, Power Electronics, Electric Drives, Control Techniques, Smart Grids, Communication Protocols, Intelligent Charging Infrastructure and Standards, machine learning, artificial intelligence, big data analytics, IoT, etc. Given the changing scenario under intelligent and sustainable power and energy, this conference aims to put together experts from these areas to disseminate their knowledge and experience for working years to come.

The ICISPES provides an international platform for all research scholars, practitioners, academicians, students and scientists to discuss and exhibit the latest research developments and discoveries in intelligent and sustainable technologies. This conference lets us share and exchange research knowledge and ideas in power and energy systems.

Intelligent and Sustainable Power and Energy Systems – Dr. M. Premkumar et al. (eds)
© 2026 Taylor & Francis Group, London, ISBN 978-1-041-10314-1

Preface

The unrelenting quest for an enhanced, healthier, and more comfortable existence has spurred innovation. This technological evolution and its implementation in practical solutions is a pivotal force propelling contemporary society forward. Current advancements in various engineering fields, including electrical, electronics, mechanical, communications, chemical, and computer fields such as AI, target developing more intelligent systems encapsulated in concepts like smart homes, smart cities, and smart societies. Furthermore, innovation in renewable energy, electric transportation, power electronics, electric drives, control techniques, smart grids, sustainability, communication protocols, and intelligent charging infrastructures and standards are intrinsically interconnected. Innovations in one sphere address specific issues and potentially facilitate solutions or foster advancements in other domains. Therefore, ongoing research in these sectors must be paired with effectively disseminating progress and findings. This approach is crucial for fostering a global society that is continuously evolving and advancing towards a future that embraces the synergistic potential of these innovations.

The International Conference on Intelligent and Sustainable Power and Energy Systems, also known as ICISPES 2024, is set to be hosted by the Department of Electrical and Electronics Engineering at the Dayananda Sagar College of Engineering, Bangalore, Karnataka, India, 13th and 14th December 2024. This pivotal event is poised to serve as a vibrant hub for the convergence of ground-breaking ideas from researchers worldwide, steering our cities and societies towards greater intelligence and integration. The ICISPES 2024 is dedicated to fostering innovation and collaboration across diverse yet interconnected engineering and technology domains, acknowledging that each field contributes vital to the betterment of society. To this end, the call for papers was crafted to encourage a substantial pool of contributions, attracting over 200 online submissions from various relevant fields. Out of this rich array of scholarly contributions, only 88 high-calibre submissions have been chosen to feature in the conference, a selection made possible through a stringent and impartial peer-review process. This review system encompassed the expertise and insights of a distinguished panel of experienced researchers, advisory members, and the technical and program committee of ICISPES 2024. The selection criteria were grounded in the significance of the contributions made by the authors, as well as the technical depth, originality, and alignment with the conference's central themes. It is worth noting that every stage of the peer-review process was facilitated seamlessly through an electronic system, ensuring efficiency and transparency.

The keynote address is a pivotal segment in any conference, facilitating the dissemination of research findings and insights before an audience of scientists and researchers from various disciplines and backgrounds. Acknowledging this, the organizers of ICISPES 2024 have exerted considerable effort to coordinate a series of keynote speeches that align perfectly with the conference's thematic focus. Furthermore, the assembly of author presentations has been meticulously orchestrated to highlight the significance and interconnectedness of the different pieces while resonating with the foundational philosophy and driving forces behind the conference. In light of this, we extend our heartfelt appreciation to the authors for choosing 'ICISPES' as the platform to share the fruits of their research endeavours. Moreover, we are profoundly thankful for the invaluable assistance rendered by each individual reviewer and the members of the Program Committee, whose contributions to the peer-review process have been instrumental in elevating the quality and substance of the conference

Acknowledgements

The editors are honoured to have facilitated the presentation of pivotal proposals and highlights of the ICISPES 2024 Conference. This event has garnered the attention and enthusiasm of academics and researchers worldwide, establishing itself as an optimal platform to unveil a myriad of innovative research discoveries. The submissions spanned various disciplines, including electrical, electronics, computation, and agricultural engineering.

We wish to express our heartfelt appreciation to all the authors who have amplified the value of the conference with their precious contributions, dedicating their time, expertise, and insightful research submissions. The seamless execution of pre- and post-conference activities and deliberations was made possible through our national and international advisory committees, timely and informed support and guidance. We convey our profound gratitude to all committee members for their invaluable input. Furthermore, we would like to acknowledge the remarkable efforts of our robust team of reviewers, who conducted thorough and critical evaluations of all submissions, offering feedback and suggestions that were vital in maintaining the high standard and quality of the conference proceedings.

A special note of thanks goes to all organising committee members, whose relentless efforts were pivotal in making this event a resounding success. We also extend our deepest gratitude to the editorial team at Taylor and Francis, SciTePress Publishing and STM Journals for their exceptional work in crafting the proceedings and journal in an innovative and intellectually stimulating manner. The ICISPES 2024 conference and its proceedings stand to receive well-deserved accolades in front of a vast assembly of attendees.

Finally, our heartfelt thanks to the Management of Dayananda Sagar College of Engineering, Karnataka, India and the Department of Electrical and Electronics Engineering faculty members for their continued support and encouragement to make the conference a huge success.

About the Editors

Dr. M. Premkumar is working as a Professor and head of the Electrical and Electronics Engineering Department at the Dayananda Sagar College of Engineering, Bengaluru, India. He has over 17 years of teaching experience. He has published over 180 technical articles in various National/International peer-reviewed journals, such as IEEE, Elsevier, Springer, and so on, with over 4500 citations and an H-index of 37. He has published/granted twelve patents by IPR, India, and IPR, Australia. He is also an Editor/Reviewer for leading journals of different publishers, such as IEEE, IET, Wiley, Taylor & Francis, Springer, MDPI, etc. He is recognised as one of the Top 2% of Scientists Worldwide based on the Study Conducted by Stanford University for the four consecutive years (2020, 2021, 2022, and 2023). His current research interests include power converters/inverters, renewable energy systems, smart grid and microgrids, PV parameter extraction, modern PV MPPTs, PV array faults, non-isolated/isolated dc-dc converters for renewable energy systems and electric vehicles, BMS for electric vehicles, and optimisation algorithms, including single-, multi-, and many-objectives for real-time power electronics and power systems problems.

Dr. Pasumarthi Usha received the B.Tech from E&EE in 1990, M.Tech in Power System with emphasis in High Voltage from J.N.T.U, College of Engineering Kakinada in 1992, and the Ph.D. in HVDC Power Transmission from Visvesvaraya Technical University in 2013, respectively. She works as a Professor at the Department of Electrical and Electronics Engineering, Dayananda Sagar College of Engineering. Her research areas are HVDC Power systems, Microgrid, and Power Electronics.

Dr. Sujit Kumar received his PhD in Electrical Engineering from MPUAT, CTAE, Udaipur, with the collaboration of IIT Roorkee with the DST Inspire Fellowship. He also worked as SRF at IIT Roorkee during the period of his PhD. He did his M. Tech from Sharda University, Greater Noida, as a Gold Medallist, and his B.E from Birla Institute of Technology and Science, Pilani (BITS Campus) as a gold medallist. During his B.E., he received a Governor of France Scholarship for his academic excellence. He is currently posted as an Assistant Professor in the Department of Electrical and Electronics Engineering at Dayananda Sagar College of Engineering, Bengaluru, India. He receives the prestigious Gold Medal in the NPTEL training course (Fuzzy Sets, Logic and Systems & amp; Applications) and exemplifies professional development and lifelong learning excellence. He secured all India ranks in GATE 2017 and GATE 2018 as 16 and 17, respectively. His expertise includes Nanotechnology, Material Science, Sustainable Energy, and Artificial Intelligence in Power Systems.

Dr. Manikanta Gopisetti is an assistant professor at Dayananda Sagar College of Engineering, Bengaluru, India. He completed his Ph.D. in Electrical Engineering from AUUP, Amity University in 2021. He received his M. Tech. degree in Power System Engineering, from SRM University Chennai, India. He graduated from Chaitanya Engineering College, Visakhapatnam, India. He has received two best paper awards in IEEE indexed conferences. He has published many research papers in reputed journals and conference proceedings. He delivered some talks on Distributed generators, Network reconfiguration and Electric Vehicles. His research interests include Distributed Generation, Network Reconfiguration, Electric Vehicles, Evolutionary Computation, and Renewable Energy Sources.

Intelligent and Sustainable Power and Energy Systems – Dr. M. Premkumar et al. (eds)
© 2026 Taylor & Francis Group, London, ISBN 978-1-041-10314-1

1

Comparative Study of Non-Isolated DC-DC Converters for Fuel Cell Electric Vehicles

Shilpa Rao H* and Pushpa Rajesh V

Department of Electrical Engineering, Jain (Deemed to be) University,
India

ABSTRACT: Depletion of fossil fuels and increased carbon emissions has led to a global movement to reduce the dependence on IC engine vehicles and move to greener solutions such as Electric Vehicles (EVs). EVs utilize battery packs to store energy. These battery packs have limited capacity and increasing the same leads to increased vehicle space utilization and charging time. Fuel cell is a credible alternative to battery packs that has several advantages such as increased efficiency and low carbon footprint. However, fuel cells have a non-linear load voltage to current feature that has to be taken care of. This paper thereby reviews the various non-isolated DC-DC converters which shall be adept for a FCEV system.

KEYWORDS: Boost converters, DC-DC converters, Electric vehicles (EVs), Fuel cell electric vehicles (FCEVs), SEPIC, Zeta converter

1. Introduction

Today, the climate is directly impacted by the carbon foot-print from fossil fuel internal combustion engine-based vehicles, which pose a major threat to the environment. [1], Electric vehicles (EVs) are among the greatest substitutes to conventional internal combustion (IC) engine vehicles, as they have the importance of mitigating global warming by reducing greenhouse gas emissions. [2], Fuel cell electric vehicles (FCEV) and battery electric vehicles (BEV) are two types of electric vehicles. Fuel cells transform chemical energy into electrical energy. [3],[4], FCEVs coupled with batteries or supercapacitors power the propulsion engine or charge the battery that powers the electric motors to propel the vehicle. Because FCEVs have limitations, such as large output voltage ranges, delayed auxiliary power sources are required to act as external energy sources, [5]. The statistics as in

---

*Corresponding author: shilpaach16@gmail.com

DOI: 10.1201/9781003654469-1

[6] presented by the International Energy Agency (IEA), reports that to ensure we reduce carbon emissions mandate rapid development of low emissions hydrogen. To increase the production of hydrogen, various challenges that impact the widespread adoption and integration into the market need to be addressed. Fuel cells are ingenious energy conversion devices that generate electricity through an electrochemical reaction.

The amount of electricity an FCEV can produce is decided by the size, number of stacks of fuel cells and kind of fuel cells, its electrolyte, and its temperature. The various types of fuel cells are proton exchange membrane fuel cells (PEMFC), solid oxide fuel cells (SOFC), molten carbon fuel cells (MCFC), phosphate-acid fuel cells (PAFC), direct methanol fuel cells (DMFC), alkaline fuel cell (AFC), regenerative fuel cell (RFC). Amongst all fuel cells, PEMFC is suitable for applications such as transportation, portable electronics, stationary power generation, etc.

The block diagram of a primary architecture for the FCEV power train is shown in Fig. 1.1. As the fuel cell stack does not meet the DC bus voltage, a boost converter is used. The converter plays a crucial role in FCEVs by facilitating efficient operation and ensuring compatibility with the vehicle's power train.

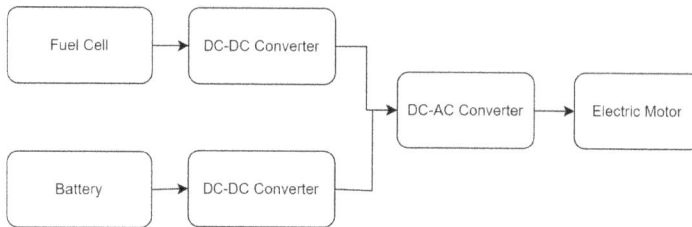

Fig. 1.1 Fuel cell power train topology

The different topologies of DC-DC converters used in FCEVs and their challenges have been reviewed in this paper. The first section compares and discusses various converter topologies. In the second section, the two most widely used converters are seen through literature review viz. SEPIC and Zeta converters are discussed followed by a comparative study of battery and fuel cell electric vehicles.

2. Literature Study

Converter topologies are divided into isolated and non-isolated, unidirectional, and bidirectional, single-stage and multistage categories. As the name suggests in a non-isolated converter the ground is shared between input and load, whereas in an isolated converter, the input and output stages have separate grounds. The advantages of an isolated converter are that the input side is protected, and thereby does not see any changes in the load conditions affecting its operation. Non-isolated converters are simpler to implement and debug, however, they lack fault resilience, gate side protection etc. The non-isolated cases are typically not used in industrial applications. Unidirectional converters convert energy from one form to another in one direction. FCEVs typically use a combination of unidirectional converters [5] for steady-state loads and bidirectional converters for battery management systems and dynamic power requirements. Single-stage converters have an inherent advantage of being highly efficient, simple and cheap to implement and debug. Until now, because of their simplicity and lower cost, these are found in various power supply designs, renewable energy systems, motor drives, and other applications. In FCEV Boost Converters are commonly used to amplify the input voltage and produce a higher output. The DC-DC converters with very high duty ratios are not practical nor economically feasible due to limitations such as voltage gain clipping due to parasitic parameters and stress on the power switches, Jayachandran, [7]. There are various topologies for boost converters in FCEVs such as SEPIC Converters, [17]. Luo Converters, Selvi, [24], Zeta Converters, [29], Coupled Inductors-Based Interleaved Boost Converters for Fuel Cell Electric Vehicles, Cascaded Buck and Boost Converter, [30].

3. Single Ended Primary Inductor Converter (SEPIC) Converters

A SEPIC [9] uses two inductors one for the input and the other for the ground, [10]-[14]. It has the properties of both a Boost converter and an inverted buck-boost converter. Being non-inverted and having the same polarity as that of the input voltage is the primary benefit of SEPIC over buck-boost converter, [15], [16]. Further, SEPIC has high gain, lower switching stress and high efficiency, [22]. A modified SEPIC Converter with a Wide step-up/step-down range is discussed in [25]. Another important advantage of using SEPIC is that there is no voltage overshoot during the turnoff process. But this converter suffers from huge power loss during switch turn-on and turn-off transition.

4. Luo Converters

Luo Converters are a new class of step-up (boost) conversion circuits. These circuits have low ripple output voltage and current, great efficiency and simple circuit structure. [26], The voltage can be increased or decreased by using these converters and are particularly helpful in situations where large output voltages are needed, Pradhan, [27] The advantage of Luo converters is that despite large voltage gain, they can have sufficiently low ripples, and are relatively simpler with a low-cost bearing. One of the prime issues though with Luo converters is the high switching losses in non-optimal regions which can overall reduce the converters efficiency. If the operation is ensured to not fall into the non-optimal regions, this can be a good choice for FCEVs, [28]

5. Zeta Converters

Zeta Converters are fourth-order DC-DC Converter topology with a high-speed MOSFET driven by a buck controller. [29] It requires a series capacitor known as flying capacitor and two inductors. Its ability to keep input and output current constant is one of its key features. [31]. Zeta converters offer a stable output response with fewer transients; however, its non-isolated nature drives challenges in case of FCEVs which needs to be countered through high input side impedance. This converter operates with a very low settling time as well, thereby offering a high efficiency output with reduced noise and EMI interference.

6. Coupled Inductor-Based Interleaved Boost Converters for Fuel Cell Electric Vehicles

[32] discusses Inductor based cascade cyclic coupled structure is beneficial in improving the power quality by reducing current ripple. Although the current ripple is reduced [20],[21], it has various technical issues that need to be carefully managed such as commutation management, inductor size etc. The converter however does offer a much-enhanced voltage ratio and also has the advantage of having a much lower ripple current effect.

7. Cascaded Buck and Boost Converter

Non-isolated hard-switching bidirectional cascaded buck and boost converter for fuel cell hybrid vehicles is proposed in [33], which uses a load-dependent switching frequency modulation by which the converter efficiency largely increases. However, the switching losses rise as the switching frequency increases but greatly reduces the inductor losses, Reddy, [23] There must be a trade-off done in such cases based on the power capacity hereby needed in FCEVs to increase its efficiency.

Table 1.1 Comparison of converters for FCEVs

Converter/Parameter	Dynamic Range	Transient Behaviour	Isolation	Cost
SEPIC	High	Poor	Good	Medium
Luo	High	Good	Medium	Medium
Zeta	High	Medium	Medium	Medium
Coupled Inductors Based Interleaved	Medium	Medium	Medium	Medium
Cascaded Buck and Boost	Medium	Medium	Medium	Medium

8. Problem Definition

The essential characteristics of DC-DC converter for FCEV's include good dynamic and transient behaviour, high Voltage Range, high efficiency, good degree of Isolation, Cost effective, modular nature of converter. To power the electric motor, fuel cell stack's DC output must be boosted to ensure compatibility with the motor's requirements. DC-DC Converters use the switching operation to maintain the required output power [8].

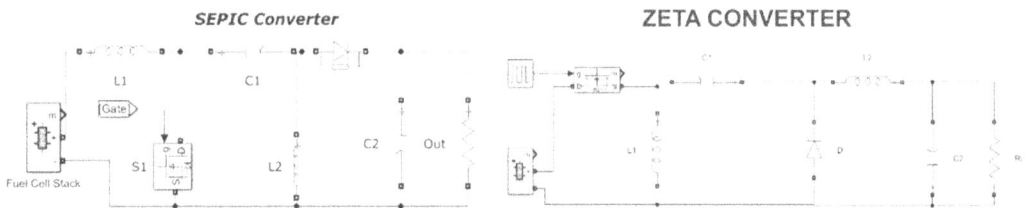

Fig. 1.2 Simulation study of SEPIC and zeta for FCEV

9. Comparative Simulation Study of SEPIC and Zeta Converters for Fuel Cell-based EVs

This section discusses the Voltage and Current characteristics of 2 popular converters (SEPIC and Zeta) used in FCEV applications. For SEPIC Converters, there needs to be a fine balance in choosing the amount of ripple to be allowed wherein the values of L and C need to be carefully chosen [18].

From the simulation results it is observed that traditional SEPIC has a high ripple content in voltage in steady state as well. In transient state, a traditional SEPIC has a higher overshoot voltage. This can be further improved by various methods. In the case of Zeta converter, the ripple content is higher than the SEPIC however the overshoot voltage is lower. Zeta converter has lower voltage gain as compared to SEPIC, [31].

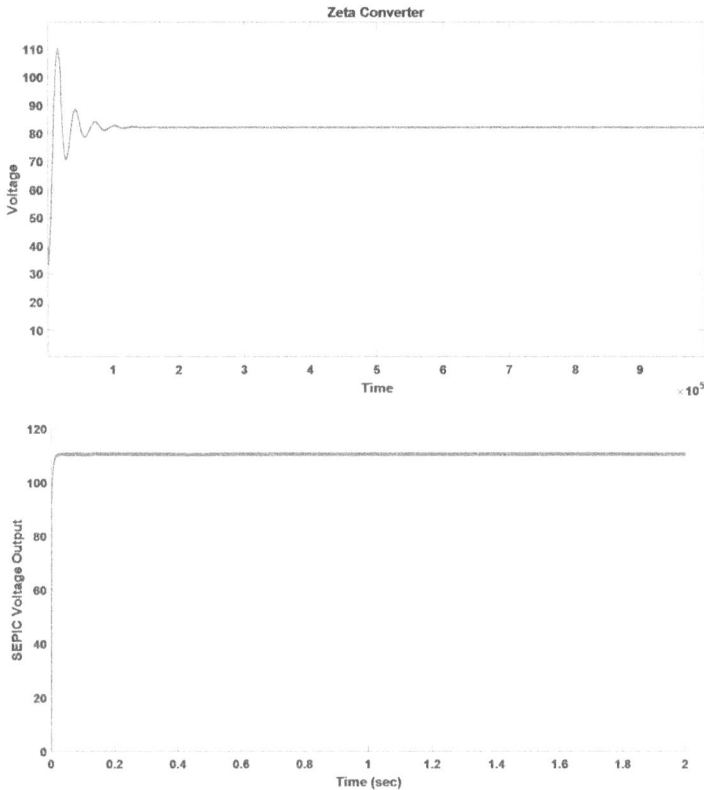

Fig. 1.3   Zeta and SEPIC converter steady state ripple and voltage

10. Comparative Study of Battery and Fuel Cell-Based EVs

Both the Battery EVs and FCEVs have similar lifecycle costs. As seen from the simulations, the Fuel Cell-based EVs are sluggish in performance meaning are slow to react to sudden rise in current demand or fall in current demand as compared to BEVs. This is happening also due to the nature of DC-DC converters being used. Better design of converters however can help greatly in changing this trend, through fast reaction and improved power quality. Multiple topologies are proposed in prior arts for overcoming the challenges such as Battery FCEVs, Super-capacitor Battery FCEVs and Bidirectional converters with battery, Ultracapacitor and FCEV.

11. Conclusion

This paper has reviewed multiple DC-DC converter topologies. The key aspects for choosing the boost converter have been DC link voltage, transient behaviour, and ripple content which are essential for FCEV operation. SEPIC converters are gaining popularity due to their output voltage regulation capacity [19], low ripple, and stable output voltage, which is ideal for unpredictable power sources such as Fuel Cells. However, there is enough work to be done to ensure the complex control in the case of SEPIC controllers is circumvented along with the fact that the converter topology is in general non-isolated. Luo converters too offer an immensely promising approach

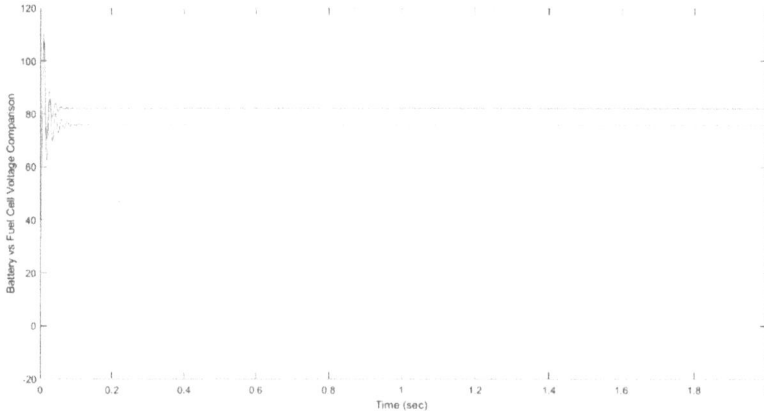

Fig. 1.4 Comparative study of FCEV and battery EV

with an optimally designed controller to ensure that the non-optimal regions of the converter are not entered. Therefore, in essence, Luo or SEPIC converters are highly suitable for FCEV operation considering that the controller is suitably designed to take care of the regions of operation. Further research in this direction can pave the way for wider adoption of FCEVs.

REFERENCES

1. Farhani, Slah et al, 2021. "Design and hardware investigation of a new configuration of an isolated DC-DC converter for fuel cell vehicle." Ain Shams Engineering Journal 12.1: 591–598..
2. Selmi, Tarek, et al, 2022. "Fuel cell–based electric vehicles technologies and challenges." Environmental Science and Pollution Research 29.52: 78121–78131.
3. Priya, Rajbala Purnima, et al, 2023. "Review of Fuel Cell Power Converter Topologies for Electric Vehicles." *IECON 2023-49th Annual Conference of the IEEE Industrial Electronics Society*. IEEE.
4. Bhaskar, Mahajan Sagar, et al, 2020. "Survey of DC-DC non-isolated topologies for unidirectional power flow in fuel cell vehicles." *Ieee Access* 8: 178130–178166..
5. Zhang, Yun, et al, 2017. "A wide input-voltage range quasi-Z-source boost DC–DC converter with high-voltage gain for fuel cell vehicles." IEEE Transactions on Industrial Electronics 65.6 : 5201–5212.
6. https://www.iea.org/reports/the-role-of-carbon-credits-in-scaling-up-innovative-clean-energy-technologies
7. Jayachandran, Divya Navamani, et al, 2022. "A comprehensive review of the quadratic high gain dc-dc converter for fuel cell application." International Journal of Electronics and Telecommunications : 299–306..
8. Pradhan, Sovit Kumar, et al , 2024. "An Extensible Non-isolated Enhanced Gain DC–DC Converter Integrating Switched Capacitor Cell for FCEV." *Arabian Journal for Science and Engineering* : 1–15..
9. Banaei, Mohamad Rezaet al, 2018. "Analysis and implementation of a new SEPIC-based single-switch buck–boost DC–DC converter with continuous input current." *IEEE transactions on power electronics* 33.12 : 10317–10325..
10. Kavyapriya, S., et al, 2020. "Modeling and Simulation of DC-DC Converters for Fuel Cell System." *Int. J. of Engg. and Adv. Tech* 9.3..
11. Munus, S. George, et al, 2023. "Modelling and Simulation of DC-DC Converters for PEM Fuel Cell Electric Vehicles." *International Conference on Microelectronics, Electromagnetics and Telecommunication*. Singapore: Springer Nature Singapore.

12. Samosir, Ahmad Saudi et al, 2011. "Simulation and implementation of interleaved boost DC-DC converter for fuel cell application." *International Journal of Power Electronics and Drive Systems* 1.2: 168–174..

13. Gopal, S. Madan, et al 2023. "Design and control of high voltage gain interleaved boost converter for fuel cell based electric vehicle applications." *J Electron Comput Netw Appl Math* 3.32 : 9–24.

14. Elsayad, Nour, et al., 2020. "A new SEPIC-based step-up DC-DC converter with wide conversion ratio for fuel cell vehicles: Analysis and design." IEEE Transactions on Industrial electronics 68.8 : 6390–6400..

15. Gumaste, Smita, et al, 2019. International Journal of Research in Engineering, Science and Management Volume-2, Issue-3.

16. Wang, Jiulong, et al, 2022. Modified SEPIC DC-DC converter with wide step-up/step-down range for fuel cell vehicles. *IEEE Transactions on Power Electronics.*

17. Balamurugan, Rangaswamy, et al, 2024. "Fuzzy logic controller-based Luo converter for light electric vehicles." *Indonesian Journal of Electrical Engineering and Computer Science* 34.1: 641–646..

18. Ansari, Sajad Arab, et al, 2018. "A novel high voltage gain noncoupled inductor SEPIC converter." *IEEE Transactions on Industrial Electronics* 66.9 : 7099–7108...

19. Wang, Jiulong, et al, 2022. "Modified SEPIC DC-DC converter with wide step-up/step-down range for fuel cell vehicles." *IEEE Transactions on Power Electronics.*

20. Wang, Ping, et al, 2017. "Input-parallel output-series DC-DC boost converter with a wide input voltage range, for fuel cell vehicles." *IEEE Transactions on Vehicular Technology* 66.9: 7771–7781..

21. Pathak, Pawan Kumar, et al, 2022. "Fuel cell-based topologies and multi-input DC–DC power converters for hybrid electric vehicles: A comprehensive review." *IET generation, transmission & distribution* 16.11: 2111–2139..

22. Wang, Hanqing, et al, 2019. "A review of DC/DC converter-based electrochemical impedance spectroscopy for fuel cell electric vehicles." Renewable Energy 141: 124–138.

23. Reddy, Bondu Pavan Kumar et al, 2024. "PV-based performance evaluation of ZETA and SEPIC topologies for EV applications." Journal of Electrical Systems 20.5s: 438–446.

24. Selvi, R. Kalai et al, 2020. "A bridgeless Luo converter based speed control of switched reluctance motor using Particle Swarm Optimization (PSO) tuned proportional integral (Pi) controller." Microprocessors and Microsystems 75: 103039..

25. Mo, Liping, et al, 2024. "A Novel Topology Derivation Method Revealed from Classical Cuk, Sepic, and Zeta Converters." *IEEE Transactions on Power Electronics.*

26. Nandish, B. M et al, 2023. "DC-DC Power Converter Topologies for Sustainable Applications." *Power Converters, Drives and Controls for Sustainable Operations*: 1–19.

27. Pradhan, Sovit Kumar, et al, 2024. "An Extensible Non-isolated Enhanced Gain DC–DC Converter Integrating Switched Capacitor Cell for FCEV." *Arabian Journal for Science and Engineering*: 1–15..

28. Jarin, T., et al, 2022. "Fuel vehicle improvement using high voltage gain in DC-DC boost converter." *Renewable Energy Focus* 43: 228–238.

29. Benzine, Meryem, et al, 2023. "Coupled inductors-based interleaved boost converters for Fuel Cell Electric Vehicles." *2023 IEEE Transportation Electrification Conference & Expo (ITEC).* IEEE.

30. Kunstbergs, Noass, et al, 2022. "Efficiency improvement of a cascaded Buck and Boost converter for Fuel Cell Hybrid Vehicles with overlapping input and output voltages." *Inventions* 7.3 : 74.

31. Pangtey, Tanya, et al, 2022. "Control and Analysis of SEPIC Topology based Boost DC-DC Converter with High Gain for Fuel Cell Fed Electric Vehicle Driving System." *2022 IEEE Students Conference on Engineering and Systems (SCES).* IEEE.

32. Singh, Shiv Prakash, et al, 2024. "A Review on DC-DC Converters for Fuel Cell Electric Vehicle Applications." *2024 IEEE Students Conference on Engineering and Systems (SCES).* IEEE.

33. Bauman, Jennifer, et al, 2008. "A comparative study of fuel-cell–battery, fuel-cell–ultracapacitor, and fuel-cell–battery–ultracapacitor vehicles." IEEE transactions on vehicular technology 57.2 : 760–769.

Note: All the figures and table in this chapter were made by the author.

Intelligent and Sustainable Power and Energy Systems – Dr. M. Premkumar et al. (eds)
© 2026 Taylor & Francis Group, London, ISBN 978-1-041-10314-1

2

Secure File Storage System to Secure the Security Algorithm and Password

Ronald Doni[1], Patibandla Sri Pranavi[2]
Department of Computer Science Engineering,
Sathyabama Institute of Science and technology,
Chennai

ABSTRACT: Development of cloud computing era with speedy improvement for unstructured records, cloud storage era is gaining more attention A cloud company lacks in phrases of increase offerings Data and facts are stored and maintained globally inside the cloud. A cloud privacy protection strategy is usually based totally on encryption. Life has many privacy safety strategies don't keep statistics inside the cloud. We advise a three-tier storage architecture counting fog can fully make use of the proposed framework Cloud storage and records privacy protection. Here we use the hash- The Salomon code set of rules is designed to divide the facts into exclusive components. Yes there are not any facts, we've lost information about the statistics. We are in this structure applying algorithms based on the card concept and keeping the information and then he can display the safety and efficacy of our software. Also, step this algorithm can calculate the distribution based on computational intelligence the ratio is stored in cloud, cloud and nearby computer respectively. Software like As a Service (SaaS): The client runs the utility in a hosted surroundings. The software can be accessed from numerous clients over the community Do now not manipulate or control the underlying cloud the usage of Infrastructure except for the confined use viable for a specific person Structural systems. 365 consists of Google Apps and Microsoft Office Examples for SaaS.

KEYWORDS: Multi stage, Cloud security, Cloud data, Amalgamate data, Cloud computing

1. Introduction

In laptop science, cloud computing refers to a form of cloud computing. Computer services are much like how power is furnished Outsourced customers can definitely use it and don't must worry

[1]doni.cse@sathyabama.ac.in, [2]patibandlasripranavi2004@gmail.com

DOI: 10.1201/9781003654469-2

about it relies upon on how the electricity is generated or transported. They need to pay each month What he spent changed into like cloud computing; The user can simply use garage, computing power, or mainly designed improvement surroundings no matter how they feature. Internal cloud computing is typically Internet-based computing. The internet cloud metaphor is primarily based on how the internet works. Described in computer network software program; what is Pythias? An acronym that covers the complex infrastructure of the Internet. This is Calculation method wherein assets are furnished "as is". Service ", whilst users get entry to the Technology Services aInternet without know-how or manipulate ("in the cloud"). Technology in the back of these farmers. Cloud computing may be observed in both massive-scale cloud structures and huge-scale structures Using statistics structure, get entry to difficulties data obj. This leads to ok first-class output Content Impact of cloud computing and fog computing on huge statistics Settings may additionally vary. However, there may be one not unusual feature that can be highlighted Limitation on proper distribution of content material - positive to be a hassle they attempt to improve accuracy by using making meters. The fog the network includes a manipulate aircraft and a statistics plane. For instance, on- The facts aircraft permits cloud computing to residence computing offerings sometimes at the brink of the community in place of the servers on the pc. Became compared to cloud computing, cloud computing emphasizes proximity to give up users. And purchaser obstacles, dense international distribution and local sources In mixture, reduce latency and shop network spine but acquire Better Quality of Service (QoS) and Edge Analytics/Stream Mining; ensuing in a higher person revel in and redundancy in case of failure. However it can additionally be used in AAL scripts. To defend user privacy, we recommend a TLS-based structure an instance of cloud computing. The TSL framework may also offer some skills to the consumer. Manage and efficaciously shield consumer privacy. As already said, countering an insider attack is difficult. Traditional methods work properly Solution for external attack, but because CSP has issues, traditional techniques. They are all invalid. Unlike the conventional approach, in our design User statistics is split into three elements of various sizes throughout the encryption manner both of them are lacking a few vital facts of life a cloud computing model has been implemented in these 3 regions in secret Data is stored at the cloud server, the cloud server and the person's neighbourhood computer. The car is made to head from big too small. In this manner Even if an attacker receives all the data, the authentic user statistics cannot be recovered from a specific employee. As for CSP, they will no longer even have useful records Information. Without data on cloud server and nearby pc, because both Cloud server and nearby machine are managed with the aid of users.

1.1 Related Work

Literature evaluation is a important step inside the software program development manner. Before growing the device, it's far vital to perceive time elements, cost financial savings and commercial enterprise robustness. Once those conditions are met, the next step is to determine the operating systems and languages used to increase the device. Once a programmer begins constructing a device, numerous styles of external help are wanted. This aid can come from advanced programmers, books or websites. Before designing the system, we enlarge the proposed tool via considering the above problems.

A primary part of the mission development branch is to cautiously examine and evaluation all requirements for undertaking improvement. For any challenge, literature evaluation is the most critical step inside the software program improvement system. Before growing equipment and associated designs, time elements, aid necessities, human sources, economics, and organizational talents ought to be determined and analysed. After these factors are met and carefully researched, the subsequent step is to determine the software program software specs to your specific PC, the working system required for your task, and the software programs required for switch. Steps like growing gear and features associated with them.

Attribute-based encryption (ABE) is a promising encryption method using the cloud can clear up many safety problems. In this article we suggest A framework for replacing metropolis facts by means of attribute For the actual use of cities anywhere cryptography, we enlarge our software to help dynamic operations. Especially from the aspect Performance analysis allows us to finish that our layout is secure and viable can withstand attack. Also, test outcomes and comparisons show our intention is exceptionally efficient computing. From the experiment Analysis, we discover that our kinetic functions require less computation 340. We believe that in time this framework can be applied to actual-world records. Sharing programs in virtually ubiquitous cities. Its features Court appropriate to protect and use personal statistics of residents; together with bioenergy consumption and fitness data Not confined to apply and amendment via sure individuals and users given to the owners. Acceleration of urbanization is increasingly more People live in towns. Dealing with huge amounts of information generated Citizens and State Public Sectors, New Information and Communication Technology is used to manner city statistics, which simplifies the system Hat cloud computing is a new computing technology. Behind the cloud Computing have become commercial and cloud technologies emerged. Affection because thru the cloud it's far supplied by means of a third celebration, i.e. The cloud Due to the character of quasi-nic cloud computing, there are numerous safety capabilities. Cloud computing in question [1].

Introduce the idea of graphs (CGs) as an information illustration device. So we present schemes (PRSCG and PRSCG-TF) based totally on CG. According to exceptional eventualities. Let's do a numerical calculation Convert unique laptop pictures to linear format with a few modifications and graphs as a vector of numbers. Second, we use a couple of keyword strategies Searching for encrypted cloud facts is considered as the idea for each hazard fashions Set up PRSCG and PRSCG-TF to cope with privacy troubles Computer Graphics-Based Semantic-Intelligent Search. Finally, we pick actual information; The Rhonchus dataset is our intention to test. Also analyse privacy and performance explains the proposed initiatives. The effects of our experiment display that Plans are useful. We put in force our method on a actual dataset to check it Efficiency and effectiveness. Because we nevertheless need to have a look at Ability to semantically search encrypted cloud information in herbal language Delivery method. Due to the scalability and excessive overall performance of the cloud Servers are a completely unstable, reasonably-priced manner to get entry to public data Company, especially in small businesses. However, statistics owners are hesitant Data privacy and existing devices use facts encryption to solve the problem of data leakage. How to execute a seek the cause of encryption is a complicated and critical hassle [2].

Vectors have cozy set of rules and picture Images are encoded the use of a recognised stream. Again, D If any authorized customers request to copy and distribute the Act We advocate using watermarks while receiving pictures from strangers A protocol to prevent such illegal distribution. In our hint-based protocol A unique footprint is immediately connected to encrypted images the use of the cloud Open the server before sending pics to the requester. So, with the incorrect image although the fake declare of the person who distributed the image can be investigated, the sample is discovered Water mark elimination. Health evaluation and checks Safety and efficacy of the proposed scheme. Considering the wrong customers SE isn't always unlawful within the schemes and protocols of watermarking Distribution of images. Generally, the capabilities of the pictures are included Cyphertext-simplest assault model: The contents of documents are covered in opposition to selective assaults. Improving plaintext assault modelling and seek performance. The image retrieval (CBIR) approach has been broadly studied. Compare with text I shop documents and pics in quite a few area. Hence his protection A common instance is cloud garage outsourcing. To Sensitive pictures together with clinical and private to preserve privacy Images need to be encrypted earlier than outsourcing, which is what CBIR does Plain textual content generation becomes vain [3].

Concretely, we develop a novel method of keyword transformation and introduce the stemming algorithm. With these two techniques, the proposed scheme is able to efficiently handle more misspelling mistake. Moreover, our proposed scheme takes the keyword weight into consideration during ranking. Like Wang et al.'s scheme, our proposed scheme does not require a predefined keyword set and hence enables efficient file update too. We also give thorough security analyses and conduct experiments on real world data set, which indicates the proposed scheme's potential of practical usage. Experiments using real-world data show that our scheme is practically efficient and achieve high accuracy. The majority of the existing techniques are focusing on multi- keyword exact match or single keyword fuzzy search. However, those existing techniques find less practical significance in real-world applications compared with the multikeyword fuzzy search technique over encrypted data. The first attempt to construct such a multi-keyword fuzzy search scheme was reported by Wang et al., who used locality-sensitive hashing functions and Bloom filtering to meet the goal of multi-keyword fuzzy search. Nevertheless, Wang's scheme was only effective for a one letter mistake in keyword but was not effective for other common spelling mistakes. Moreover, Wang's scheme was vulnerable to server out-of-order problems during the ranking process and did not consider the keyword weight.

In this overview, we explored exceptional sorts of cryptographic communication. Cloud is one of the simplest solutions for health and cozy information switch The data owner have to encrypt his information before storing it within the cloud Knowledge included via the facts cloud service issuer and Attacking others. Also mathematical contraptions, cryptographic machines be greater cell and frequently upload a couple of key Used then it looks at a way to compress non-public keys into public keys. Cryptosystems that guide the transmission of mystery keys for unique ciphers Our approach to instructions in the cloud is extra bendy than a hierarchical one of the very best solutions for comfortable information switch over the internet is the cloud It encrypts its information earlier than it's far stored within the cloud Knowledge blanketed by means of the facts cloud carrier company and Attacking others. Also mathematical contraptions, cryptographic machines they have become greater mobile and often have a couple of key Used then it looks at the way to compress private keys into public keys. Cryptosystems that assist the transmission of mystery keys for one-of-a-kind ciphers our method to training inside the cloud is bandier than a hierarchical one,this article offers a top-level view and take a look at of cryptography. Methods for relaxed and green records transfer inside the cloud [5].

1.2 Existing System

Cloud computing technology have advanced in current years. With the explosive growth of unstructured records, cloud garage era He gets greater interest and better development.Computer era is developing swiftly. This is cloud computing It step by step came together thru many tries. In a coin storage gadget, consumer data is completely stored in the cloud. Servants. If the user loses manage over records and privacy dangers. Privacy protection schemes are typically primarily based on encryption. Such modes of existence cannot efficiently counter assaults from out of doors Internal a part of cloud server.

Disadvantages of Existing System
- Changes in expertise of risk from growth sometimes within the cloud.
- Low latency and region focus.

2. Requirement Analysis

2.1 Evaluation of the Rationale and Feasibility of the Proposed System

The suggested structure permits the utilization of distributed archives while maintaining information security. Here we use Hash-Solomon code evaluation to divide the data into different parts. If any part is missing, the data information will be lost. In this system, we collect information and demonstrate the safety and accuracy of our design by calculating on the basis of containers. In addition, considering statistical data, this assessment can calculate the degree of loss in the cloud, fog and virtual machines available as a guide (SaaS): it is provided by the customer his request to work with the weather, it can be from him fulfilled. by form. From different vendors by software vendors.

2.2 Proposed System

Fully occupy cloud storage platform to shield privacy Information. This cloud computing is getting loads of attention from numerous resources Department of Society. Three tier clouds shop 3 distinctive regions of records. If a part of the elements statistics is missing, we have misplaced statistics approximately the data. In it a proposed framework the use of algorithms based totally on bucket idea.

A proposed method for a secure file storage system often involves hybrid cryptography. This approach combines both symmetric and asymmetric encryption techniques to enhance security. Here's a brief overview of how it works:

1. Symmetric Encryption: Algorithms like AES (Advanced Encryption Standard) are used to encrypt the actual data files. Symmetric encryption is fast and efficient, making it suitable for large datasets2.
2. Asymmetric Encryption: Algorithms like RSA (Rivest-Shamir-Adleman) are used to encrypt the symmetric keys. Asymmetric encryption provides a robust mechanism for key distribution, ensuring that the symmetric keys are securely transmitted2.
3. Multi-layered Security: The data is first encrypted using a symmetric key, which is then encrypted with the recipient's public key using asymmetric encryption. This ensures that both the data and the encryption keys are secure2.
4. Steganography: Techniques like Least Significant Bit (LSB) steganography can be used to hide key information within the encrypted data, adding an extra layer of security.
5. File Segmentation: The file can be divided into segments and stored in separate locations to further enhance security.

Steps for Secure File Storage System

1. User Authentication:
   - User logs in with credentials.
   - System verifies user identity.

2. File Encryption:
   - User uploads a file.
   - The file is encrypted using a symmetric encryption algorithm (e.g., AES).
3. Key Encryption:
   - The symmetric key is encrypted using the user's public key (asymmetric encryption, e.g., RSA).
4. File Upload:
   - Encrypted file and encrypted key are uploaded to the storage server.
5. Storage:
   - File segments may be distributed across multiple locations for added security.
6. File Access Request:
   - User requests access to a stored file.
   - System verifies user identity.
7. Key Decryption:
   - The user's private key decrypts the symmetric key.
8. File Decryption:
   - The symmetric key decrypts the file.
9. File Download:
   - The decrypted file is made available to the user.

Advantages OF Proposed System

- We use CART idea in our device which minimizes records losses and reduce the procedure time.
- We use BCH code set of rules. It is highly elastic.
- The BCH code is used in lots of verbal exchange programs and has a small range termination of employment.

BLOCK DIAGRAM

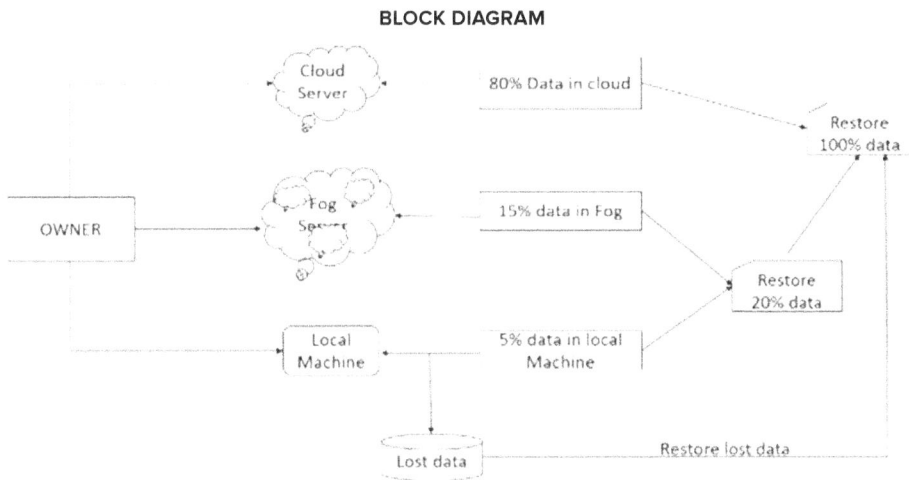

Fig. 2.1 Block diagram of proposed system [6]

Figure 2.1 shows the Block Diagram

1. User Interface (UI) Layer:
   - Authentication Module: For user login and identity verification.
   - Dashboard: Interface for users to upload, download, and manage files.
2. Encryption/Decryption Layer:
   - Encryption Module: Encrypts files using secure algorithms (e.g., AES, RSA).
   - Key Management System: Handles storage and retrieval of encryption keys.
3. Storage Layer:
   - File Storage Module: Secure storage of encrypted files, possibly in a database or cloud storage.
   - Access Control Module: Manages permissions for file access.
4. Security Layer:
   - Intrusion Detection System: Monitors for unauthorized access or security breaches.
   - Audit Logs: Maintains records of user actions for tracking and accountability.

Figure 2.2 shows the Flow Chart

1. User Authentication:
   - Start
   - User enters credentials
   - Validate credentials
     - If valid, proceed to Dashboard
     - If invalid, prompt user to re-enter credentials
2. File Upload Process:
   - User selects file to upload
   - Encryption Module encrypts file using the encryption algorithm
   - Key Management System stores the encryption key securely
   - Store encrypted file in File Storage Module
   - Display success message to user
3. File Download Process:
   - User requests file download
   - Access Control Module checks permissions
     - If authorized, proceed
     - If unauthorized, deny access

FLOW CHART

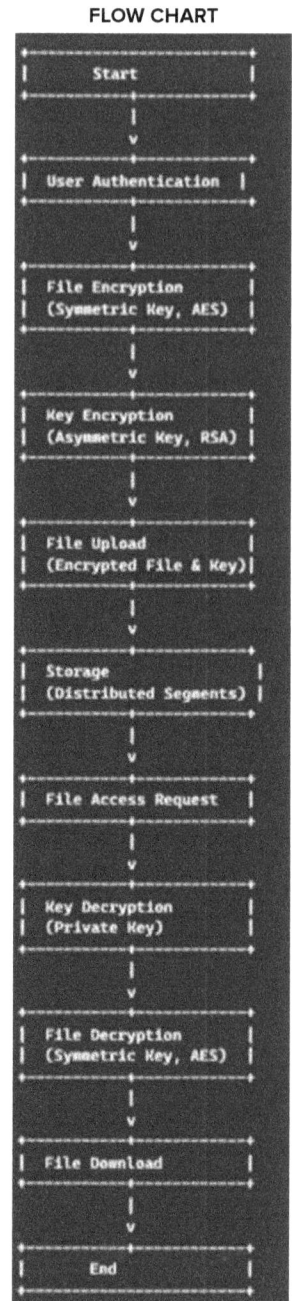

Fig. 2.2   Flowchart
Source: Author

- Retrieve encrypted file from File Storage Module
- Decryption Module decrypts the file using the corresponding key
- Deliver decrypted file to user
- Log the download action in Audit Logs

4. Security Monitoring:
- Continuously monitor system for unauthorized access
- Trigger alerts for any suspicious activities
- Log details in Audit Logs

3. System Modules

1. Login
2. Registration
3. Storage Scheme
4. Recovery Scheme

3.1 Modules Description

1. Login

It is the primary motion that opens with the internet site consumer. Person User need to offer correct touch variety and password" Entered in registration to enter the website. If records User-supplied records matches the database desk and user Successful login to the website online, otherwise a message approximately failed login will seem is displayed and the person must re-input the perfect records. Connection Registration approaches for registering new users also are supplied.

2. Registration

A new user who desires to get entry to the website have to sign in first. Register with the aid of clicking the Register button in the course of the login procedure earlier than logging in. Action is open. Enter your complete name and password and register as a new user. And get in touch with no. The person need to affirm the password once more. Password text field for affirmation. When entering consumer facts In all textual content fields, whilst you press the report button, the data will be modified Both the database and the user are prompted to log in once more. Then the registered user you want to go into an internet web page to get right of entry to it. Use validations all text fields for correct functioning of the page. As a symptom each text field should imply every text field or call, contact; Password affirmation at some point of registration is invalid. Yes such an area isn't empty. The utility will display a message through the facts. In each text subject. You have to additionally offer records within the Password and Confirm Password fields. According to registration. Another affirmation touch no. It must be robust and 10 inches lengthy. If such test is violated then the registration will fail and then the consumer will should sign up again if one of the fields is empty, a message is displayed at the web page. If all such facts is correct, the user will be redirected to the account Log in to the website.

3. Storage Scheme

In this module, the user can keep his documents on 3 special servers. Once the records is uploaded to the cloud the owner has no manage over it. In this module, the original record is divided into 3 tiers. Data may be encrypted at any level the usage of distinctive cryptographic methods.

4. *Recovery Scheme*

In this module, the consumer can retrieve their files from three exceptional storages. A cloud server is a server that could be a cloud server and a nearby pc. We are here One way to divide the data is to apply Solomon's subtraction notation Different elements. If part of the facts is missing, we've got lost records about the facts.

4. System Methodologies

4.1 Cloud Computing

Pay-per-use access to computing resources, such as physical or virtual servers, data storage, networking capabilities, application development tools, software, and AI-powered analytic tools, is known as cloud computing. When compared to conventional on-premises infrastructure, customers benefit from greater scalability and flexibility through the cloud computing model. When we play a cloud-hosted video game, stream a movie on Netflix, or access a cloud application like Google Gmail, cloud computing plays a crucial role. Additionally, cloud computing has emerged as a necessity for businesses of all sizes, from start-ups to multinational corporations.

Its numerous business applications incorporate empowering remote work by making information and applications open from anyplace, making the structure for consistent omnichannel client commitment and giving the immense registering power and different assets expected to exploit state of the art advances like generative simulated intelligence and quantum figuring. A cloud administrations supplier (CSP) oversees cloud-based innovation administrations facilitated at a far-off server farm and commonly makes these assets accessible for a pay-more only as costs arise or month to month membership expense.

Tools

1. Cloud Storage Services: Services like Google Drive, Dropbox, and OneDrive offer secure cloud storage solutions. They provide encryption for data at rest and in transit, ensuring that your files are protected.
2. Network Attached Storage (NAS): Solutions like Synology NAS offer robust security features, including multi-factor authentication and device management.

Programming Languages

1. Python: Widely used for its simplicity and extensive libraries, Python has libraries like cryptography for encryption and secure data handling.
2. Go (Golang): Known for its performance and scalability, Go is a good choice for cryptographic operations.
3. C++: Offers low-level control over system resources and has libraries like Crypto++ for cryptographic functions.
4. Java: Popular for its portability and security features, Java has built-in support for encryption and secure coding practices4.

Frameworks

1. Spring Boot: A Java-based framework that provides comprehensive security features, including OAuth2 and JWT for secure authentication.

2. Django: A high-level Python web framework that includes security features like CSRF protection and secure password handling.
3. Express.js: A Node.js framework that can be used with various security middleware to protect your application.

Figure 2.3, Fig. 2.4, Fig. 2.5, Fig. 2.6, and Fig. 2.7 explains us about the modules of the user interface

SCREEN SHOTS FOR MODULE DESCRIPTION

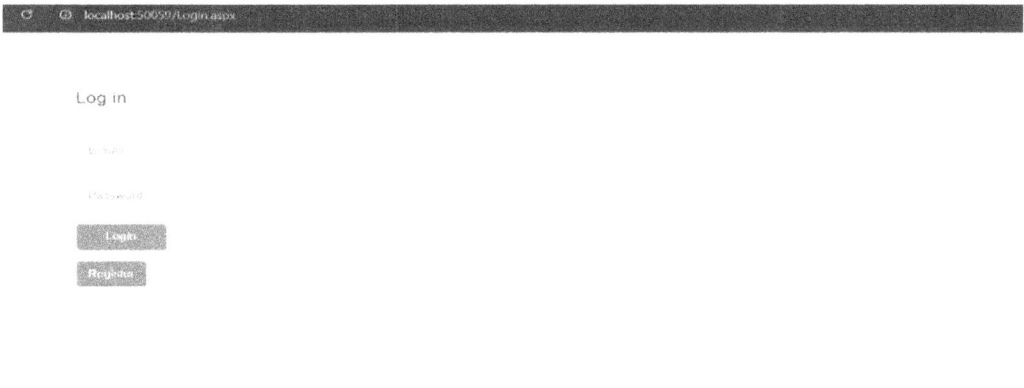

Fig. 2.3 Authentication *Module*

Source: Author

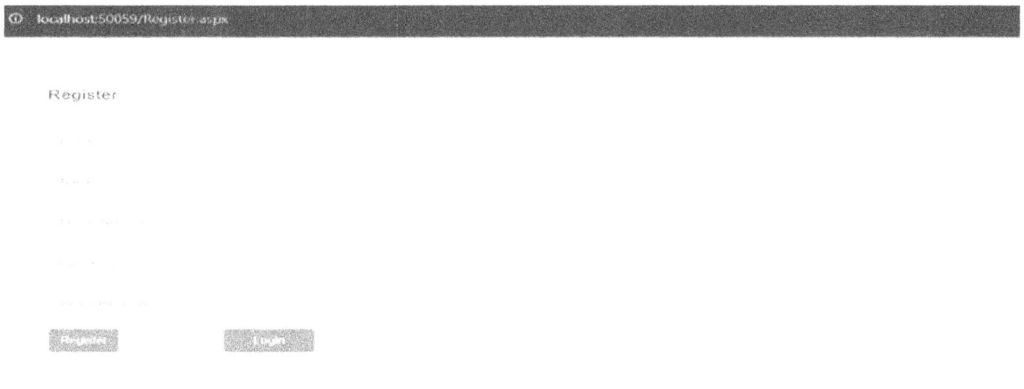

Fig. 2.4 *Security layer*

Source: Author

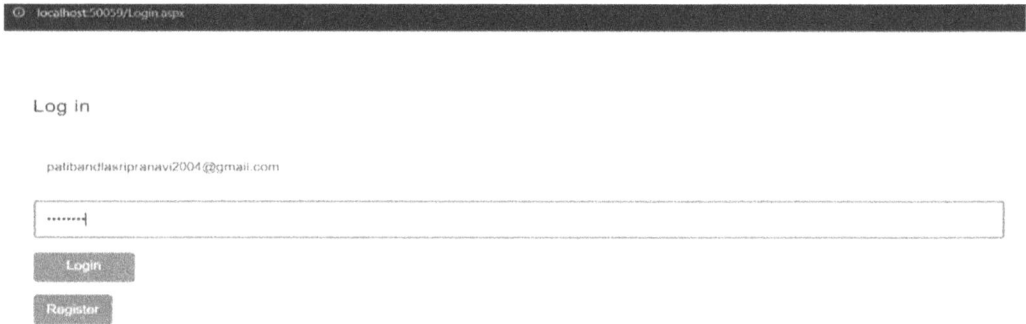

Fig. 2.5 Login *Module and Dashboard*

Source: Author

User Interface (UI) Layer

1. Authentication Module:
   - Purpose: Ensures that only authorized users can access the system.
   - Functions: Verifies user credentials (username and password), handles login attempts, and provides error messages for invalid credentials.

2. Dashboard:
   - Purpose: Acts as the main interface for users to manage their files.
   - Functions: Allows users to upload, download, and delete files. Displays file statuses and provides access to account settings.

Encryption/Decryption Layer

3. Encryption Module:
   - Purpose: Protects files by converting them into an unreadable format.
   - Functions: Uses encryption algorithms (e.g., AES, RSA) to encrypt files before storage. Ensures data privacy by preventing unauthorized access.

4. Key Management System:
   - Purpose: Manages encryption keys securely.
   - Functions: Stores, retrieves, and rotates encryption keys. Ensures that keys are only accessible to authorized components of the system.

Storage Layer

5. File Storage Module:
   - Purpose: Stores encrypted files securely.
   - Functions: Manages file storage in databases or cloud storage. Provides secure access and retrieval mechanisms.

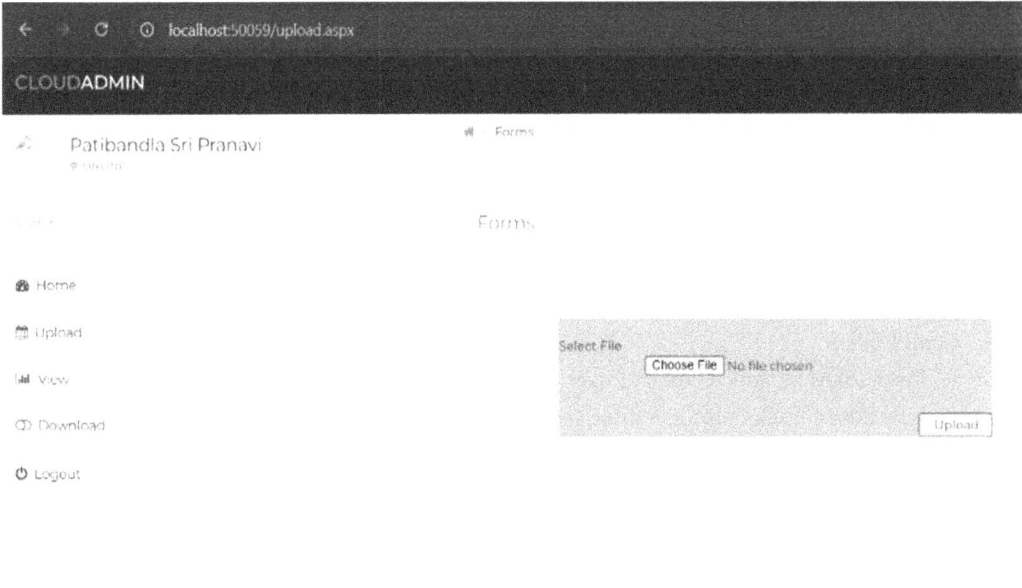

Fig. 2.6 Encryption layer

Source: Author

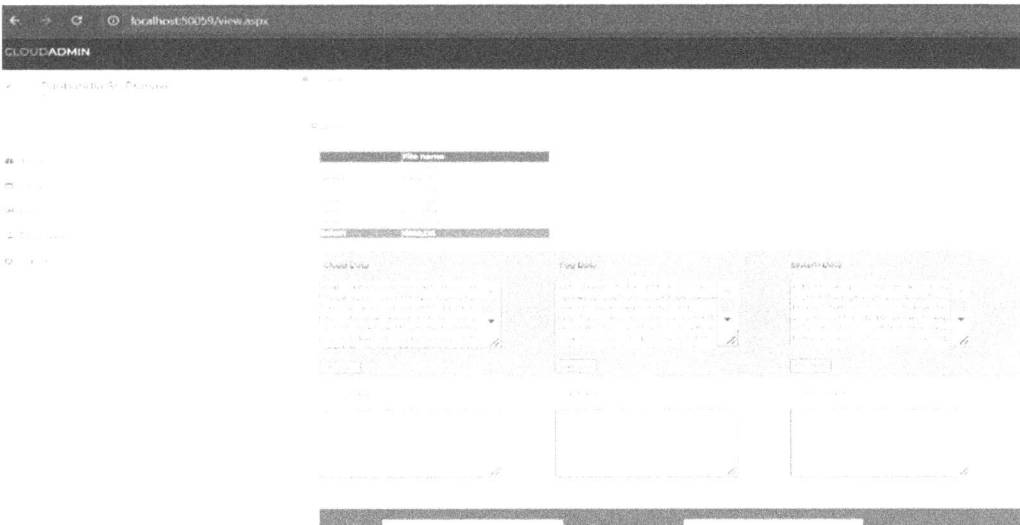

Fig. 2.7 Storage layer

Source: Author

6. Access Control Module:
   - Purpose: Controls permissions for file access.
   - Functions: Determines user permissions for uploading, downloading, and deleting files. Prevents unauthorized access.

Security Layer

7. Intrusion Detection System:
   - Purpose: Monitors the system for security breaches.
   - Functions: Detects and alerts on unauthorized access attempts. Logs suspicious activities for further analysis.

8. Audit Logs:
   - Purpose: Maintains records of user actions.
   - Functions: Tracks user activities such as file uploads and downloads. Provides a trail for accountability and troubleshooting.

5. Conclusion

Many agencies and agencies use cozy garage And security will gain from this choice Takes care of safety for documents stored inside the cloud is solved the usage of the method The complete record is a totally risky cookie. With handiest one cloud server, this possibility will increase Attack and the attacker receives the entire board. Yes, secondly when this method is added, growth occurs secure files are saved on distinct servers that shop them inside the cloud and encrypt them. The cloud uses numerous storage ideas. Along with the concept of encryption for increased protection Content within the cloud, in place of storing the whole report in a single the cloud gadget divides the document into multiple components, the hash Encrypt them and any other cloud Servants. This mode of trade permits companies to take advantage of -degree encryption in your files Adopted here with the introduced gain of information integrity.

REFERENCES

1. P. Mell and T. Grance, "The NIST definition of cloud computing," Nat. Inst. Stand. Technol., vol. 53, no. 6, pp. 50–50, 2009.
2. H. T. Dinh, C. Lee, D. Niyato, and P. Wang, "A survey of mobile cloud computing: Architecture, applications, and approaches," Wireless Commun.MobileComput., vol. 13, no. 18, pp. 1587–1611, 2013.
3. J. Chase, R. Kaewpuang, W. Yonggang, and D. Niyato, "Joint virtual machine and bandwidth allocation in software defined network (sdn) and cloud computing environments," in Proc. IEEE Int. Conf. Commun., 2014, pp. 2969–2974.
4. H. Li, W. Sun, F. Li, and B. Wang, "Secure and privacy-preserving data storage service in public cloud," J. Comput. Res. Develop., vol. 51, no. 7, pp. 1397–1409, 2014.
5. Y. Li, T.Wang, G.Wang, J. Liang, and H. Chen, "Efficient data collection in sensor-cloud system with multiple mobile sinks," in Proc. Adv. Serv. Comput., 10th Asia-Pac. Serv. Comput. Conf., 2016, pp. 130–143.
6. A. Bhardwaj, F. Al-Turjman, M. Kumar, T. Stephan, and L. Mostarda, ''Capturing- theinvisible (CTI): Behavior-based assaults acknowledgment in IoT-situated modern control frameworks," IEEE Access, vol. 8, pp. 104956–104966, 2020.

7. M. Kumar, A. Rani, and S. Srivastava, ''Image legal sciences upheld lighting assessment,'' Int. J. Picture Graph., vol. 19, no. 3, Jul. 2019, Art. no. 1950014.
8. A. Bhardwaj, F. Al-Turjman, M. Kumar, T. Stephan, and L. Mostarda, "Capturing-the-invisible (CTI): Behavior-based attacks recognition in IoT-oriented industrial control systems," IEEE Access, vol. 8, pp. 104956–104966, 2020.
9. M. Kumar, A. Rani, and S. Srivastava, "Image forensics based on lighting estimation," Int. J. Image Graph., vol. 19, no. 3, Jul. 2019, Art. no. 1950014.
10. Industry 4.0 and big data innovations, G. Li, J. Tan, S. S. Chaudhry, Enterprise Information Systems, 13 (2) (2019) 145–1.

Intelligent and Sustainable Power and Energy Systems – Dr. M. Premkumar et al. (eds)
© *2026 Taylor & Francis Group, London, ISBN 978-1-041-10314-1*

3

Optimal Allocation of Distributed Generators in Distribution System with Various Load Models

Babita Gupta[1]

Department of Electrical and Electronics Engineering
BVRIT HYDERABAD College of Engineering
for Women, India

School of Electronics and Electrical Engineering,
Lovely Professional University,
Phagwara, Punjab, India

Suresh Kumar Sudabattula[2],
Sachin Mishra[3]

School of Electronics and Electrical Engineering,
Lovely Professional University,
Phagwara, Punjab, India

Nagaraju Dharavat[4]

School of Computer Science and Artificial Intelligence,
SR University, Warangal, Telangana, India

ABSTRACT: Reduced power losses, improved voltage profile, increased voltage stability index, minimized total voltage deviation, and greater energy savings can be achieved by implementing an advanced solution for the optimal allocation of banks of shunt capacitors (SCs) and distributed generators (DGs) in a radial distribution system (RDS). This research proposes an integrated strategy utilizing the VSI and the Multi-Objective Evolution of the Dragon Fly (DA) and Whale Optimization Algorithm (WOA) algorithm to address the aforementioned issue. Within the context of determining the best location and capacity of DGs, the DA and WOA algorithm technique has been used for different load models using IEEE 69-bus system. Various load models like CP, IND, RES, COM, CI, and CZ are utilized to evaluate the effectiveness of the techniques. The optimal positions are identified using the VSI approach, while the sizes corresponding to these places are determined using DA & WOA technique. The outcome indicate that the suggested method can substantially improve the voltage profile, reduce P_{Loss}, and provide savings from

[1]guptababita151@gmail.com, [2]suresh.21628@lpu.co.in, [3]sachin.20444@lpu.co.in, [4]nagaraju.dnr98@gmail.com

DOI: 10.1201/9781003654469-3

P_{Loss} minimization. DA algorithm is used to compare the efficiency of optimizing P_{Loss} and voltage profile improvement with the other algorithm WOA. The DA algorithm outperforms the WOA approach in terms of power loss reduction and voltage profile enhancement, as shown in the simulation results.

KEYWORDS: DA & WOA algorithm, Distributor generators, Power loss, Voltage profile, VSI

1. Introduction

The DS accounts for considerable P_{Loss}, often ranging from 10% to 13% of the generated output. Significant distribution losses indicate inefficiency and inadequate regulation of system voltage. Locally deployed DGs and SCs banks integrated into the network may enhance its capacity and power quality. In distributed generation technology, small production units (from 1 kW to 50 MW) are connected near the consumer end [1]. The DGs may comprise both conventional and unconventional energy sources [2]. Non-conventional distributed generators have emerged as a significant alternative owing to the constraints of fossil fuels. DGs can be strategically placed to reduce both actual P_{Loss} and reactive power loss Q_{Loss} as well as enhance the VSI, diminish TVD and revamp the consistency of the network. To leverage the benefits of DG technology, it is essential to ascertain the best dimensions and locations of the DGs in the distribution system (DS). Failure to do so may result in adverse impacts, including heightened losses, diminished voltage profiles, and escalated costs. The incorporation of DGs into the DS is a vital decision, as it provides enhanced efficiency, increased dependability, superior power quality, and a decrease in greenhouse gas emissions [3]. The identification of the appropriate dimensions and locations of DGs and SCs inside the RDS has garnered significant attention from researchers over the past decade. The primary advantages of DG-based Electrical Distribution Networks include an enhanced voltage profile, reduce transmission line losses, improved efficiency, increased energy output with lower emissions, better system protection and stability, decrease transmission and distribution congestion and augmented power quality. Determining the proper capacity and placement of distributed generation units in distribution systems is essential for maximizing their beneficial effects. Investigations have indicated that an inappropriate selection of DG location and size might result in increased active power losses and voltage variations within the network [4]. Moreover, numerous scientific investigations have endeavored through diverse methodologies to address the appropriate positioning and capacity of distributed generation units in distribution networks. The Sine Cosine Algorithm (SCA) is employed to address the optimization problem of size and locating DG units in various studies as outlined below. Authors in [5] introduced a novel Pareto-based Multi-objective (MOSCA) approach for optimal DG allocation in RDS. This method aims to enhance VSI, minimize total active power loss, reduce gas emissions, and decrease annual energy loss costs, while incorporating four distinct voltage-dependent load models and adhering to the system's operational constraints and DG limitations. To assist network operators in making prompt and suitable decisions, the authors have produced an optimal Pareto set. The suggested methodology is implemented on 69-bus test systems. Nevertheless, in this research, the authors have not incorporated the supplementary expenses associated with the new DGs units as a component of the objective function.

The penetration level of DG is estimated to range from 10% to 80% of the system's demand, potentially necessitating the replacement and enhancement of system components, deemed a non-economic option; furthermore, the operating power factor of the DG units has not been considered. The authors utilized capacitor banks (CBs) [6] to enhance system reliability and optimize cost efficiency. The ideal locations and dimensions of the CBs are established using two procedures; the initial strategy involves selecting suitable buses based on the loss sensitivity factor. Subsequently, the SCA ascertains the ideal dimensions of CBs. In the second technique, SCA identifies the ideal locations and sizes of the CBs. The suggested technique is evaluated on the IEEE 33 and 69-bus RDS, incorporating various single and multi-CBs. In this work, the authors focused exclusively on one sort of distributed generation, specifically capacitor banks, without addressing voltage stability enhancement. This research was further developed in [7] to achieve cost minimization, enhancement of voltage profiles, and reduction of system losses. Authors in [8] advocate for a specific solution that involves the integration of two types of distributed generation technologies: Photovoltaic (PV) systems and DSTATCOMs, into realistic radial distribution networks (RDNs). The best positions and ratings of the DGs are ascertained by the Improved Sine Cosine approach (ISCA) to evaluate the efficacy of this approach. The fitness function encompasses the optimization of power losses, the minimizing of voltage variations, and the enhancement of voltage stability. The research was conducted with and without DSTATCOMs at photovoltaic penetration levels of 60%, 70%, 80%, and 90%. The findings obtained demonstrated enhanced performance when integrating both D STATCOMs and PV units, validating the efficacy of the proposed optimization technique. This study, however, examined only two types of distributed generation, excluded solution cost as a target, and did not account for the constraints of the operational system. Authors in [9] sought to enhance system performance by utilizing a methodology based on SCA to determine the ideal size and placement of single and dual distributed generators in radial distribution systems. The proposed method is evaluated on IEEE 15, 33, 69, and 85 RDS. The research aims to reduce power loss and enhance the overall voltage profile. The results indicated the viability of the proposed technology and its beneficial impact on reducing power loss and enhancing the voltage profile.

2. Problem Formulation and Objective

The objective of allocating DGs and SCs in RDS simultaneously is to minimize P_{Loss} and enhance the voltage profile and VSI.

A. Minimizing the Active P_{Loss} - In the RDS, the shallow voltage profile results in greater losses than those experienced in the transmission system. The following equation computes the active P_{Loss} in the RDS.

$$P_{Loss} = R_{ij} * \frac{P_j^2 + Q_j^2}{|V_j|2} \tag{1}$$

Where, the resistance between buses i and j is represented by R_{ij}. P_j and Q_j denote the actual and reactive power flowing in the nodes j. V_j indicates the voltage at bus j.

The sum of all losses at different nodes yields the system overall real P_{Loss} (P_{TL}) and is expressed as

$$P_{TL} = \sum_{n=1}^{N} P_{Loss}$$

Where N represents the number of nodes

So, the OF of the active PLoss is given by $OF_{AL} = min(P_{TL})$

B. Voltage Stability Index-

$$VSI = 2*|V_1^2|*|V_2^2|-|V_2^4|-2*|V_2^2|*(P_2R_{12}+Q_2X_{12})+Z_{12}^2*(P_2^2+Q_2^2) \qquad (2)$$

Where V1 & V2 are the voltages at bus1 & bus2, respectively, Z_{12} is the line impedance between buses1 and 2.

C. Minimizing the Reactive Q_{Loss}. The equation for determining the reactive power loss (Q_{Loss}) of any branch is

$$Q_{Loss} = I_{ij}*\frac{P_j^2+Q_j^2}{|V_j|2} \qquad (3)$$

The sum of all losses at different nodes yields the system overall rreactive Q_{Loss} (Q_{TL}) and is expressed as: $Q_{TL} = \sum_{n=1}^{N} Q_{Loss}$

Where, N represents the number of buses So, the reactive Q_{Loss} is given below:

$$OF_{RL} = min(Q_{TL})$$

D. Constraints - The bus voltage must be specified within the range of ±5% ensuring that it does not exceed the maximum and lowest voltages. The voltage at each system bus must be specified requirements:

$0.95 \leq V_n \leq 1.05$, Where V_n is the voltage at any node n.

The power limits of DGs and thermal limits of line current shall be within the limits as given below.

$$P_{DG}^{min} \leq P_{DG} \leq P_{DG}max; Q_C^{min} \leq Q_C \leq Q_C^{max} \ \& \ I_m \leq I_m^{max}$$

E. The key assumptions considered prior to the problem formulation are defined below.

1. The source bus is not considered as the location for the installation of distributed generators and synchronous condensers.
2. Loads are classified as constant power factor (PQ) loads.
3. The upper and lower limits of DGs and SC banks are presented in Table 3.1.

Table 3.1 Maximum and minimum limits of DGs and SCs

Cases	DG$_{min}$ (MW)	DG$_{max}$ (MW)	SC$_{min}$ (MVAr)	SC$_{max}$ (MVAr)
Case-I	0.5	1.5	0.2	1.5
Case-II	0.5	1.5	0.2	1.5
Case-III	0.5	2	0.2	1.5
Case-IV	0.5	2.5	0.2	1.5

Source: Author

3. Location of DGs and SCs using VSI

A radial distribution system's stability can be evaluated with the use of the VSI.The VSI is an important indicator that provides the information of voltage collapse occurring in a PS. Finding vulnerable busses and gauging a system's voltage stability margins are the goals of this index. Furthermore, VSI can figure out where the Distributed Energy Resources (DERs) are best suited to be installed. If the stability index indicates poor performance, the system cannot handle an overload and requires immediate action. A higher value is preferable as it enables the bus to accommodate heavier loads without compromising its stability.

4. Integration of DGs into RDS

The integration of DGs into the RDS is achieved by implementing the proposed technique on the IEEE 69 bus system. Research on electrical power distribution networks can benefit from the test systems produced by IEEE 69. Analysis of load flows, investigations of voltage stability, optimization, and the integration of distributed generation and renewable energy sources. This system operates at a voltage level of 12.66kV and has a total load demand of 3.802 MW and 2.694 MVAr. The purpose of this implementation is to verify the resilience and efficacy of the distribution network. Three types of distributed generation (DG) technologies is presented to ascertain the ideal combination necessary to validate the goal function about the location and sizing problem [10].

a) First Type: DGs that produce only active power include things like fuel cells and photovoltaic systems. These DGs are typically linked to the power grid via inverters and converters that work at a unity power factor.

b) Second Type: DGs with a zero-power-factor operation, like synchronous compensators and capacitor banks, supply only reactive power.

c) Third Type: A variety of DGs such as gas turbine synchronous machines, which can provide both reactive as well as active power.

5. Algorithm

5.1 DA Algorithm

A new population-based meta-heuristic approach is the Dragonfly technique (DA). Idealized dragonfly hunting behavior (sometimes called static swarm feeding) and movement patterns are mostly impacted by DA. Dragonflies, when found in nature, often gather in clusters to forage for food. The hunting mechanism describes this process. The migratory mechanism allows large groups of dragonflies to move more easily since they all fly in the same direction. The dragonflies swarming behaviour is explained below by following points

1. The search agents are kept apart in the neighborhood by the separation mechanism. Eq. (1) shows the mathematical model of the separation behavior.

$$S_i = \sum_{i=1}^{N} X - X_i \tag{1}$$

2. Alignment shows how a search agent's speed is matched to that of other search agents in the area. As indicated in Eq. (2), the alignment behavior can be mathematically modeled.

$$A_i = \frac{\sum_{j=1}^{N} V_j}{N} \tag{2}$$

where V_j represent the speed of the jth neighbour.

3. First, cohesiveness shows how people swoosh from their local region to the hub. It describes how people prefer to swoop down into the nearby mass center. Equation (3) presents the mathematical model of the Cohesion behavior.

$$C_i = \frac{\sum_{j=1}^{N} x_j}{N} - X \tag{3}$$

4. The term "attraction" describes the way in which a food supply draws in birds. In Eq.4 and Eq.5, we can see the mathematical representation of this behavior.

$$F_i = F_{loc} - X \tag{4}$$

$$F_i = F_{loc} + X \tag{5}$$

5. The human propensity to flee from an aggressor is an example of distraction. In Eq. (6), we can model the adversary and the i-th solution quantitatively.

$$f_3 = Minimise \sum_{i=1}^{N} |(V_n - V_i)| \tag{6}$$

5.2 WOA Algorithm in DN

Introduced in 2016, WOA is a new algorithm that relies on demographic data (Mirjalili & Lewis), 2016. The social behavior of humpback whales is mimicked by this program. At each iteration, WOA uses three principles to improve and refine the positions of these solutions: surrounding prey, spiral updating of positions, and hunting for prey. This is similar to other population-based algorithms. The following are their descriptions:

1. Generate the intial population X_i (i= 1,2,………NP)
2. The fitness of each potential solution in Xi should be evaluated.
3. X= the best candidate solutions
4. While the halting criterion is not staisified do
5. for i=1 to NP do
6. Update a,A,C,l and P;
7. For j=1 to NP do
8. If P< 0.5 then
9. If (|A|<1)
10. D = |C.X - Xi|
11. Xi(j) = X*(j) -A.D
12. Else if (|A|≥1)
13. Select a random individual Xrand
14. D = |C.X$_{rand}$ - X$_i$|

5.3 Load Flow Analysis

Figure 3.1 illustrates a single line diagram (SLD) of a basic RDS, where i represents the transmitting end node and j denotes the receiving end node. Yi and Yj correspond to the shunt conductance of nodes i and j, respectively, indicating leakage current between conductance's. Shunt conductance is regarded as negligible in calculations due to the minimal capacitance of the distribution system [11,13]. The multi-objective function is examined by load flow analysis of RDS, utilizing the backward-forward sweep approach [12]. This approach is utilized due to its little memory need, simplicity, robustness, enhanced convergence rate, and efficiency

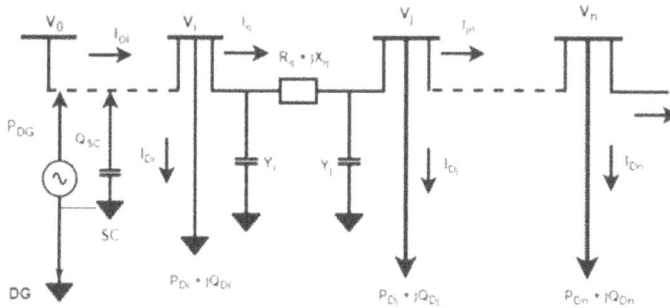

Fig. 3.1 Single line diagram of RDS [12]

6. Results and Discussion

The proposed method is assessed on IEEE 69 bus systems to ascertain its effectiveness and validity. The primary objective function encompasses the reduction of active/reactive power loss, enhancement of voltage profile, annual cost savings, and total operating cost reduction. The DA & WOA assesses the efficacy of the RDS. The suggested DA was evaluated against WOA optimization methods, while considering various load models such as CP, IND, RES, COM, CI, and CZ. Electrical loads categorized as CP are those that, despite fluctuations in voltage and current, sustain a constant power consumption. It is present in various power electronics devices that possess a designated power rating for operation. Due to their consistent power, it aids in maintaining the reliability of the PS. The entire electricity consumption of an average household is referred to as a RES. The RES load models will exhibit varying daily consumptions, including lighting, heating, and television usage, among others. IND loads refer to the electrical power consumption of heavy machinery and equipment. Their power necessitates diverse and sophisticated control and protection systems to prevent equipment damage. The COM load models pertain to loads utilized in commercial settings, including restaurants, retail establishments, and non-industrial applications. The CI load models are characterized by continuous current supply regardless of voltage fluctuations, while loads with constant impedance, unaffected by variations in voltage and frequency, are referred to as CZ loads.

7. Simulation of IEEE 69 Test System

The effectiveness of the DA & WOA is implemented and validated using the IEEE 69 test system. The single-line diagram of the 69-bus system, including the locations of the DGs and SCs, is illustrated in Fig. 3.1. Various metrics, including active and reactive power loss, are examined for three distinct load types. In reference A. Dogan et al (2019), we find the 69-test system's line and bus statistics. VSI is calculated and weak buses are identified based on the simulation result. DGs & SCs are placed on the weak buses. The best locations are 11,17 & 61 and sizes of DGs at these locations are found using DA & WOA algorithms. The results obtained for different cases are analyzed to evaluate the power losses and voltage profile and also the best optimization technique out of the two are mentioned in this paper. The results are tabulated for different cases.

7.1 Base Case

This scenario does not take into account the integration of DGs and SCs. In the most basic scenario, the system voltage is 13.8 kV, and the active and reactive power demands for CP, IND, RES, COM,

CI, and CZ are 3801.49 kW and 2694.60 kVAr, 3767.41 kW and 2046.78 kVAr, 3632.30 kW and 2226.22 kVAr, 3530.10 kW and 2290.92 kVAr, 3618.16 kW and 2564.43 kVAr, and 3448.76 kW and 2444.18 kVAr, respectively. The values given for the VSI_{min} (p.u) and V_{min} (p.u) are 0.6822 and 0.9172, respectively. The simulated results are shown in Table 3.2.

7.2 Case 1: Integration of Single DG and Single SC

Utilizing VSI, buses 11, 17, and 61 are designated as the ideal sites for the installation of DGs and SCs within the IEEE 69 bus test system. To validate and confirm the superiority of the suggested technology, a single DG and SC are positioned at bus site 61. Applying DA technique, we were able to ascertain the DG and SC dimensions. The simulated outcomes are displayed in Table 3.2. The minimum observed values for the COM load model are 19.93 kW and 12.978 kVAr, whereas the maximum values for the CP load model are 23.17 kW and 14.418 kVAr. The enhanced VSI_{min} (p.u) and the observed V_{min} (p.u) are recorded as 0.8942 and 0.9724, respectively. The suggested approaches significantly improved the RDS system's voltage profile and stability by reducing PLoss compared to other methodology.

7.3 Case 2: Integration of Two DGs and Two SCs

In this case two DGs and two SCs are included into the renewable distribution system (RDS) at bus positions 11 and 61, respectively. The dimensions of the DGs and SCs were determined using DA methodology. DGs introduce real power into the system, whereas SCs provide reactive power. Table 3.2 presents the outputs of the minimum voltage profiles and VSI values for all load models. The minimal values of P_{Loss} are 8.71 kW and Q_{Loss} 8.047 kVAr for COM and maximum values for CP load are 10.11kW and 8.68 KVAr. The enhanced VSI_{min} (p.u) and the observed V_{min} (p.u) are recorded as 0.9429 and 0.9854, respectively the study result clearly indicates that the incorporation of two DGs together with two SCs diminishes P_{Loss} and enhances the voltage profile and VSI values. Furthermore, Case-2 markedly surpasses Case-1result for P_{Loss} reduction, voltage profile, and VSI.

7.4 Case 3: Integration of Three DGs and Three SCs

In this instance, three DGs and three SCs are placed at positions 11, 17, and 61. These locations are optimal locations determined by VSI. By applying DA methods, the DG and SC dimensions were obtained. After three DGs generators and three SCs were installed into the RDS, the system's performance was much improved. The results also show that the system's performance is improved when DGs are placed at multiple locations. The minimum values are 3.83 kW and 6.43 kVAr for COM, while the maximum values are 4.29 kW and 6.76 kVAr for the CP load model. The enhancements are 0.9588 pu and 0.9943 pu for VSImin and Vmin respectively. Table 3.1 illustrates the simulation outcomes for multiple scenarios. It is clear that P_{Loss} without DGs is 202.671 kW, and the minimum voltage is 0.9131p.u. After locating the DGs at optimal place, the P_{Loss} are reduced, VSI increased in all cases. The maximum P_{Loss} reduction and best voltage profile improvement is observed in case3 of type3 power reduction and minimum voltage of 0.9677 p.u. is obtained in the two DGs allocations for the type 1 scenario. There is a drastic loss reduction of 64.74% for three DGs allocations with a minimal voltage of 0.9679 p.u. is obtained. i.e. placement of three DGs. The result shows a 41.95 % reduction in P_{Loss} and a minimum voltage of 0.9362 p.u. in the single DG allocation for the type 1 case 57.54%.

Table 3.2 Results of the IEEE 69 bus system using the DA algorithm

Algorithm/ Cases	Item/Load	DG size (kW)	SC size (kVAr)	P_{loss}(kW)	Q_{loss} (kVAr)	V_{min} (p.u)	VSI_{min} (p.u)
Base Case	CP			225.00	102.1983	0.9090	0.6822
	IND			171.40	79.1125	0.9196	0.7136
	RES			164.87	76.3614	0.9217	0.7211
	COM			156.92	72.956	0.9242	0.7289
	CI			188.67	86.6196	0.9172	0.7073
	CZ			158.79	73.773	0.9246	0.7304
Single DGs & Single SCs	CP	1930.69	680.42	23.17	14.4189	0.9724	0.8942
	IND	1830.34	444.94	20.41	13.1759	0.9738	0.8993
	RES	1689.00	896.57	20.29	13.1323	0.9741	0.9003
	COM	1614.07	801.09	19.93	12.978	0.9745	0.9017
	CI	1763.70	649.20	21.54	13.6897	0.9735	0.8983
	CZ	1584.13	723.42	20.04	13.0347	0.9746	0.9021
Two DGs & Two SCs	CP	2659.268	1087.387	10.11	8.6802	0.9854	0.9429
	IND	2528.27	1121.074	8.95	8.1397	0.9861	0.9457
	RES	2524.446	533.0946	8.87	8.1227	0.9862	0.9461
	COM	2473.32	504.0946	8.71	8.0476	0.9865	0.9471
	CI	2383.368	1681.224	9.39	8.3614	0.986	0.9452
	CZ	2563.485	861.7355	8.75	8.0694	0.9886	0.9476
	CP	2551.423	1481.609	4.29	6.764	0.9944	0.9593
	IND	2519.205	1178.059	3.89	6.4746	0.9944	0.9592
Three DGs & Three SCs	RES	2389.744	1016.441	3.87	6.4735	0.9944	0.9593
	COM	2548.026	321.3011	3.83	6.4326	0.9944	0.9594
	CI	2747.779	250.2844	4.08	6.616	0.9943	0.9590
	CZ	2400.976	621.6403	3.85	6.4536	0.9943	0.9588

Source: Author

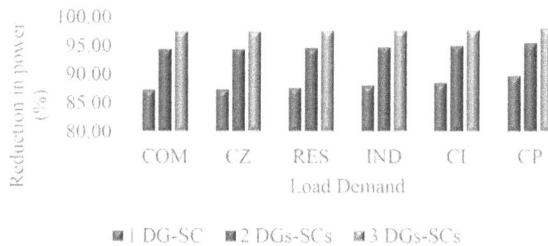

Fig. 3.2 Reduction in power loss by considering different loads

Source: Author

Table 3.3 Results of IEEE 69 bus system simulation results under various load conditions using WOA algorithm

Algorithm	Cases	Load Type	DG size (kW)	SC size (kVar)	P_{Loss} (kW)	Q_{Loss} (kVar)	V_{min} (P.U)	VSI_{min} (P.U)
WOA	Base Case	COM			156.92	72.96	0.9242	0.7287
		CZ			158.79	73.77	0.9246	0.7304
		RES			164.87	76.36	0.9217	0.7199
		IND			171.40	79.11	0.9196	0.7112
		CI			188.67	86.62	0.9172	0.7072
		CP			225.00	102.20	0.9090	0.6821
WOA	Single DGs & Single SCs	COM	1811.88	130.17	19.936	12.97	0.9745	0.9018
		CZ	1722.61	324.17	20.044	13.07	0.9745	0.9017
		RES	1791.77	324.17	20.293	13.12	0.9742	0.9006
		IND	1841.31	388.60	20.412	13.13	0.9741	0.9003
		CI	1679.58	1075.60	21.538	13.66	0.9737	0.8989
		CP	2098.77	269.74	23.169	14.41	0.9725	0.8943
WOA	Two DGs & Two SCs	COM	2241.11	1220.83	8.94	8.16	0.9858	0.9445
		CZ	2336.09	878.94	9.02	8.27	0.9861	0.9456
		RES	2506.96	1028.25	9.10	8.18	0.9880	0.9527
		IND	2676.45	302.84	9.21	8.23	0.9862	0.9460
		CI	2726.17	422.40	9.43	8.36	0.9867	0.9479
		CP	3108.58	28.02	10.13	8.70	0.9853	0.9425
WOA	Three DGs & Three SCs	COM	2345.29	737.20	4.68	6.74	0.9943	0.9545
		CZ	2403.19	532.04	4.80	6.81	0.9943	0.9544
		RES	2363.44	1146.43	4.92	6.78	0.9944	0.9545
		IND	2456.07	452.82	5.28	6.95	0.9944	0.9546
		CI	2190.50	1221.27	5.44	7.24	0.9943	0.9590
		CP	2716.50	733.80	5.57	7.24	0.9943	0.9588

Source: Author

8. Conclusion

Worldwide, there was a dramatic spike in the demand for electrical energy. Further, proper distribution of DGs in DS fulfill the demand and generation gap and minimizes the losses and enhancement of voltage profile. In this paper, an effective methodology such as VSI and different optimization technique is proposed to address this issue. Various scenarios are considered while testing the created approach on the IEEE 69 bus test system. Various cases such as only DG, 1DG & 1SC and multiple DGs and multiple SCs are considered and simulated results are projected. Type3 shows the most improvement in voltage stability and voltage profile, as well as the biggest reduction in P_{Loss}, according to the results. Type3 reached a minimum VSI of 0.9640 and a minimum voltage of 0.9921 per unit, resulting in a maximum loss reduction of 94.25%. DA algorithm is used to compare the efficiency of optimizing P_{Loss} and voltage profile improvement with WOA algorithm. According to the simulated results, the DA algorithm show the superiority compared to WOA method with respect to amount of power loss reduction and improved in the voltage profile.

REFERENCES

1. Ackermann, T., Andersson, G., & Söder, L. (2001). Distributed generation: a definition. Electric power systems research, 57(3), 195–204.
2. Ghosh, S., Ghoshal, S. P., & Ghosh, S. (2010). Optimal sizing and placement of distributed generation in a network system. International Journal of Electrical Power & Energy Systems, 32(8), 849–856.
3. Singh, B., Mukherjee, V., & Tiwari, P. (2015). A survey on impact assessment of DG and FACTS controllers in power systems. Renewable and Sustainable Energy Reviews, 42, 846–882.
4. Farh, H. M., Al-Shaalan, A. M., Eltamaly, A. M., & Al-Shamma'A, A. A. (2020). A novel crow search algorithm auto-drive PSO for optimal allocation and sizing of renewable distributed generation. IEEE Access, 8, 27807–27820.
5. Raut, U., & Mishra, S. (2021). A new Pareto multi-objective sine cosine algorithm for performance enhancement of radial distribution network by optimal allocation of distributed generators. Evolutionary Intelligence, 14(4), 1635–1656.
6. Abdelsalam, A. A., & Mansour, H. S. (2019). Optimal allocation and hourly scheduling of capacitor banks using sine cosine algorithm for maximizing technical and economic benefits. Electric Power Components and Systems, 47(11–12), 1025–1039.
7. Abdelsalam, A.A., 2020. Optimal distributed energy resources allocation for enriching reliability and economic benefits using Sine-cosine algorithm. Technol. Econ. Smart Grids Sustain. Energy 5 (1), 1–18
8. Ramadan, A., Ebeed, M., & Kamel, S. (2019, February). Performance assessment of a realistic egyptian distribution network including PV penetration with DSTATCOM. In 2019 International Conference on Innovative Trends in Computer Engineering (ITCE) (pp. 426–431). IEEE.
9. Ang, S., Leeton, U., Chayakulkeeree, K., & Kulworawanichpong, T. (2018). Sine cosine algorithm for optimal placement and sizing of distributed generation in radial distribution network. GMSARN International Journal, 12(4), 202–212.
10. Satyanarayana, C., Rao, K. N., Bush, R. G., Sujatha, M. S., Roja, V., & Nageswara Prasad, T. (2019). Multiple DG placement and sizing in radial distribution system using genetic algorithm and particle swarm optimization. Computational Intelligence and Big Data Analytics: Applications in Bioinformatics, 21–36.
11. Dogan, A., & Alci, M. (2019). Simultaneous optimization of network reconfiguration and DG installation using heuristic algorithms. Elektronika Ir Elektrotechnika, 25(1), 8–13.
12. Bompard, E., Carpaneto, E., Chicco, G., & Napoli, R. (2000). Convergence of the backward/forward sweep method for the load-flow analysis of radial distribution systems. International journal of electrical power & energy systems, 22(7), 521–530.
13. Dogan and M. Alci, "Simultaneous optimization of network reconfiguration and DG installation using heuristic algorithms," Elektron. Elektrotech., vol. 25, no. 1, pp. 8–13, 2019. DOI: 10.5755/j01.eie.25.1.22729.

Intelligent and Sustainable Power and Energy Systems – Dr. M. Premkumar et al. (eds)
© 2026 Taylor & Francis Group, London, ISBN 978-1-041-10314-1

4

Power Quality Issues and Improvement Techniques

Bhumika Shrimali[1],
Vikramaditya Dave[2], Megha Sen[3]
and Kamlesh Jat[4]
Department of Electrical Engineering,
College of Technology and Engineering, MPUAT,
Udaipur, India

ABSTRACT: In the rapidly evolving landscape of modern power systems, maintaining high power quality has become paramount due to the increasing reliance on sensitive electronic equipment and the integration of renewable energy sources. This paper explores the critical issues associated with power quality, including voltage sags, swells, harmonic distortion, flickers, and transients, while providing a comprehensive analysis of various mitigation techniques. Through case studies and simulations, the effectiveness of traditional methods, such as passive and active filters, voltage regulators, and advanced solutions like Custom Power Devices (CPDs) and Flexible AC Transmission Systems (FACTS) is examined. The research highlights the significant impact of these technologies on improving power quality in both industrial and renewable energy systems. Additionally, the dissertation discusses the challenges in power quality management, particularly in large-scale and renewable energy applications, while exploring future trends such as next-generation custom power devices. The findings underscore the necessity for effective power quality improvement strategies to enhance system reliability and operational efficiency.

KEYWORDS: Active filters, Custom power devices, FACTS, Harmonic distortion, Power quality, Power quality mitigation, Renewable energy, Voltage sags

1. Introduction

Power quality is an electrical power source's ability to generate a steady, distortion-free voltage waveform. As solar and wind energy become more integrated into electrical infrastructure, power

[1]bhumikashrimali1997@gmail.com, [2]vdaditya1000@gmail.com, [3]meghasen257@gmail.com, [4]kamleshjat386@gmail.com

DOI: 10.1201/9781003654469-4

quality is more crucial than ever [1]. Our increased dependence on fragile electrical and technological gadgets makes high-quality electricity more critical [2]. Voltage swells and sags—temporary voltage decreases and rises—are the most common. Events caused by sudden load shifts, switching, or short circuits can damage sensitive electronics and industrial operations [3]. Harmonics can cause overheating, device failure, and system inefficiency. Voltage variations can disrupt manufacturing. Small deviations from nominal voltage can cause equipment failure, output losses, and greater maintenance costs [4].

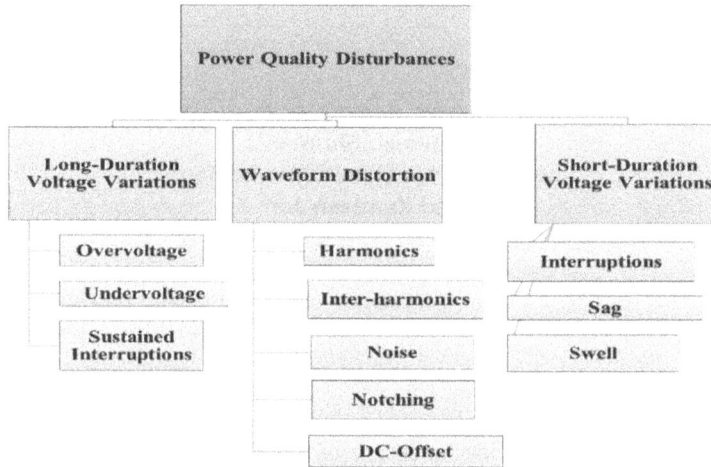

Fig. 4.1 Typical power quality disturbances [5]

Power quality issues include flicker, or rapid voltage fluctuations, and transients, or brief disturbances caused by switching activities or external sources like lightning strikes [6]. Power quality is getting harder to maintain in modern power networks, especially those using distributed generation (DG) and renewable. This paper aims to identify and analyse the biggest power quality issues in modern power systems, with a focus on renewable energy sources. Our main objective is to evaluate power quality mitigation solutions. To do this, we must decide which filters, custom power devices (CPDs), and flexible AC transmission systems (FACTS) are best for transients, harmonic distortion, and voltage sags. The Distribution Static Compensator (D-STATCOM), Dynamic Voltage Restorer (DVR), and Unified Power Quality Conditioner (UPQC) have shown promise in improving power quality [7]. This paper explains power quality enhancement research by examining power quality issues in present power systems and evaluating creative solutions [8].

2. Power Quality Issues

When discussing electrical systems, "power quality" describes how well the power supply maintains clean, consistent voltage and current. Pure sinusoidal power waves have constant voltage and frequency [9]. The severity and duration of voltage variations affect electrical equipment. Overvoltage and sags can damage or disable equipment. Voltage swell is a brief surge in voltage above 110% of its nominal level [10].

The equation that defines harmonic frequencies is as follows:

$$f_h = n \times f_{fund} \tag{1}$$

Table 4.1 The key characteristics of voltage variations

Voltage Variation	Definition	Duration	Causes	Impact
Voltage Sag	Reduction in voltage below 90%	0.5 cycles to 1 minute	Motor startups, faults, short circuits	Equipment tripping, malfunction
Voltage Swell	Increase in voltage above 110%	0.5 cycles to 1 minute	Load reduction, switching events	Equipment damage, reduced lifespan
Long-duration Variation	Sustained deviation from nominal voltage	>1 minute	Load demand changes, equipment failures	Overheating, energy inefficiency

Source: Author

Where: f_h is the harmonic frequency, n is the harmonic order, f_{fund} is the fundamental frequency (e.g., 50 Hz or 60 Hz) [11].

When voltage changes quickly, flicker occurs, causing light brightness to vary [12]. These systems cause voltage swings and harmonic distortion, especially when connected to the grid via inverters [13]. The power quality parameters (THD, voltage sag, swell, flicker etc.) their acceptable limits, and causes are shown below.

Table 4.2 The key power quality parameters, their acceptable limits, and typical causes

Parameter	Acceptable Limit	Typical Causes
THD (Voltage)	≤ 5%	Non-linear loads, inverters, rectifiers
Voltage Sag	90% – 100% of nominal	Motor startups, short circuits
Voltage Swell	110% – 120% of nominal	Load shedding, sudden disconnection of large loads
Flicker (Short-term)	≤ 1.0 Pst	Voltage fluctuations due to large load changes
Frequency Deviation	±0.1 Hz	Load/generation imbalance, grid disturbances
Voltage Unbalance	≤ 2%	Uneven loading in three-phase systems

Source: Author

3. Power Quality Improvement Techniques

Due to the widespread usage of sensitive electronic devices power systems require power quality for efficient and reliable operation. Traditional methods for improving power quality include filters, voltage regulators, and other devices that stabilise voltage [14]. Complex power systems that use renewable energy sources require advanced mitigation methods like Custom Power Devices (CPDs) and Flexible AC Transmission Systems solution [15]. Traditional procedures are helpful, but not enough. The Dynamic Voltage Restorer (DVR), Unified Power Quality Conditioner (UPQC), and Distribution Static Compensator D-STATCOM improve system stability and dependability through real-time disturbance adjustment [16]. The Distribution Static Compensator (D-STATCOM) tackles reactive power imbalances and harmonic distortion in distribution networks [17]. The D-STATCOM injects compensating currents to reduce harmonic currents and improve power factor. Power quality issues like harmonic distortion can cause systems to malfunction, equipment to overheat, and efficiency to drop. Most harmonic distortion reduction filters are active or passive [18].

A continuous power supply requires voltage management, especially in dynamic load systems [19]. Flexible AC Transmission Systems (FACTS) demonstrate how current technology has enhanced power transmission system efficiency and reliability. The static synchronous compensator (STATCOM) and static VAR compensator (SVC) are FACTS devices that improve power quality, especially in renewable energy systems [20].

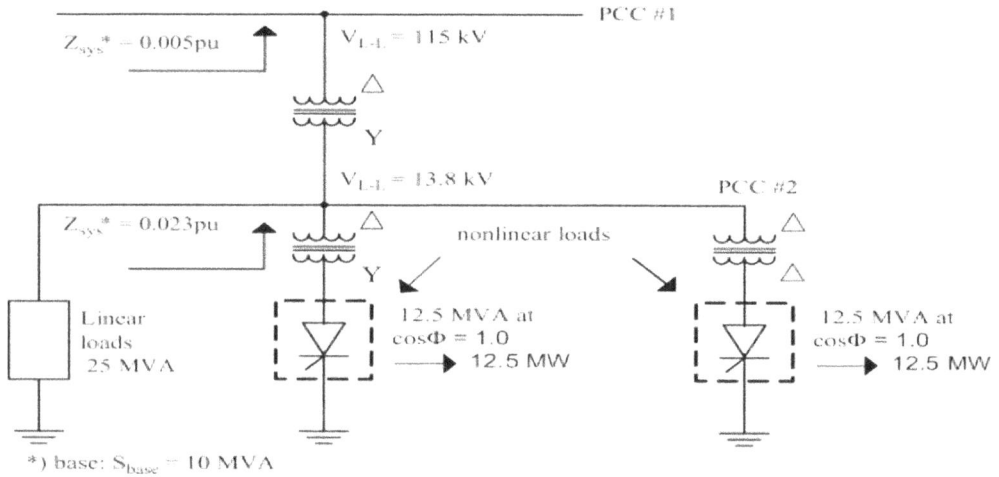

Fig. 4.2 A schematic representation of an active harmonic filter and its interaction with the power grid to cancel out harmonic currents [19]

3.1 Case Studies and Simulation Analysis

Case Study 1: Industrial Power System

This case study examines an industrial power system with numerous power electronic devices and non-linear variable frequency drives (VFDs). During the initial assessment, the current's total harmonic distortion (THD) exceeded IEEE 519 limitations, causing transformer overheating, conductor losses, and possibly equipment failure [21]. To mitigate these harmonics, the power system had active filters and a D-STATCOM. By adding compensatory currents, the D-STATCOM can reduce harmonic currents from non-linear loads. Active filters targeted certain harmonic frequencies, lowering THD.

Simulation Results: The simulation results showed a significant reduction in harmonic distortion after the installation of the D-STATCOM and active filters. The THD of the current was reduced from over 12% to below 5%, which is well within the IEEE limits. The overall system performance improved markedly, with reduced heating in transformers and better operational efficiency for the equipment.

Case Study 2: Renewable Energy System

This power quality case study at a wind farm addresses voltage stability and harmonic distortion mitigation issues caused by inverter technology and wind generation unpredictability Kumar, C. & Mishra, M.K. (2014). A Static Synchronous Compensator (STATCOM) and other Custom Power Devices in Flexible AC Transmission Systems (FACTS) eased these issues. STATCOM supported dynamic reactive power to stabilise voltage during quick wind variation. Utilising power devices like active filters reduced inverters' harmonic distortion.

Simulation Results: The simulations indicated a notable improvement in voltage stability and a reduction in harmonic distortion following the installation of the STATCOM and active filters. Voltage levels were maintained within acceptable limits, even during high variability conditions, and the harmonic distortion was reduced from 8% to less than 3%. The combined approach of using

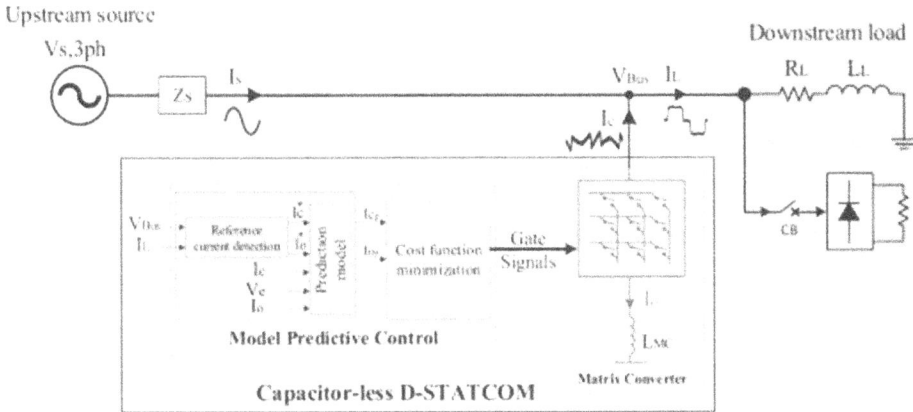

Fig. 4.3 D-STATCOM in an industrial power system [22]

FACTS and custom power devices proved effective in enhancing the reliability and performance of the wind farm.

3.2 Comparative Analysis

The comparative analysis of the two case studies reveals valuable insights into the effectiveness of different power quality improvement techniques. In both industrial and renewable energy systems, the implementation of D-STATCOMs, active filters, and FACTS devices demonstrated significant improvements in mitigating power quality issues.

Table 4.3 Comparative analysis of the two case studies

Technique	Case Study 1: Industrial	Case Study 2: Renewable	Cost-Effectiveness	Implementation Complexity
D-STATCOM	Reduced THD from >12% to <5%	Stabilized voltage	Moderate initial cost	Moderate
Active Filters	Targeted specific harmonics	Reduced harmonics to <3%	High initial cost	Moderate to complex
FACTS (STATCOM)	N/A	Enhanced voltage stability	Moderate initial cost	Moderate

Source: Author

D-STATCOM and FACTS devices increased system dependability and reduced power quality maintenance costs in all case studies, making them cost-effective. Active filters cost more upfront but paid for themselves in longer equipment lifespans and less downtime. The difficulty of integrating these technologies into existing systems varied.

4. Discussion

Increasing energy consumption and the inclusion of renewable sources make power quality disturbances more critical than ever to maintain electrical system reliability and efficiency. This research examined the effectiveness of different power quality solutions. Results show standard and novel strategies to eliminate disruptions. Filters and voltage regulators are traditional methods, while CPDs and flexible AC transmission systems are advanced. Filters, active or passive, are known to reduce harmonic distortion. Passive filters are cheap and quick to set up, but they can't adjust to harmonic levels and can create resonance. Active filters detect and modify harmonic distortions in real time, making them more sensitive. Custom Power Devices (CPDs) like Distribution Static Compensators (D-STATCOM) and Dynamic Voltage Restorers can solve specific power quality issues, notably in industrial and renewable energy systems. The DVR injects corrective voltages to maintain supply levels, saving sensitive equipment from power outages caused by voltage dips and spikes. D-STATCOMs reduce total harmonic distortion (THD) and increase electrical system operating reliability in non-linear loads. FACTS devices like STATCOMs and SVCs regulate voltage and reduce reactive power fluctuations in renewable energy systems. FACTS devices stabilise voltage levels by providing dynamic reactive power support when wind and solar generation fluctuates rapidly. Through case studies, this research shows that FACTS devices increase power quality and system reliability. These mitigation measures work, but they require careful planning, design, and connection with existing systems. The best answer for any power system is usually a mix of ideas with pros and cons.

5. Conclusion

This study examined power quality concerns, their effects on electrical systems, and effective approaches to reducing them. Research shows that transients, harmonic distortion, voltage sags, and other power quality disruptions can affect renewable energy and industrial power system reliability and efficiency. The results demonstrate the importance of power quality management plans to reduce downtime and boost efficiency. Case studies showed how active filters, D-STATCOMs, and FACTS devices may improve power quality. In the industrial power system case study, D-STATCOM and active filters reduced harmonic distortion, improving efficiency and protecting sensitive equipment. In a renewable energy case study, STATCOMs stabilised voltage levels and reduced inverter harmonic output, illustrating the importance of cutting-edge technology in renewable energy. Analysis has helped us understand how different mitigation techniques work in different conditions. FACTS systems and specialist power devices are improving power quality, according to this study. This study's findings suggested many research options. First, next-gen tailored power gadgets need more design and optimising study. These devices must be more efficient, adaptable, and inexpensive to solve additional power quality issues. Additionally, renewable energy source integration into power systems has significant challenges that require ongoing research. Future research should focus on hybrid solutions that use many technologies to improve power quality in renewable energy-heavy systems. Developing real-time electricity quality monitoring methods is also vital. Given the worldwide nature of energy markets and the rising interconnectivity of power systems, a single power quality management strategy could improve solutions and reliability across regions.

REFERENCES

1. Afonso, J. L., Tanta, M., Pinto, J. G. O., Monteiro, L. F., Machado, L., Sousa, T. J., & Monteiro, V. (2021). A review on power electronics technologies for power quality improvement. Energies, 14(24), 8585.
2. Abas, N., Dilshad, S., Khalid, A., Saleem, M. S., & Khan, N. (2020). Power quality improvement using dynamic voltage restorer. IEEE access, 8, 164325–164339.
3. Rao, S. N. V. B., Kumar, Y. V. P., Pradeep, D. J., Reddy, C. P., Flah, A., Kraiem, H., & Al-Asad, J. F. (2022). Power quality improvement in renewable-energy-based microgrid clusters using fuzzy space vector PWM controlled inverter. Sustainability, 14(8), 4663.

4. Mohammed, A. B., Ariff, M. A. M., & Ramli, S. N. (2020). Power quality improvement using dynamic voltage restorer in electrical distribution system: an overview. Indonesian Journal of Electrical Engineering and Computer Science (IJEECS), 17(1), 86–93.

5. Bagdadee, A. H., Khan, M. Y. A., Ding, H., Cao, J., Zhang, L., & Ding, Y. (2020). Implement industrial super-dynamic voltage recovery equipment for power quality improvement in the industrial sector. Energy Reports, 6, 1167–1175.

6. Choudhury, S., & Sahoo, G. K. (2024). A critical analysis of different power quality improvement techniques in microgrid. e-Prime-Advances in Electrical Engineering, Electronics and Energy, 100520.

7. Gade, S., Agrawal, R., & Munje, R. (2021). Recent trends in power quality improvement: Review of the unified power quality conditioner. ECTI Trans. Electr. Eng. Electron. Commun, 19(3), 268–288.

8. Goud, B. S., Reddy, C. R., Bajaj, M., Elattar, E. E., & Kamel, S. (2021). Power quality improvement using distributed power flow controller with BWO-based FOPID controller. Sustainability, 13(20), 11194.

9. Siddula, S., Achyut, T. V., Vigneshwar, K., & Teja, M. U. (2020, October). Improvement of power quality using fuzzy based unified power flow controller. In 2020 International conference on smart technologies in computing, electrical and electronics (ICSTCEE) (pp. 55–60). IEEE.

10. Sahoo, B., Alhaider, M. M., & Rout, P. K. (2023). Power quality and stability improvement of microgrid through shunt active filter control application: An overview. Renewable Energy Focus, 44, 139–173.

11. Prabhu, M. H., &Sundararaju, K. (2020). Power quality improvement of solar power plants in grid connected system using novel Resilient Direct Unbalanced Control (RDUC) technique. Microprocessors and Microsystems, 75, 103016.

12. Awaar, V. K., Jugge, P., Kalyani, S. T., & Eskandari, M. (2023). Dynamic voltage restorer–a custom power device for power quality improvement in electrical distribution systems. In Power Quality: Infrastructures and Control (pp. 97–116). Singapore: Springer Nature Singapore.

13. Sharma, A., Indliya, J. N., Swami, R. K., and Gupta, S. (2023) Generation Forecasting of Solar PV in Distribution Network using ANN. 2023 2nd International Conference on Automation, Computing and Renewable Systems (ICACRS), IEEE, Dec. 2023, 156–161.

14. Singh, B., Chandra, A., & Al-Haddad, K. (2015). Power quality: problems and mitigation techniques. John Wiley & Sons.

15. Shah, S. K., Hellany, A., Nagrial, M., & Rizk, J. (2014). Power quality improvement factors: An overview. 2014 11th Annual High Capacity Optical Networks and Emerging/Enabling Technologies (Photonics for Energy), 138–144.

16. Teke, A., Saribulut, L., & Tumay, M. (2011). A novel reference signal generation method for power-quality improvement of unified power-quality conditioner. IEEE Transactions on power delivery, 26(4), 2205–2214.

17. Gupta, N., Swarnkar, A., & Niazi, K. R. (2014). Distribution network reconfiguration for power quality and reliability improvement using Genetic Algorithms. International Journal of Electrical Power & Energy Systems, 54, 664–671.

18. Liang, X. (2016). Emerging power quality challenges due to integration of renewable energy sources. IEEE Transactions on Industry Applications, 53(2), 855–866.

19. Mahela, O. P., & Shaik, A. G. (2016). Topological aspects of power quality improvement techniques: A comprehensive overview. Renewable and Sustainable Energy Reviews, 58, 1129–1142.

20. Urquizo, J., Singh, P., Kondrath, N., Hidalgo-León, R., & Soriano, G. (2017, October). Using D-FACTS in microgrids for power quality improvement: A review. In 2017 ieee second ecuador technical chapters meeting (etcm) (pp. 1–6). IEEE.

21. Kar, B., & Halder, B. (2016, December). Comparative analysis of a Hybrid active power filter for power quality improvement using different compensation techniques. In 2016 International Conference on Recent Advances and Innovations in Engineering (ICRAIE) (pp. 1–6). IEEE.

22. Kumar, C., & Mishra, M. K. (2014). A voltage-controlled DSTATCOM for power-quality improvement. IEEE transactions on power delivery, 29(3), 1499–1507.

Intelligent and Sustainable Power and Energy Systems – Dr. M. Premkumar et al. (eds)
© 2026 Taylor & Francis Group, London, ISBN 978-1-041-10314-1

5

Advanced Control Strategies for Distributed Generation in Microgrids: Enhancing Efficiency, Reliability and Renewable Energy Integration

Birudala Venkatesh Reddy[1]
SV College of Engg., Tirupati, India

Mamidala Vijay Karthik[2]
EEE Department, CMREC, Hyderabad, India

Budidha Nagaraju[3]
EEE Department, Vagdevi College of Engineering,
Warangal, India

N. Kalpana[4]
EEE Department, Matrusri Engineering College,
Hyderabad, India

CH. Hariprasad[5]
EEE Department, Bapatla Engineering College,
Bapatla, India

Abbaraju Himabindu[6]
EEE Department, Annamacharya Inst. Tech &
Sci-Boyanapalli, Rajampet, India

G. Madhusudhana Rao[7]
Department of EEE, CMRIT, Hyderabad, India

ABSTRACT: Recent advances in power electronics and control systems have made microgrids key for managing Distributed Energy Resources (DERs). This paper discusses control techniques for distributed generation units in microgrids, focusing on their integration with utility grids, enhancing efficiency, reliability, and power sharing. The review examines hybrid AC/DC microgrids models, focusing on the effectiveness of droop control techniques, particularly the

[1]venkateshreddy.b@svce.edu.in, [2]mvk291085@gmail.com, [3]nagaraju_b@vaagdevi.edu.in, [4]kalpana@matrusri.edu.im, [5]hariprasad777@gmail.com, [6]ahimabindu.eee@gmail.com, [7]gmrgurrala@cmritonline.ac.in

DOI: 10.1201/9781003654469-5

quasi-proportional resonance (PR) controller, compared to traditional PI controllers. Simulations confirm improvements in grid synchronization and reduced transients. The paper also explores the integration of Energy Storage Systems (ESS) to support grid and islanded modes, addressing unbalanced conditions caused by single-phase loads. A novel Multi-Functional Power Converter (MFPC) is proposed to resolve three-phase current and reactive power issues. Additionally, decentralized energy storage control, based on state of charge (SoC), dynamically adjusts droop coefficients. The research proposal highlights the role of smart grid technologies in improving synchronization and energy management, offering key insights for future microgrid development

KEYWORDS: Control techniques, Distributed generation, Droop control, Energy storage systems, Microgrids, Quasi-proportional resonance controller, Renewable energy integration, Voltage and frequency regulation

1. Introduction

Micro networks are crucial for supplying electricity and promoting central energy networks. Renewable energies are evolving to reduce CO_2 emissions, but they face challenges due to diverse energy sources are described by [1]. Microgrid converters must maintain voltage and frequency during island mode service. Advanced strategies for microgrid management, electrical, and power are examined. The study focuses on estimating alternate secondary layer regulation in contactless microgrids, analyzing control systems' benefits and disadvantages, determining network efficiency, and analyzing AC simulations. Grid-forming converters (GFCs) are becoming the foundation for future power systems are clearly explained [2-5].

The proposal presents a comprehensive review of control techniques for microgrid management, focusing on efficiency, reliability, and power-sharing improvements. It explores hybrid AC/DC microgrids and their effectiveness in droop control, comparing quasi-proportional resonance (PR) controllers and traditional PI controllers. The research also addresses Energy Storage Systems (ESS) integration for both grid-connected and islanded modes, addressing unbalanced conditions from single-phase loads [6]. The paper proposes a Multi-Functional Power Converter (MFPC) to address three-phase current imbalances and reactive power problems, and introduces a decentralized SoC-based droop control strategy. The research also highlights the role of smart grid technologies in synchronization and energy management shown by [7-8].

This paper discusses the complexities of DC power systems and compares traditional methods with four approaches to power control, current limitation, and grid-forming regulation. It suggests that connecting small and medium-sized enterprises to low-voltage grids through Micro Sources (MS) can increase productivity and improve resource management [9]. The study examines microgrid operations using two independent inverter control systems and primary energy sources, demonstrating the feasibility of MG insulation production through PV generators and battery storage systems connected to the LV grid. The transition of a microgrid to island mode requires a structured monitoring scheme to ensure the non-linear Voltage Controller sliding mode can improve power management and network convergence in microgrids by reducing DC link voltage power, thereby reducing potential instability and delays in islanding detection by [10-13].

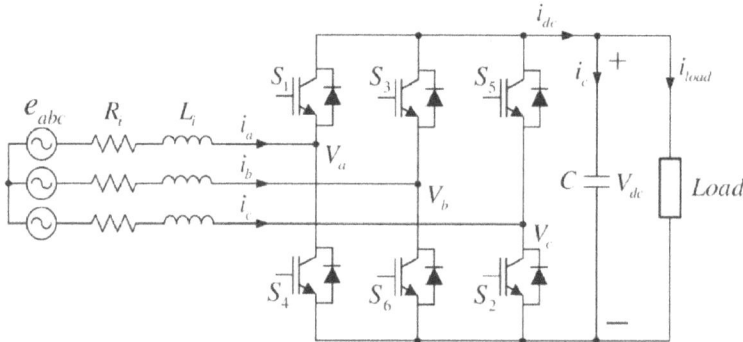

Fig. 5.1 Configuration of a three-phase Ac/Dc converter [14]

The primary objective of this work is to enhance the operational efficiency, reliability, and power-sharing capabilities of hybrid AC/DC microgrids by advancing control techniques for Distributed Generation Units (DGUs) and Energy Storage Systems (ESS) referred in [14]. The study introduces a comparison between traditional PI controllers and Quasi-Proportional Resonance (PR) controllers, showcasing their impact on grid synchronization, transient response, and harmonic reduction. A novel Multi-Functional Power Converter (MFPC) is proposed to address persistent challenges like reactive power issues and three-phase current imbalances. This article suggests controlling drop control on both systems using the DC voltage loop, which maintains DC link voltage's dignity and offers unintentional isolation for smooth mode transitions and stabilizes micro grids. Hierarchical administration is recommended for dynamic transitions between different modes of operation. Microgrids are a potential solution for integrating Distributed Energy Resource units into renewable energy networks. They manage volatility, reduce emissions, and enhance efficiency. Modern microgrid architectures include smaller units, shared energy storage systems, and automated control features. Transformers regulate power flows, creating resilient and sustainable low-to-medium-voltage transmission networks. Advances in ICT enable autonomous reactive voltage control, reducing power exchange are referred [15].

2. Literature Review

The study in [16] explores the article critiques a study on renewable energy integration, highlighting its complexity, limitations, and shortcomings. It criticizes the study for overlooking grid variability, scalability, and cyber security risks, and calls for a more balanced approach to integrating renewable energy technologies. Authors in [17] presents the predictive control in power electronics and drives can be enhanced by integrating local networks with distributed renewable generation, such as microgrids. Research on these systems can reduce outages, controlled energy supply. Key approaches include regulatory frameworks, automatic control systems, and energy storage systems. The paper presents by Katleho Moloi (2020), a fault diagnostic algorithm using wavelet packet transform and support vector machine, achieving 99% fault detection accuracy in the Eskom 90-bus electrical system. However, the high accuracy may not account for varying fault conditions or extreme weather events, and its generalizability may be limited. Future work should focus on testing under diverse operational conditions. The study in [18] explores This article introduces a decentralized adaptive droop-based control for active power sharing in autonomous microgrids, improving transient performance and demonstrating real-time operation using MATLAB and FPGA-based hardware setups. Future research should validate control techniques under diverse operating conditions. Ref. [19] explores the evolution of distribution networks due to distributed energy resources (DERs) and digital transformation at grid edges. It presents a concept of grid-edge control, which combines traditional grid control systems with autonomous DER control. However, it lacks in-depth analysis of practical implications and scalability, and lacks concrete examples or case studies to illustrate its application. The paper highlights the importance of different control methods but lacks detailed examination of challenges, limitations, and the impact of regulatory frameworks and market dynamics on their deployment.

3. Methodology

This research introduces Dynamic Droop Control, improves stability under dynamic conditions, bridges grid-connected and islanded modes, and proposes MFPC Innovation for effective reactive power and unbalanced load conditions. The DC standard for a microgrid consists of variable loads, power generation, and energy storage devices such as photovoltaic (PV) panels. It includes a DC/DC converter, a battery-operated photovoltaic monitor, and sensors for electric vehicles or household usage. As power fluctuates rapidly, control actions are needed to either generate or absorb energy to maintain stable voltage levels.

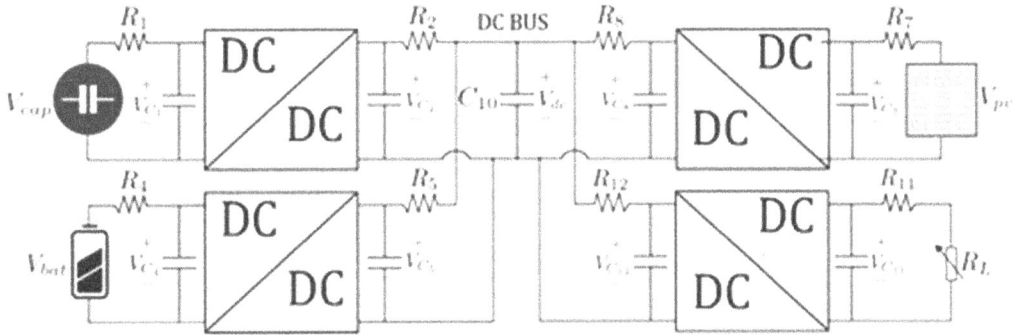

Fig. 5.2 DC micro grid integrating with two energy storage systems operating at different time scales [17]

The microgrid uses a battery, super condenser, PV series, battery, and super capacitor for charging. Local controllers manage devices, while secondary controllers coordinate electricity flow. Economic benefits, local demand, and CO_2 emissions reduction are prioritized. Utilizing the microgrid research approach, super capacitors can contribute to voltage stability within the system.

$$Vc_1 = \frac{1}{R_1 C_1} Vs - \frac{1}{R_1 C_1} Vc_1 - \frac{1}{C_1} IL_3 \tag{1}$$

$$Vc_2 = \frac{1}{R_2 C_2} V_{dc} - \frac{1}{R_2 C_2} Vc_2 + \frac{1}{C_2} IL_3 (1 - u_1) \tag{2}$$

$$IL_3 = \frac{1}{L_3} Vc_1 - \frac{1}{L_3} Vc_2 (1 - u_1) - \frac{R_{01}}{L_3} IL_3 \tag{3}$$

Super capacitors improve energy efficiency by responding quickly to grid oscillations. They have longer life cycles and higher power densities, but typically have lower energy density. They can maintain voltage stability in microgrids. Wind energy generation requires grid management, addressing stability, uncertainty, and marginal costs. Efficiency depends on wind turbine rotational speed. Photovoltaic units rely on solar and thermal radiation, and their performance depends on power production ratio and solar irradiance input. Modular pulses convert energy into electrical energy using AC and DC converters or DC/AC converters is mentioned in [17].

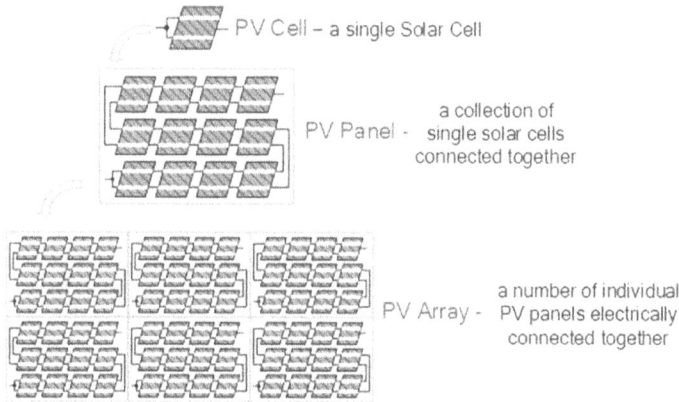

Fig. 5.3 PV cell configuration from solar cell to solar array [22]

This problem develops a robust framework for hybrid AC/DC microgrids, integrating Distributed Generation Units (DGUs), Energy Storage Systems (ESS), and advanced control techniques to enhance grid synchronization, stability, and power sharing. By comparing traditional PI controllers with Quasi-Proportional Resonance (PR) controllers, the research highlights improved transient response and harmonic reduction. A novel Multi-Functional Power Converter (MFPC) addresses reactive power and current imbalances, while a decentralized SoC-based droop control strategy ensures optimal ESS utilization. The paper presents a power regulation scheme for grid-connected distributed energy units using iterative learning control, demonstrating its fast, stable, and high-quality performance through simulations and comparisons will be from [20]. To prevent mechanical damage, the wind turbine automatically shuts down when wind speeds reach 22 m/s and 45 m/s.

$$P = \frac{1}{2}\rho_a C_P A_S V^3 \tag{4}$$

Where – kg / m^3 air density, CP, AS the wind turbine rotor slip area in m^2, and V wind speed in m/S. where – m^3. The rotor AS 'swept zone can be extended:

The CP coefficient of rotor power can be expressed as follows:

$$C_p = \frac{P_{rotor}}{P_{wind}} \tag{5}$$

According to the maximum power that can be extracted from wind can be calculated as:

$$P = \frac{16}{27}\frac{\rho_a}{2}V^3 A_S \tag{6}$$

Rotor diameter, tip-speed ratio, rotor design, and wind intensity all impact power generation in wind turbines. Higher TSR leads to efficiency and reduced noise. Blade pitch control optimizes power coefficient, and mechanical torque is influenced by wind speed. This research study explores the use of DC sources in microgrid topology, specifically PV and battery-based outlets. Inverters regulate voltages using a pi-controller, while MPPT tracks output current. The SC controls battery energy modifications and external SOC controllers. The ESS module controls grid-linked AC current sources, activating DC voltage control loops. Storage devices supply or absorb power when frequency gradient deviates from zero or microgrid operates outside its range. Monitoring techniques address frequency fluctuations. Two approaches: secondary local control and secondary centralized control. Power adjustments target frequency deviations. Employing lower switching frequencies helps minimize errors in these filtering systems. The role of the LC philtre transmission is:

$$G_f(s) = \frac{u_0(s)}{V_{(s)}} = \frac{R_f + 1/(j\omega C_f)}{j\omega L_f + R_f + 1/(j\omega C_f)} \tag{9}$$

$$= \frac{j\omega\varepsilon\omega_0 + \omega_0{}^2}{(j\omega)^2 + j\omega\varepsilon\omega_0 + \omega_0{}^2} \tag{10}$$

Then, the resonant frequency fc can be calculated as follows:

$$f_c = \frac{1}{2\pi\sqrt{L_f C_f}} \tag{11}$$

It satisfies the following constraint condition:

$$10 f_n \le f_c \le f_s/10 \tag{14}$$

4. Results Analysis

Fast Fourier Transform (FFT) analysis is an essential tool for evaluating the performance of control strategies in microgrids, especially when assessing their impact on power quality, synchronization, and transient behavior. Fast Fourier Transform (FFT) analysis is a crucial tool for evaluating control strategies in microgrids, particularly in assessing their impact on power quality, synchronization, and transient behavior. FFT analysis can provide insights into Harmonic Distortion, Transient Analysis during Mode Transitions, Unbalanced Load Analysis, Energy Storage System Dynamics, and Grid Synchronization Performance. By assessing Total Harmonic Distortion (THD), Transient Analysis during Mode Transitions, Unbalanced Load Analysis, Energy Storage System Dynamics, and Grid Synchronization Performance, researchers can identify and quantify the effectiveness of proposed control strategies in improving grid synchronization and power quality. The results of FFT analysis can provide a quantitative basis for implementing these strategies in future microgrid development.

System Performance at Grid-connected Mode

The machine operates in grid mode for 15 seconds, with a ramp mounted on the battery and a 1 MW active energy output. The PV scale is increased from 1000W/m^2 to 2000W/m^2, and the active power output varies with charging ramp. The microgrid faces voltage and frequency drops, and the controller transitions to autonomous mode after 30 milliseconds. The study examines normal and inefficient machinery, controller output, and active power injector optimization. The PV station's performance is analyzed based on changes in solar irradiation, with a plasma array surface temperature set at 25°C. Graph 10 shows active power injection and DC/C converter controller's effectiveness in managing current amplitude. Figure 5.5 illustrates wind farms operating at different

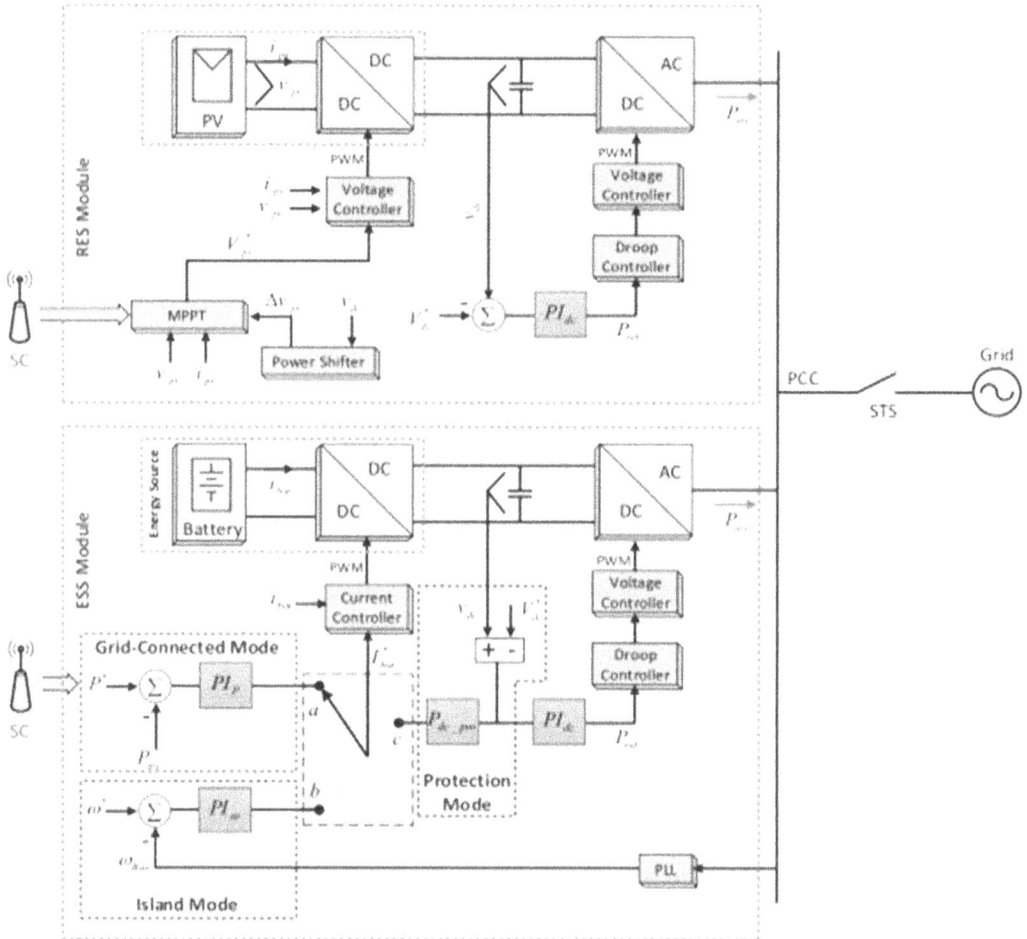

Fig. 5.4 Proposed microgrid control topology [21]

wind speeds, ranging from 10 to 15 meters per second. The active energy produced by the wind farm fluctuates with wind intensity, while the Maximum Power Point Tracking (MPPT) controller regulates the maximum rotational speed of the turbines. The Rotor Side Converter (RSC) controller oversees the current amplitude, and the Grid Side Converter (GSC) controller ensures a stable DC bus voltage of 1150 V.

The wind turbine can hold up to 150ms when voltage falls to 0% of the nominal value, and quick disconnections are enabled if voltage drops to 20%. Various approaches have been proposed for controlling voltage and reactive control, including I- pitch angle power-based, DC-Link condenser size-based solutions, crowbar-based, additional energy storage approach, and kinetic method to rotor inertia-based energy storage. The microgrid controller maintains power balance between load and generation, while batteries sustain the energy balance on the microgrid. Mat lab TM Simulink simulates the H-bridge grid tying inverter with built controllers, dual current control loops, and external PR control improved by harmonic compensation. The estimated efficiency is 94.9%.

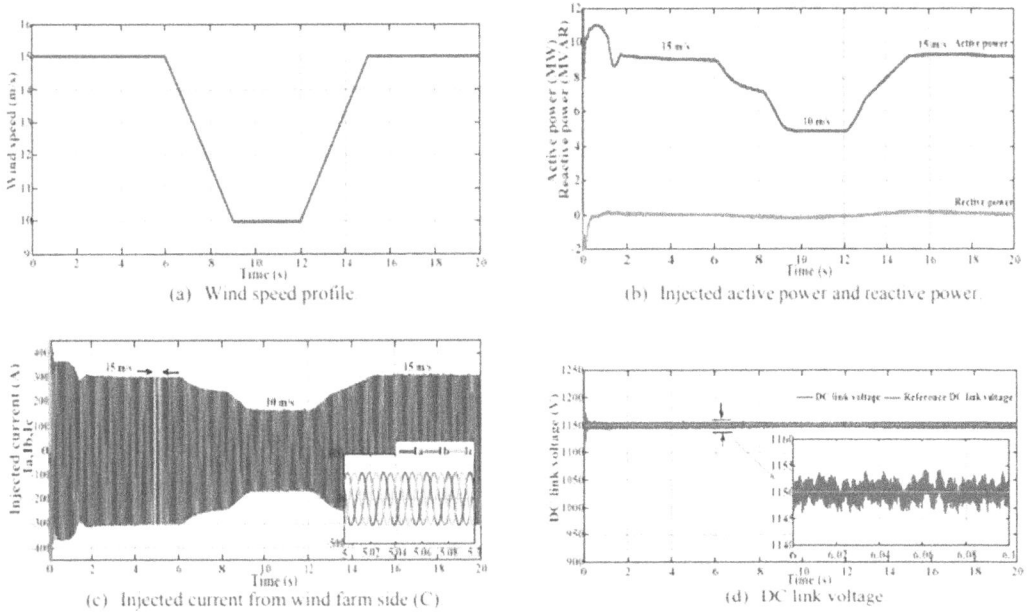

Fig. 5.5   Performance of wind farm

Source: Author

Figure 5.6 shows the working power output of a PHV generator and battery, with 75 kW of MPP power and near-zero battery power. The power system network is modelled in SIMULINK and simulated using STATCOM and DFIG behavior analysis. Wind speed patterns are shown, and the MPPT algorithm is used to optimize rotor speed for maximum power harvest.

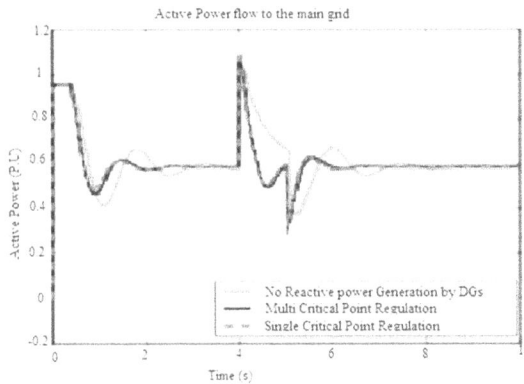

Fig. 5.6   Voltage profile at PCC without reactive power compensation

Source: Author

Summary of Key Advantages of the Proposed Methods:

- **PR Controllers** offer faster, more accurate synchronization and superior transient response compared to traditional PI controllers.
- **Dynamic SoC-Based Droop Control** enhances ESS efficiency and power-sharing flexibility, reducing energy wastage and improving system performance.

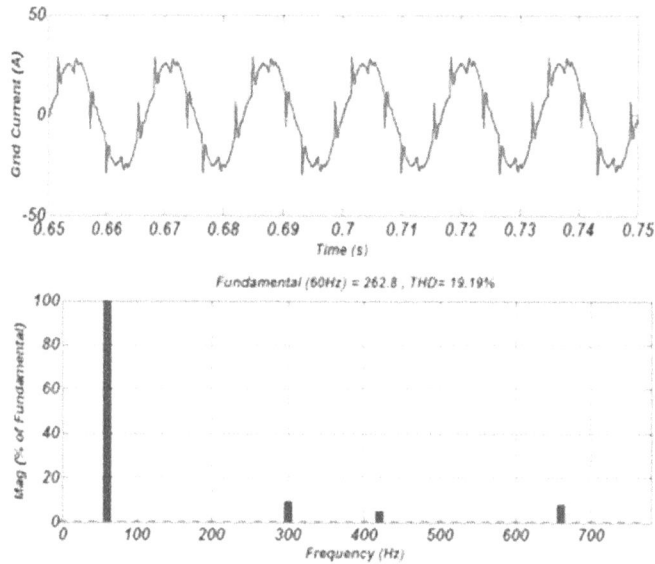

Fig. 5.7 Grid current and its FFT analysis when proposed technique is applied

Source: Author

- **MFPC** addresses both reactive power compensation and three-phase current imbalance, ensuring optimal grid quality under varying load and generation conditions.
- **Decentralized ESS Control** ensures more efficient, real-time energy management across the microgrid, enhancing scalability and performance.

Overall, this work provides significant advancements in microgrid control, improving stability, power quality, and energy management over conventional methods.

5. Conclusion

A modern microgrid control network integrates a photovoltaic power source with an energy storage system (ESS) battery to regulate the DC voltage from the inverter. The control system has been validated through MATLAB/Simulink experiments, yielding promising results that demonstrate the effectiveness of the proposed control scheme in achieving microgrid stability and facilitating the transition from grid-connected to standalone operation. This article explores various aspects and configurations of DC microgrids, emphasizing the significance of implementing diverse control policies and transforming topologies for enhanced efficiency. The study on the hydrogen-dependent solar delivery network of the Hybrid Microgrid (HPVHS) successfully accommodates both active and reactive loads. Simulation results indicate that forced microgrid isolation can be effectively executed under varying conditions of electricity import and export, highlighting the necessity for the medium-voltage (MV) network to rapidly address faults to ensure a seamless transition to independent service. Additionally, the research examines the power-sharing capabilities of control devices, focusing on maximizing energy distribution through a proportional power-sharing mechanism. As the demand for self-sufficient energy solutions grows among consumers and the public sector, microgrids and their associated electrical and electronic systems are becoming increasingly vital for strengthening existing grid networks.

REFERENCES

1. Abeyasekera, T.; Johnson, C.M.; Atkinson, D.J.; Armstrong, M. Suppression of line voltage related distortion in current controlled grid connected inverters. IEEE Trans. Power Electron. 2005, 20, 1393–1401.

2. Abusara, M.A.; Sharkh, S.M.; Guerrero, J.M. Improved droop control strategy for grid-connected inverters. Sustain. Energy Grids Netw. 2015, 1, 10–19.

3. Ahmed, K.; Massoud, A.; Finney, S.; Williams, B. Optimum selection of state feedback variables PWM inverters control. In Proceedings of the 4th IET Conference on Power Electronics, Machines and Drives, York, UK, 2–4 April 2008.

4. Anto Joseph, A Review of Power Electronic Converters for Variable Speed Pumped Storage Plants: Configurations, Operational Challenges, and Future Scopes, IEEE Journal of Emerging and Selected Topics In Power Electronics, VOL. 6, NO. 1, Issn no: 2168–6785, MARCH 2018.

5. Barklund, E.; Pogaku, N.; Prodanovic, M.; Hernandez-Aramburo, C.; Green, T.C. Energy management in autonomous microgrid using stability-constrained droop control of inverters. IEEE Trans. Power Electron. 2008, 23, 2346–2352.

6. Bharath.k.r, A Review on DC Microgrid Control Techniques, Applications and Trends, International Journal Of Renewable Energy Research, Vol.9, No.3, September, 2019.

7. Bhaskara, S.N.; Chowdhury, B.H. Microgrids A review of modeling, control, protection, simulation and future potential. In Proceedings of the 2012 IEEE Power and Energy Society General Meeting, San Diego, CA, USA, 22–26 July 2012; IEEE: Piscataway, NJ, USA, 2012; pp. 1–7.

8. Bhuiyan, F.A.; Yazdani, A. Multimode control of a DFIG-based wind-power unit for remote applications. IEEE Trans. Power Deliv. 2009, 24, 2079–2089.

9. Bidram, A.; Davoudi, A. Hierarchical structure of microgrids control system. IEEE Trans. Smart Grid 2012, 3, 1963–1976.

10. Blaabjerg, F.; Teodorescu, R.; Liserre, M.; Timbus, A.V. Overview of control and grid synchronization for distributed power generation systems. IEEE Trans. Ind. Electron. 2006, 53, 1398–1409.

11. Bottrell, N.; Prodanovic, M.; Green, T.C. Dynamic stability of a microgrid with an active load. IEEE Trans. Power Electron. 2013, 28, 5107–5119.

12. Bhavana Pabbuleti, A Review On Hybrid Ac/Dc Microgrids: Optimal Sizing, Stability Control And Energy Management Approaches, Journal of Critical Reviews, ISSN- 2394–5125, Vol 7, Issue 1, 2020.

13. Chen, C.L.; Wang, Y.; Lai, J.S.; Lee, Y.S.; Martin, D. Design of parallel inverters for smooth mode transfer microgrid applications. IEEE Trans. Power Electron. 2010, 25, 6–15.

14. Kang, Jin-Wook, et al. "A study on stability control of grid connected DC distribution system based on second order generalized integrator-frequency locked loop (SOGI-FLL)." *Applied Sciences* 8.8 (2018): 1387.

15. Chen, Z.; Luo, A.; Wang, H.; Chen, Y.; Li, M.; Huang, Y. Adaptive sliding-mode voltage control for inverter operating in islanded mode in microgrid. Int. J. Electr. Power Energy Syst. 2015, 66, 133–143.

16. Maqbool, Rashid, et al. "Identifying the critical success factors and their relevant aspects for renewable energy projects; an empirical perspective." *Journal of Civil Engineering and Management* 24.3 (2018): 223–237.

17. Perez, Filipe, et al. "DC microgrid voltage stability by dynamic feedback linearization." *2018 IEEE international conference on industrial technology (ICIT)*. IEEE, 2018.

18. Prakash, Surya, Vaibhav Nougain, and Sukumar Mishra. "Adaptive droop-based control for active power sharing in autonomous microgrid for improved transient performance." *IEEE Journal of Emerging and Selected Topics in Power Electronics* 9.3 (2020): 3010–3018.

19. Mai, Tam T., et al. "An overview of grid-edge control with the digital transformation." *Electrical Engineering* (2021): 1–19.

20. Delghavi, Mohammad B., Amirnaser Yazdani, and Abdollah Alizadeh. "Iterative learning control of dispatchable grid-connected distributed energy resources for compensation of grid current harmonic distortions." *International Journal of Electrical Power & Energy Systems* 131 (2021): 107064.

21. Issa, Walid R., et al. "Control strategy for uninterrupted microgrid mode transfer during unintentional islanding scenarios." IEEE Transactions on Industrial Electronics 65.6 (2017): 4831-4839.

22. www.alternative-energy-tutorials.com/photovoltaics/photovoltaic-array.html.

Intelligent and Sustainable Power and Energy Systems – Dr. M. Premkumar et al. (eds)
© *2026 Taylor & Francis Group, London, ISBN 978-1-041-10314-1*

6

A Novel Nine-Level Switched Capacitor Multilevel Inverter Topology with Common Ground Configuration

Md. Ashraf Ali Khan[1],
Sameer Alam[2], Yusra Wahab[3]
Department of Electrical Engineering, Aligarh Muslim University,
Aligarh, India

ABSTRACT: This research presents a novel nine-level common ground multilevel inverter (MLI) topology with quadruple voltage boosting, designed for transformer-less PV converters to address current leakage issues impacting power quality and safety. The proposed single-phase MLI employs a DC power source, 4 diodes, 4 capacitors, and 12 switches to generate nine output voltages while achieving four-fold voltage amplification through a switched capacitor approach, eliminating the need for additional components. By alternating capacitor charging and discharging, the topology generates the desired waveform with an impressive efficiency of 96.95%. Leveraging Level Shifted Modulation, its performance is analyzed in MATLAB and PLECS, demonstrating improvements in voltage gain, leakage current reduction, and reduced switch count, supported by qualitative comparison studies.

KEYWORDS: Multilevel inverter (MLI), Common ground (CG) configuration, Switched capacitors, Level shifted modulation, Solar photovoltaic (PV) applications

1. Introduction

Power electronics advancements have led to the development of sophisticated inverter technologies for converting direct current (DC) to alternating current (AC) in high-power applications. Among these, the switched capacitor multilevel inverter (SCMI) stands out for its advantages over conventional inverters. Unlike traditional designs, SCMIs utilize switched capacitor networks to generate multiple voltage levels, enabling high-quality AC waveforms with improved harmonic profiles, reduced switch stress, and minimized electromagnetic interference (EMI) [1-4].

Corresponding author: [1]khanashraf8738@gmail.com, [2]amustudent0@gmail.com

DOI: 10.1201/9781003654469-6

SCMIs are highly versatile, capable of addressing both low and high voltage needs. Their significance in renewable energy systems and transmission settings highlights their efficiency in handling large power loads in high-power sectors. At the same time, their adaptability benefits low-power units, enhancing efficiency and control across various smaller-scale applications. This broad applicability positions SCMIs as tailored and reliable solutions for diverse power demands in modern power electronics [4-5].

While traditional multilevel inverters (MLIs) have demonstrated solid performance, their reliance on numerous components results in higher production costs and complex assembly [5]. Recent topological advancements aim to streamline designs, reduce costs, and simplify conventional MLI architectures. Techniques such as flying capacitor converters, diode-clamped converters, and cascaded inverters (CMIs) have been integral in minimizing component count [6]. Alongside these, emerging strategies continue to optimize device configurations for efficiency, cost-effectiveness, and practical application in contemporary power electronics.

In switched-capacitor (SC) MLIs, series and parallel capacitor arrangements enable unique charging and discharging mechanisms. During discharging, the capacitors exhibit an increase in output voltage, demonstrating SC-MLIs' capability for higher voltage outputs at specific operating instances.

SC-MLIs have been extensively studied, with the common ground multilevel inverters (CG-MLIs) emerging as a promising milestone, particularly for solar photovoltaic (PV) applications. CG-SC-MLIs are vital in addressing leakage current issues inherent to PV systems caused by parasitic capacitance between PV panels and the grid. Leakage current reduces PV panel lifespan and negatively impacts system performance [7]. By employing a common ground configuration, CG-SC-MLIs effectively curb leakage currents, enabling stable and efficient PV operation. This innovation promotes reliability and longevity in PV systems, making CG-SC-MLIs a focal point in advancing PV system efficiency and durability.

This research introduces an innovative 9-level SC-MLI topology that offers several advantages over existing designs:

1. The inverter utilizes a single DC source and achieves a lower total standing voltage per unit.
2. It features an inherent fourfold voltage boosting capability and ensures self-capacitor voltage balancing.
3. The design incorporates a common-ground configuration, effectively eliminating common-mode leakage currents.

The structure is as follows: Section 1 reviews prior research, Section 2 presents a comprehensive literature review, Section 3 explains the schematic and operational principles, Section 4 provides a comparative analysis, Section 5 discusses simulation results, Section 6 examines power loss analysis, and Section 7 concludes the findings.

2. Literature Review

Extensive research has been conducted to achieve high voltage gain at the load using standard switched-capacitor approaches. Several nine-level inverter configurations have been proposed, each with distinct features and limitations. For instance, Siddique et al. [8] designed a nine-level inverter using 11 power switches and 3 capacitors, achieving a voltage boost of two. Similarly, [9] proposed a nine-level inverter employing 10 switches and 2 capacitors with two-times voltage boosting. Another configuration by Siddique et al. [10] used 12 switches and 3 capacitors, also providing a two-times voltage gain. However, the topology in [11] demonstrated four-times voltage gain with nine output levels but required 17 switches, 5 diodes, and 4 capacitors, significantly increasing system cost. Alternatively, the topology in [12] utilized 8 semiconductors, 2 capacitors, and 1 diode for double voltage boosting. While it reduced component count compared to prior designs, uniform capacitor charging was not achieved, limiting voltage gain to two. Similarly, Varesi et al. [13] achieved nine levels with quadruple boosting using a single source, 13 semiconductors, 2 diodes, and 2 capacitors, but the high component count resulted in lower efficiency. Another design [14] achieved nine levels with two-times voltage gain using 11 semiconductors but operated with lower efficiency. Reference [15] utilized 12 switches and 2 capacitors to produce nine levels with two-times voltage gain, while the setup in [16] used 12 switches and 3 capacitors for nine output levels without voltage boosting capability.

A more advanced configuration [17] achieved quadruple voltage boosting using 14 switches and 5 capacitors in a common ground setup, similar to the proposed topology but with higher component requirements. Recent studies [18–25] continue to explore switched-capacitor-based multilevel inverters (SC-MLIs), aiming to maximize voltage gain while minimizing essential components such as diodes, capacitors, and IGBTs.

3. Proposed Topology of 9-level SC-MLI

The circuit arrangement in Fig. 6.1 illustrates a nine-level switched capacitor multilevel inverter (SC-MLI) with a common ground configuration. This novel topology comprises 12 IGBT switches, 4 diodes (D1 to D4), and 4 capacitors (C1 to C4). Among the switches, 8 operate complementarily, simplifying control. The output voltage levels generated are +4Vdc, +3Vdc, +2Vdc, +1Vdc, 0Vdc, -1Vdc, -2Vdc, -3Vdc, and -4Vdc.

To address current spikes during capacitor charging, which cause electromagnetic interference (EMI) and reduce capacitor lifespan [26], a quasi-resonant inductor-diode combination (L-Dch) is included.

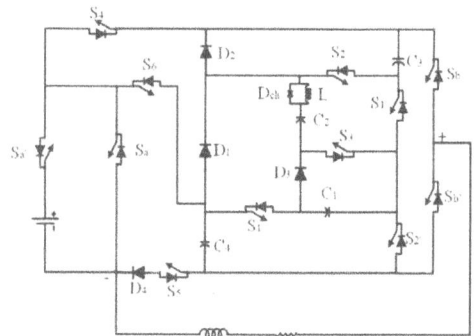

Fig. 6.1 Proposed 9-level SC-MLI with common ground configuration

This component, shown in Fig. 6.1, features an inductor (L) in series with capacitor C2 to mitigate these effects and enhance capacitor longevity.

3.1 Operating Modes of the Topology

The proposed topology operates in nine modes, producing nine voltage levels across the load, as depicted in Fig. 6.2(a-i). These figures illustrate current flow paths for both positive and negative voltage generation and capacitor charging. The generation of different load voltage levels is detailed below:

1. $+4V_{DC}$ Mode: When switches Sa', Sb, S1, S2, S3, S5, and S6 are on, capacitors C1, C2, and C3 discharge in series with Vdc, while C4 charges. Diodes D1 to D3 are non-conducting, and D4 conducts (Fig. 6.2a).

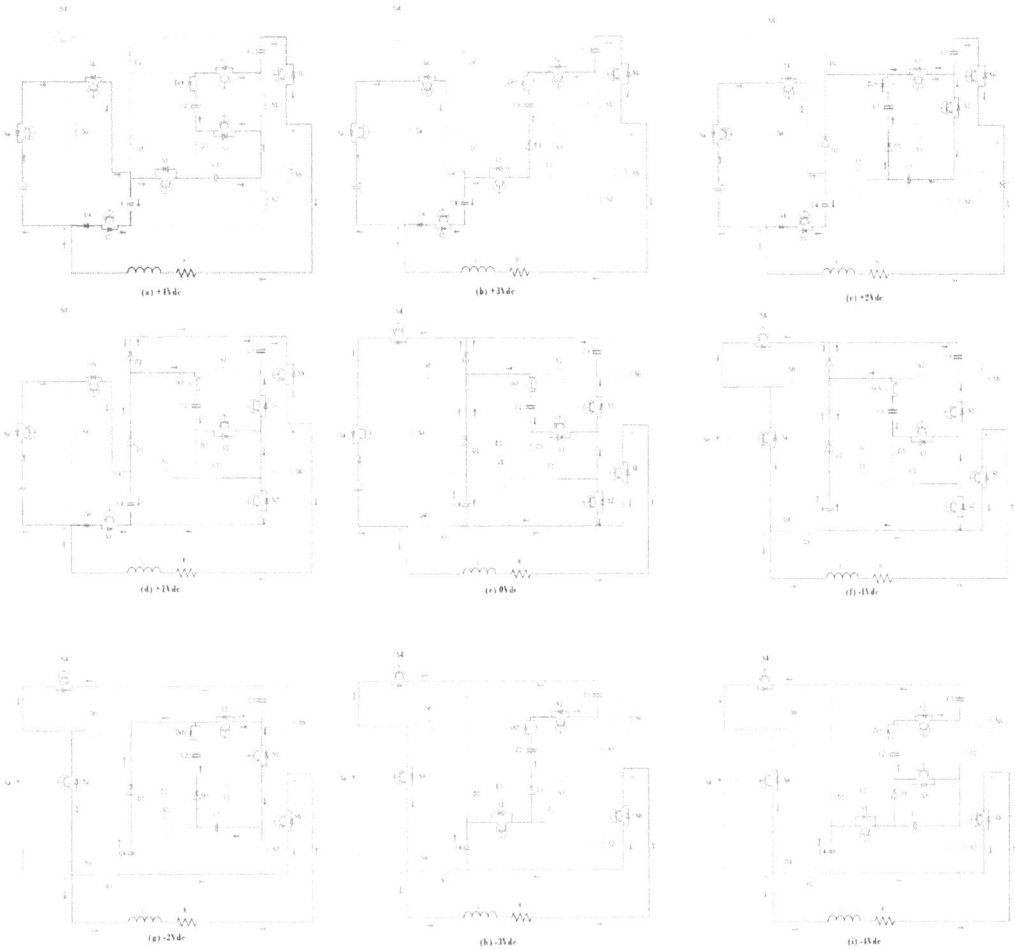

Fig. 6.2 Operating stages of proposed 9L-MLI

2. +3V$_{DC}$ Mode: Activating Sa', Sb, S1, S2, S5, and S6 links C2 and C3 in series with Vdc, discharging them while C1 remains unchanged and C4 charges. Diodes D3 and D4 conduct, and D1 and D2 do not (Fig. 6.2b).

3. +2V$_{DC}$ Mode: When Sa', Sb, S1', S2, S5, and S6 are on, C1 and C4 charge while C2 and C3 discharge. Diodes D1, D3, and D4 conduct, while D2 remains non-conducting (Fig. 6.2c).

4. +1V$_{DC}$ Mode: With Sa', Sb, S1', S2', S3, S5, and S6 active, C2, C3, and C4 charge, and C1 remains unchanged. Diodes D1, D2, and D4 conduct, and D3 is blocked (Fig. 6.2d).

5. 0V$_{DC}$ Mode: Turning on Sa', Sb', S1', S2', S3, and S4 results in 0V across the load. C4 discharges while C2 and C3 charge. Diodes D1 and D2 conduct, and D3 and D4 are non-conducting (Fig. 6.2e).

6. -1VDC Mode: With Sa, Sb', S1', S2', S3, and S4 on, C4 discharges, C1 remains unchanged, and C2 and C3 charge. Diodes D1 and D2 conduct, and D3 and D4 are non-conducting (Fig. 6.2f).

7. 2VDC Mode: Activating *S*a, Sb', S1', S2', and S4 produces -2Vdc. C4, C2, and C3 discharge, and C1 charges. Diodes D1 and D3 conduct, while D2 and D4 do not (Fig. 6.2g).

8. -3VDC Mode: Turning on *S*a, Sb', S1, S2, and S4 results in -3Vdc. C4, C2, and C3 discharge, and C1 remains unchanged (Fig. 6.2h).

9. -4VDC Mode: Activating *S*a, Sb', S1, S2, S3, and S4 discharges all capacitors, generating -4Vdc (Fig. 6.2i).

Table 6.1 outlines the switching patterns and charging/discharging sequences for attaining nine voltage levels. In Fig. 6.2, red lines represent load current paths, while purple lines indicate capacitor charging trajectories, clearly illustrating load and charging paths for each voltage state.

Table 6.1 Switching states and charging-discharging of capacitors for proposed 9-level topology

Voltage Levels	Switches State												Capacitors			
Vo	Sa	Sa'	Sb	Sb'	S1	S1'	S2	S2'	S3	S4	S5	S6	C1	C2	C3	C4
+4Vdc	0	1	1	0	1	0	1	0	1	0	1	1	DCH	DCH	DCH	CH
+3Vdc	0	1	1	0	1	0	1	0	0	0	1	1	NC	DCH	DCH	CH
+2Vdc	0	1	1	0	0	1	1	0	0	0	1	1	CH	DCH	DCH	CH
+1Vdc	0	1	1	0	0	1	0	1	1	0	1	1	NC	CH	CH	CH
0Vdc	0	1	0	1	0	1	0	1	1	1	0	0	NC	CH	CH	DCH
-1Vdc	1	0	0	1	0	1	0	1	1	1	0	0	NC	CH	CH	DCH
-2Vdc	1	0	0	1	0	1	1	0	0	1	0	0	CH	DCH	DCH	DCH
-3Vdc	1	0	0	1	1	0	1	0	0	1	0	0	NC	DCH	DCH	DCH
-4Vdc	1	0	0	1	1	0	1	0	1	1	0	0	DCH	DCH	DCH	DCH
1-: Switch ON; 0-: Switch OFF; CH-: Charging; DCH-: Discharging; NC-: No Change																

3.2 Modulation Technique

For generating switching signals, various modulation methods such as selective harmonic elimination pulse width modulation (SHE-PWM), phase-disposed pulse width modulation (PD-PWM), and level-shifted PWM are available. This study employs the level-shifted PWM technique to generate control signals for switches. This method compares triangular carrier waves (Vcr1 to Vcr8) with a sine-wave reference signal (Vref).

The carrier waves have a frequency of 3 kHz, while the reference wave operates at 50 Hz. Using this approach, nine voltage levels are produced across the load, as shown in Fig. 6.3A. In Fig. 6.3A, the carrier waves (Vcr1 to Vcr8) are compared with Vref in a comparator circuit, producing the appropriate switching signals for each power switch (S1, S2, S1', S2', S3, S4, S5, S6, Sa, Sa', Sb, and Sb'). Figure 6.3B illustrates the switching logic, showing how a complete switching cycle is generated for the proposed nine-level converter.

The modulation index (M) is calculated as the ratio of the reference wave amplitude (Vref) to eight times the amplitude of the carrier wave (Vcr), as given by:

$$M = \frac{V_{ref}}{8 \times V_{cr}} \tag{1}$$

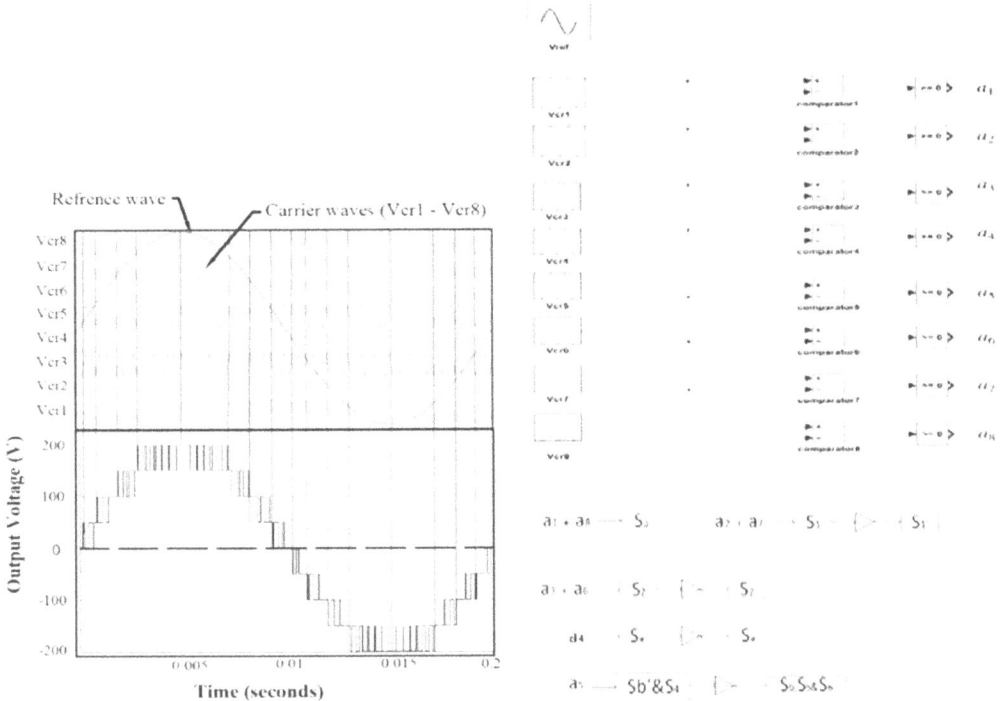

Fig. 6.3 a) Level shifted PWM scheme waveform b) Switching logic

3.3 Capacitor Design

In switched capacitor topologies, calculating optimal capacitance values is critical, influenced by factors like ripple voltage and maximum discharge time (MDT). The MDT or Longest Discharge Time (LDT) for each capacitor is used to determine appropriate capacitance. The charging and discharging cycles for the proposed topology are outlined in Table 6.1, while Fig. 6.4 illustrates MDT for capacitors C1, C2, C3, and C4. Capacitors discharge to supply current to the load while maintaining steady Vdc and recharge when connected in parallel with the source. For the proposed topology, the MDT for capacitor C1 occurs during the voltage levels +4Vdc and -4Vdc. For C2 and C3, MDT spans +2Vdc to +4Vdc (positive cycle) and -2Vdc to -4Vdc (negative cycle). Capacitor C4's MDT persists throughout the negative cycle. The capacitance is calculated by evaluating the allowable voltage ripple during the longest discharge phases across a full cycle. The charge stored by a capacitor is given by (2):

$$Qc = \int_{t_{DCH}}^{t_{CH}} I_0 \cdot \sin(2\pi f_0 t)dt \tag{2}$$

Where, $(t_{DCH} - t_{CH})$ represents the maximum discharge duration, Io is the load current, and f_0 is the fundamental frequency.

The optimal capacitance (C) is derived as (3):

$$C = \frac{Q_c}{\Delta V} = \frac{\int_{t_{DCH}}^{t_{CH}} I_0 \cdot \sin(2\pi f_0 t)dt}{\Delta V} \tag{3}$$

Capacitances (C_1, C_2, C_3, and C_4) are calculated while maintaining a ripple voltage $\Delta V < 10\%V$. [27]

$$C_1 = \frac{I_o}{\pi \times f_o \times \Delta V} \times \Delta T_1 \tag{4}$$

$$C_2 = \frac{I_o}{\pi \times f_o \times \Delta V} \times \Delta T_2 \tag{5}$$

$$C_3 = \frac{I_o}{\pi \times f_o \times \Delta V} \times \Delta T_3 \tag{6}$$

$$C_4 = \frac{I_o}{\pi \times f_o \times \Delta V} \times \Delta T_4 \tag{7}$$

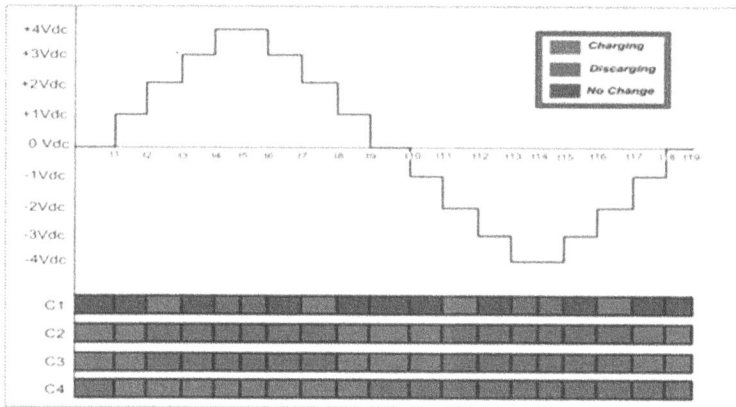

Fig. 6.4 Capacitor charging/discharging pattern with output voltage

Where, the durations $\Delta T_1 = t_5 - t_4$, $\Delta T_2 = t_7 - t_2$, $\Delta T_3 = t_7 - t_2$, and $\Delta T_4 = t_{18} - t_9$ represent the maximum discharging periods for capacitors C_1, C_2, C_3, and C_4 respectively.

3.4 Charging Inductor Design

To prevent overvoltage on capacitors, reduce electromagnetic interference, and limit fault currents, the charging inductor is designed according to the guidelines in [26].

$$L \geq \frac{1}{\left(4\pi f\right)^2 C_t} \tag{8}$$

4. Comparative Analysis

A detailed comparison with other nine-level inverters is presented to highlight the advantages and limitations of the proposed topology. Table 6.2 outlines key parameters such as the number of capacitors (Nc), diodes (N_D), switches (N_{SW}), drivers (N_{GD}) and DC sources (Ndc), switching loops (Ncl), voltage gain (A), common-ground (CG) feature, Per-unit Total Standing Voltage (TSV p.u.), cost function (CF), and efficiency (Eff).

Table 6.2 Comparison table of proposed topology with other 9-level topologies

Refs.	NSW	NGD	ND	NCL	Ndc	NC	Gain	CG	TSV (p.u)	CF	Efficiency (%)
15	12	11	0	4	1	2	Double	No	5.5	3.8	96.47
8	11	10	0	3	1	3	Double	No	5	3.7	92.2
10	12	11	0	2	1	3	Double	No	5.5	3.7	97.45
16	12	12	0	6	1	3	Unity	No	5	4.2	95.73
17	12	12	2	4	2	4	Double	No	8	9.3	95.64
18	14	12	0	6	1	5	Quadruple	Yes	7.5	4.4	97.4
19	9	9	2	3	1	3	Double	No	5.75	3.5	NA
20	8	8	2	5	1	2	Quadruple	No	6.25	3.4	95.2
21	10	9	1	5	1	2	Quadruple	No	6.25	3.7	NA
22	10	5	3	3	1	3	Quadruple	No	5.5	3.3	96
Proposed	12	5	4	3	1	4	Quadruple	Yes	5.5	3.7	96.95

The cost function (CF) is calculated using the formula derived from Siddique et al. [10]:

$$CF = \frac{[N_{sw} + N_D + N_{GD} + N_C + N_{CL} + TSV_{p.u}\} \times xN_{dc}}{Number\ of\ Levels} \tag{9}$$

Here, (x) denotes the total number of DC sources used.

Compared to other topologies, the proposed design achieves a nine-level output with an equal or fewer number of switches than most counterparts, except those in [8, 19, 20, 21, 22]. However, it is noteworthy that circuits [8, 19, 20, 21, 22] lack a common-ground feature, and [8, 19] provide only two-times voltage boosting. Circuit [16], on the other hand, does not include any voltage boosting capability. Among the studied designs, only [18] and the proposed topology offer a common-ground configuration. However, the (TSV) p.u. and cost function of [18] are significantly higher than those of the proposed design. Moreover, [18] uses more switches, further increasing its cost and complexity.

The proposed topology demonstrates competitive efficiency, achieving the same or lower (TSV) p.u. compared to most circuits, except those in [8, 16]. Additionally, it has a reduced cost function, except for the designs in [19, 20, 22]. This highlights the cost-effectiveness and efficiency of the proposed topology relative to other designs, making it a compelling choice for nine-level inverters.

5. Results and Discussion

The performance of the proposed 9-level, 4-times boost Switched Capacitor Multilevel Inverter (SC-MLI) topology was validated through simulations using MATLAB 2021. The simulations covered a range of operating conditions, including fixed and dynamic scenarios, to assess the inverter's behaviour and robustness. The load voltage and current waveforms under various conditions are presented in Figs. 6.5 and 6.6. For a power factor of 0.85, corresponding to a series circuit impedance of 50 ohms resistance and 100 millihenries inductance, Figure 6.5a illustrates the voltage and current waveforms. With an inductive load, the current waveform is nearly sinusoidal, lagging behind the voltage. At a power factor of 0.95, with impedance of 50 ohms resistance and 50 millihenries inductance, Figure 6.5b shows a similarly sinusoidal current waveform lagging the

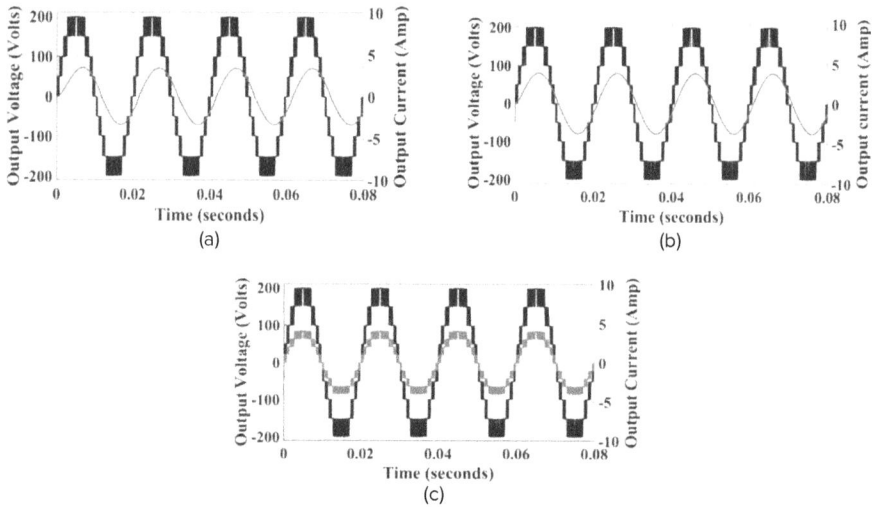

Fig. 6.5 Output voltage and current for (a) 0.85 p.f , (b) 0.95 p.f, and (c) unity p.f

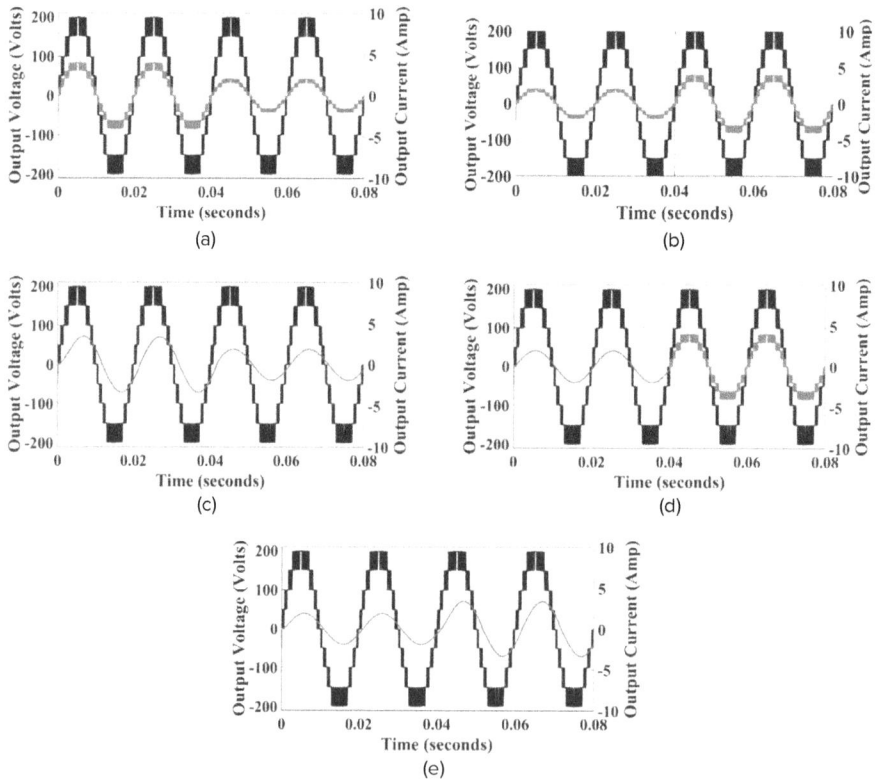

Fig. 6.6 Output voltage and current for dynamic loads (a) R=50Ω to 100Ω, (b) R=100Ω to 50Ω, (c) R=50Ω, L=100mH to R=100Ω, L=100mH, (d) R=100Ω,L=100mH to R=50Ω,L=100mH, and (e) R=100Ω,L=100mH to R=50Ω

voltage. Figure 6.5c, on the other hand, displays the waveforms of a purely resistive load with R = 50 ohms. In the case of pure resistance, both the voltage and current waveforms are in phase.

The dynamic performance of the proposed topology is analyzed under varying load conditions, as shown in Fig. 6.6. In Fig. 6.6a, the load impedance transitions from 50 ohms to 100 ohms and then reduces back to 50 ohms, while Fig. 6.6b shows the resulting current variations, confirming stable operation during load changes. Figures 6.6c and 6d depict transitions from an inductive load (50 ohms, 100 millihenries) to a higher inductive impedance (100 ohms, 100 millihenries) and back, maintaining consistent performance. A unique case is presented in Fig. 6.6e, where the load changes from an inductive configuration (100 ohms, 100 millihenries) to a purely resistive load (50 ohms). The output current variations under these dynamic conditions demonstrate the proposed topology's effectiveness and reliabilityacross diverse load profiles. This highlights its robustness and suitability for practical applications involving fluctuating operating states.

6. Power Loss Analysis

Power losses in the proposed topology are categorized into conduction loss (Pc), switching loss (Psw), and ripple loss (P_R), as shown in Equation (10). These losses, caused by energy dissipation in power electronic switches and capacitors, were analyzed using PLECS software to enhance the performance and efficiency of the proposed Switched Capacitor Multilevel Inverter (SC-MLI).

$$P_L = P_C + P_{SW} + P_R \tag{10}$$

6.1 Conduction Loss

Conduction losses arise due to the internal resistance and voltage drop in switches during their conduction periods. They are calculated using Equations (11) and (12):

$$P_{C(IGBT)} = \sum_{n=1}^{12}(V_{ON(IGBT)} \times I_{avg(IGBT)} + R_{ON(IGBT)} \times I_{rms}^2 \tag{11}$$

$$P_{C(DIODE)} = \sum_{n=1}^{4}(V_{ON(DIODE)} \times I_{avg(DIODE)} + R_{ON(DIODE)} \times I_{rms}^2 \tag{12}$$

Here, (VON) is the voltage drop across the switch in its ON-state, (RON)is the on-state resistance, and (Irms) and (Iavg)are the root mean square (RMS) and average currents, respectively.

6.2 Switching Loss

Switching losses occur during the ON-OFF transitions of switches due to non-ideal behaviour, as calculated using Equations (13) and (14). To make calculations easier, it has been assumed that switch voltage and current are linear over the course of the switching period. The voltage and current waveforms overlap during transitioning, which causes switching losses because of the non-ideal behaviour of the switches.

$$P_{SW} = P_{SW(ON)} + P_{SW(OFF)} \tag{13}$$

$$P_{SW} = \left[\sum_{n=1}^{12}\left\{\sum_{in\frac{1}{f_o}}\left(\frac{V_{ON}i_{ON}t_{ON}}{6} + \frac{V_{OFF}i_{OFF}t_{OFF}}{6}\right)\right\}\right] \times f \tag{14}$$

6.3 Ripple Loss

Ripple loss, unique to SC-MLIs, occurs due to the Equivalent Series Resistance (ESR) of capacitors. It is expressed as:

$$P_R = \sum_{n=1}^{4} C \times \Delta V_C^2 \times \frac{f}{2} \tag{15}$$

An ESR of 0.01Ω is assumed for each capacitor to estimate ripple losses.

The thermal modelling of the proposed topology using PLECS software and IGBT datasheets (IKB40N65ES5) enabled loss calculations under various load levels. Figure 6.7a illustrates the efficiency curve, comparing the proposed topology with references [22] and [23]. At 1000 W output power, the proposed topology achieves 96.95% efficiency, outperforming the topology in [22] (95%) and closely matching [23] (96%).

Figures 6.7b and 6.7d show the power loss distribution among components at 1000 W, highlighting the influence of ripple losses caused by the capacitors' ESR. Figure 6.7c presents conduction and switching losses under different load levels, showing a decrease in conduction losses with increasing load. The analysis was conducted at a constant ambient temperature of 25°C, and heat sink performance was evaluated. Conduction and switching losses were quantified using PLECS

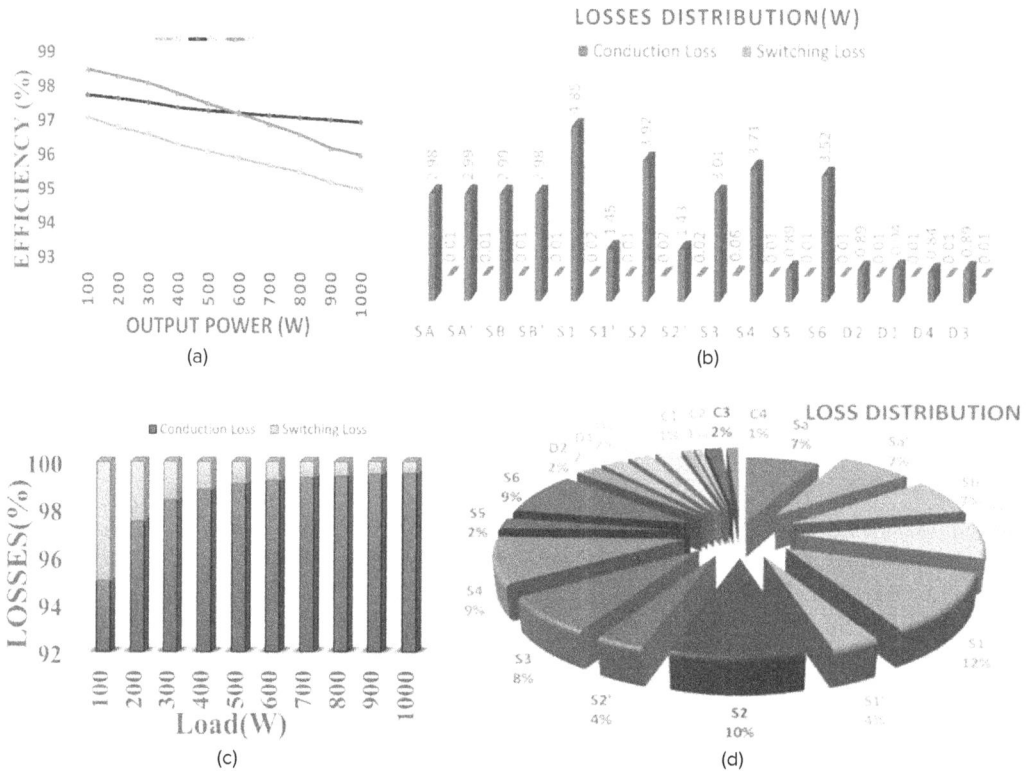

Fig. 6.7   Results extracted from PLECS. (a) Efficiency curve. (b) Conduction and switching loss across each switch and diodes. (c) Switching loss and conduction loss vs load. (d) Power loss distribution at 1KW

probes and methods like periodic averaging and impulse calculation. This comprehensive analysis demonstrates the proposed topology's superior efficiency and effective loss management under various operating conditions

7. Conclusion

This research introduces a novel nine-level Multilevel Inverter (MLI) featuring a common ground configuration and quadruple voltage boosting capability, achieving four times the input voltage at the output stage. Comparative analysis with other nine-level inverters highlights its significant advantages, including the use of only one DC source, which ensures natural capacitor balancing without additional devices. The inverter's feasibility and efficiency were validated through simulations on a 1 kW prototype, achieving an efficiency of 96.95%. Various operating conditions were simulated and analyzed in detail. Additionally, loss analysis conducted using PLECS software confirmed the topology's robustness under diverse scenarios. Development of a hardware prototype for the proposed inverter is planned for future work.

REFERENCES

1. K. K. Gupta, A. Ranjan, P. Bhatnagar, L. K. Sahu, and S. Jain, "Multilevel Inverter Topologies with Reduced Device Count: A Review," in IEEE Transactions on Power Electronics, vol. 31, no. 1, pp. 135–151, Jan. 2016. DOI: 10.1109/TPEL.2015.2405012.

2. A. Salem, H. Van Khang, K. G. Robbersmyr, M. Norambuena, and J. Rodriguez, "Voltage Source Multilevel Inverters With Reduced Device Count: Topological Review and Novel Comparative Factors," in IEEE Transactions on Power Electronics, vol. 36, No. 3, pp. 2720–2747, March 2021. DOI: 10.1109/TPEL.2020.3011908.

3. P. R. Bana, K. P. Panda, R. T. Naayagi, P. Siano, and G. Panda, "Recently developed reduced switch multilevel inverter for renewable energy integration and drives application: Topologies, comprehensive analysis and comparative evaluation," IEEE Access, vol. 7, pp. 54888–54909, 2019.

4. M. Trabelsi, H. Vahedi, and H. Abu-Rub, "Review on Single-DC-Source Multilevel Inverters: Topologies, Challenges, Industrial Applications, and Recommendations," in IEEE Open Journal of the Industrial Electronics Society, vol. 2, pp. 112–127, 2021. DOI:10.1109/OJIES.2021.3054666.

5. T. Roy and P. K. Sadhu, "A Step-Up Multilevel Inverter Topology Using Novel Switched Capacitor Converters With Reduced Components," in IEEE Transactions on Industrial Electronics, vol. 68, no. 1, pp. 236–247, Jan. 2021. DOI: 10.1109/TIE.2020.2965458.

6. Debela T, Singh J, Sood VK. "Evaluation of a grid-connected reduced-component boost multilevel inverter (BMLI) topology," International Journal of Circuit Theory and Applications, 2022;2022:1–33. DOI: 10.1002/cta.3253.

7. Sarwer Z, Anwar MN, Sarwar A. "A nine-level common ground multilevel inverter (9L-CGMLI) with reduced components and noosting ability," International Journal of Circuit Theory and Applications, 2023;51(8):3826–3840. DOI: 10.1002/cta.3550.

8. M. D. Siddique et al., "A New Single Phase Single Switched-Capacitor Based Nine-Level Boost Inverter Topology With reduced Switch Count and Voltage Stress," in IEEE Access, vol. 7, pp. 174178–174188, 2019. DOI:

9. Oorappan, G.M., Pandarinathan, S., & Arumugam, J. "A new nine-level switched-capacitor-based multilevel inverter with low voltage stress and self-balancing," Electric Engineering, 105, 867–882 (2023). DOI: 10.1007/s00202-022-01703-4.

10. M. D. Siddique et al., "Single-Phase Boost Switched-Capacitor-Based Multilevel Inverter Topology With Reduced Switching devices," in IEEE Journal of Emerging and Selected Topics in Power Electronics, vol. 10, no. 4, pp. 4336–4346, Aug. 2022. DOI: 10.1109/JESTPE.2021.3129063.

11. H. Khoun Jahan, M. Abapour and K. Zare, "Switched-Capacitor-Based Single-Source Cascaded H-Bridge Multilevel Inverter featuring Boosting Ability," in IEEE Transactions on Power Electronics, vol. 34, no. 2, pp. 1113–1124, Feb. 2019, doi:10.1109/TPEL.2018.2830401.

12. B. S. Naik, Y. Suresh, J. Venkataramanaiah, and A. K. Panda, "A Hybrid Nine-Level Inverter Topology With Boosting capability and Reduced Component Count," in IEEE Transactions on Circuits and Systems II: Express Briefs, vol. 68, no. 1, pp.316–320, Jan. 2021. DOI:10.1109/TCSII.2020.2998496

13. Varesi K, Esmaeili F, Deliri S, Tarzamni H. "Single-input quadruple-boosting switched-capacitor nine-level inverter with self-balanced capacitors," IEEE Access, 2022;10:70350-70361. DOI:10.1109/ACCESS.2022.318700

14. Bhatnagar P, Agrawal R, Dewangan NK, Jain SK, Gupta KK. "Switched-capacitors 9-level module (SC9LM) with reduced device count for multilevel DC to AC power conversion," IET Electrical Power Applications, 2019;13(10):1544–1552.

15. S. S. Lee, "Single-Stage Switched-Capacitor Module (S3CM) Topology for Cascaded Multilevel Inverter," in IEEE Transactions on Power Electronics, vol. 33, no. 10, pp. 8204–8207, Oct. 2018. DOI: 10.1109/TPEL.2018.2805685.

16. N. Sandeep and U. R. Yaragatti, "Operation and Control of an Improved Hybrid Nine-Level Inverter," in IEEE Transactions on Industry Applications, vol. 53, no. 6, pp. 5676–5686, Nov.-Dec. 2017, doi: 10.1109/TIA.2017.2737406.

17. R. Barzegarkhoo, S. S. Lee, S. A. Khan, Y. P. Siwakoti, and D.-C. Lu, "A novel generalized common-ground switched-capacitor multilevel inverter suitable for transformerless grid-connected applications," IEEE Transactions on Power Electronics, 2021;36(9):10,293–10,306.

18. Ali, M. Tayyab, A. Sarwar, and M. Khalid, "A Low Switch Count 13-Level Switched-Capacitor Inverter With Hexad Voltage-Boosting for Renewable Energy Integration," *IEEE Access*, vol. 11, pp. 36300–36308, 2023, doi:10.1109/ACCESS.2023.3265467.

19. Liu H, Wu J, Zeng J, Guo H. "A novel nine-level inverter employing one voltage source and reduced components as high-frequency AC power source," in IEEE Transactions on Power Electronics, 2017;32(4):2939–2947.DOI:10.1109/TPEL.2016.2582206.

20. Sathik MA, Siddique MD, Sandeep N, et al. "Compact quadratic boost switched-capacitor inverter," in IEEE Transactions on Industrial Applications, 2022;58(4):4923–4931. DOI: 10.1109/TIA.2022.3172235.

21. Hosseini SA, Varesi K. "Hybrid switched-capacitor 9-level boost inverter," in 2021 12th Power Electronics, Drive Systems, and Technologies Conference (PEDSTC), 2021;1–4. DOI: 10.1109/PEDSTC52094.2021.9405886.

22. Debela Awas T, Singh J. High-gain nine-level switched-capacitor multilevel inverter featuring less number of devices and leakage current. Int J Circ Theor Appl. 2023;1–28. Doi:10.1002/cta.3603

23. A. Srivastava and J. Seshadrinath, "A New Nine-Level Highly Efficient Boost Inverter for Transformerless Grid-Connected PV Application," in *IEEE Journal of Emerging and Selected Topics in Power Electronics*, vol. 11, no. 3, pp. 2730–2741, June 2023, doi: 10.1109/JESTPE.2022.3210512.

24. V. Anand and V. Singh, "A 13-Level Switched-Capacitor Multilevel Inverter With Single DC Source," *IEEE J Emerg Sel Top Power Electron*, vol. 10, no. 2, pp. 1575–1586, Apr. 2022, doi: 10.1109/JESTPE.2021.3077604

25. K.-M. Kim, J.-K. Han, and G.-W. Moon, "A High Step-Up Switched-Capacitor 13-Level Inverter With Reduced Number of Switches," *IEEE Trans Power Electron*, vol. 36, no. 3, pp. 2505–2509, Mar. 2021, doi: 10.1109/TPEL.2020.3012282.

26. Jahan, H. K., Shotorbani, A. M., Abapour, M., Zare, K., Hosseini, S. H., Blaabjerg, F., & Yang, Y. (2021). Switched capacitor based cascaded half-bridge multilevel inverter with voltage boosting feature. *CPSS Transactions on Power Electronics and Applications (CPSS TPEA)*, *6*(1), 63–73. [Article ID: 9399336](https://doi.org/10.24295/CPSSTPEA.2021.00006)

27. Iqbal A, Siddique MD, Prathap Reddy B, Maroti PK. (2021). Quadruple boost multilevel inverter (QB-MLI) topology with reduced switch count. *IEEE Transactions on Power Electronics*, *36*(7).

Note: All the figures and tables in this chapter were made by the author.

Intelligent and Sustainable Power and Energy Systems – Dr. M. Premkumar et al. (eds)
© 2026 Taylor & Francis Group, London, ISBN 978-1-041-10314-1

7

Optimised EV Charging Infrastructure: A Grid-Integrated Solar and Battery Storage System with Triple-Loop Control

Umasankar Loganathan[1],
Ramana Krishnan N.[2], Raffik H.[3], Venkatesan K.[4]
Department of Electrical and Electronics Engineering,
S. A. Engineering College, Thiruverkadu,
Chennai

ABSTRACT: This paper presents an energy-efficient electric vehicle (EV) charging station that integrates a photovoltaic (PV) array and battery energy storage (BES) within a grid-connected framework, employing a triple-loop control strategy. As EV adoption grows, the need for sustainable charging infrastructure is critical. Our proposed system addresses this by harnessing renewable solar energy to reduce grid dependency and improve environmental sustainability. The charging station's hierarchical control design features an outer loop for stabilising grid voltage and frequency, a middle loop for managing BES charge cycles, and an inner loop for controlling the DC-DC converter with maximum power point tracking (MPPT). This arrangement maximises PV energy usage and allows seamless switching between grid-connected and autonomous modes based on PV availability and battery charge levels. Simulation results demonstrate that the system meets dynamic load requirements, maintains power quality, and reliably adapts to varying solar and grid conditions. This model offers a scalable and resilient solution for EV charging infrastructure that supports energy independence and reduces peak grid loads, contributing to the broader goals of a low-carbon energy transition.

KEYWORDS: Battery energy storage, Electric vehicles, Grid integration, Photovoltaic systems, Triple-loop control

[1]drumasankar@saec.ac.in, [2]ramanakamu@gmail.com, [3]raffik488@gmail.com, [4]savenky21@gmail.com

DOI: 10.1201/9781003654469-7

1. Introduction

The international shift toward electric vehicles (EVs) is driven by the desire to reduce greenhouse gas emissions and reliance on fossil gasoline. However, the demanding circumstances of EV charging infrastructure, including grid dependence and peak load demands, prevent large-scale adoption. Integrating renewable energy, especially photovoltaic (PV) structures, into electric vehicle charging stations can provide sustainable solutions by reducing grid strength dependence and carbon emissions [1, 2]. Combining PV structures with battery energy storage (BES) can complement grid balancing and cargo control. Stored solar energy can be utilised through low generation or excess demand, increasing machine resiliency and reducing energy outages [3]. In addition, electric vehicles charged through power stations supported by renewable energy can provide ancillary services such as height regulation and voltage stabilisation, positioning such power stations as valuable grid assets [4]. This study proposes a three-loop control method with a grid-connected electric vehicle charging station structure. The outer loop stabilises the grid voltage and frequency, the middle loop manages the BES rate cycle, and the inner loop optimises photovoltaic power generation utilisation through Maximum Power Point Tracking (MPPT). This multi-level control enhances charging stability and energy even under grid fluctuations [5, 6]. The proposed machine minimises dependence on the grid, optimises the use of renewable energy, and supports sustainable electric vehicle charging. This paper is prepared as follows: Section 2 reviews relevant literature, Section 3 introduces the machine layout, Section 4 explains the three-ring manipulation method, Section 5 provides experimental results, and Section 6 summarises the direction of fate research.

2. Related Research Work

The integration of renewable energy, especially photovoltaic (PV) systems, with electric vehicle (EV) charging systems has been extensively studied to improve efficiency and sustainability through stable, independent charging (reducing grid dependence, reducing peak demand, and reducing carbon emissions). Continuity. Recent research has explored advanced technologies and control algorithms for more efficient EV charging. For example, [7] demonstrated the advantages of grid-connected solar photovoltaic systems in improving power reliability, while [8] introduced an adaptive control algorithm to handle network disturbances dynamically. [9] highlighted the increased capacity of photovoltaic-based EV charging stations connected to the three-phase grid, stabilising the performance. Active wireless charging systems are also receiving attention. [10] studied onboard DC-DC converters to improve wireless charging efficiency. [7] developed the best implementation scheme for integrating PV-BES systems to support the grid and effectively manage energy. [11] Reviewed maximum power point tracking (MPPT) algorithms and highlighted their importance in large-scale PV deployments. [12] described three methods for BES control in renewable energy-integrated power plants, consistent with the method adopted in this study. [13] Emphasised the role of advanced control methods in ensuring the reliable operation of renewable hybrid electric vehicle charging systems. This paper proposes a triple-loop control technology to optimise power flow, stabilise grid interaction, and achieve green and reliable electric vehicle charging under various conditions.

3. Proposed Method

This study proposes an EV charging station with a photovoltaic (PV) array, battery energy storage (BES), and grid connectivity controlled by a hierarchical triple-loop strategy. The triple-loop control includes an outer loop for grid stability, a middle loop for BES management, and an inner loop for PV optimisation. The charging station consists of the following components: PV Array: Provides renewable energy to power EV chargers and charge the BES. Battery Energy Storage (BES) Stores surplus PV energy for later use and discharges when solar availability is low. DC-DC Converter with MPPT: Optimizes PV output through maximum power point tracking (MPPT). Inverter: Converts DC to AC for grid interface. Triple-Loop Controller: Manages energy distribution among PV, BES, and the grid for system stability and efficiency. The block diagram shown in Fig. 7.1 below outlines the major system components and their interactions:

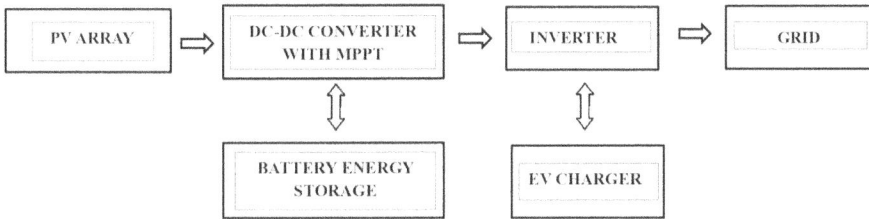

Fig. 7.1 Block diagram of the proposed EV charging system

Source: Author

The triple-loop control approach dynamically allocates energy from the PV array and BES to ensure grid stability and efficient EV charging. The outer loop monitors and adjusts voltage and frequency, maintaining stable grid interactions even under variable PV generation [14, 15]. The Proportional-Integral (PI) controller modulates power flow, discharging from BES when demand increases or reducing power input during surplus to stabilise grid conditions. The middle loop monitors the BES's state of charge (SOC) to maintain optimal charge levels, switching between charging and discharging as required [16]. It charges the BES with excess PV energy during periods of low EV demand and discharges during high demand or low PV availability, reducing grid dependency. The inner loop uses the Perturb and Observe (P&O) algorithm for MPPT, ensuring maximum power output from the PV array [17]. Rapid adjustments in response to irradiance and temperature changes allow efficient energy capture, reducing grid reliance by utilising more solar energy [18, 19].

3.1 Flow Chart and Algorithm

The flowchart in Fig. 7.2 outlines the operation of an energy management system integrating photovoltaic (PV) generation, grid supply, and battery energy storage (BES) to meet electric vehicle (EV) charging demand while ensuring grid stability. The steps are as follows.

Algorithm

The following algorithm outlines the triple-loop control: Initialize system parameters (grid voltage, frequency, PV power, BES SOC, EV load) [19, 20]. Monitor grid stability. Adjust power flow between PV, BES, and grid based on voltage and frequency targets. Monitor BES SOC [20]. Charge or discharge BES depending on PV power availability and EV demand. Track PV output. Adjust the DC-DC converter duty cycle for MPPT to maintain maximum PV output.

The triple-loop control system manages power flow within an EV charging station with a photovoltaic (PV) array and battery energy storage (BES). This approach optimises energy use and stabilises the interaction with the grid. Each control loop in this hierarchical system plays a unique role. The outer loop maintains grid voltage and frequency, which are essential for seamless integration with the grid.

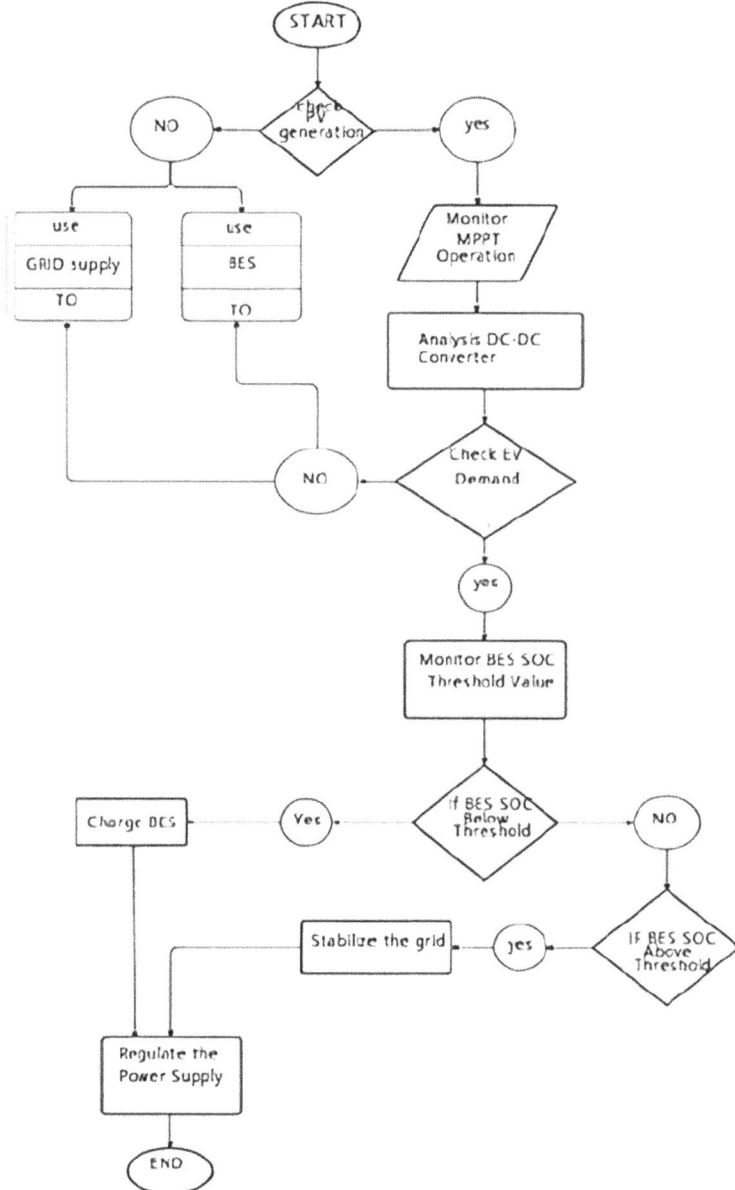

Fig. 7.2   Flowchart of the triple-loop control strategy

Source: Author

It also manages power flow to stabilise the grid, especially under fluctuating solar output. The inner loop controls the DC-DC converter using Maximum Power Point Tracking (MPPT) to maximise PV output. Rapid adjustments help the PV system stay efficient under varying irradiance levels. Through MPPT, the system achieves up to 95% energy efficiency, while BES storage supports load demands, reducing grid dependency by up to 50% during peak PV production.

4. Result and Discussion

4.1 Modelling and Simulation of PV-BES System with a Boost Converter and MPPT

Figure 7.3 shows the Matlab/Simulink circuit of the PV-BES System with Boost Converter and MPPT. Power rating input from the user = 2.00 kW, Minimum number of panels required per string = 8, Maximum number of panels connected per string without reaching maximum voltage = 10, Minimum power rating of the solar PV plant = 1.80 kW, Maximum power possible per string without reaching maximum DC voltage = 2.25 kW, Actual number of panel per string = 9, Number of strings connected in parallel = 1, Actual solar PV plant power = 2.03 kW.

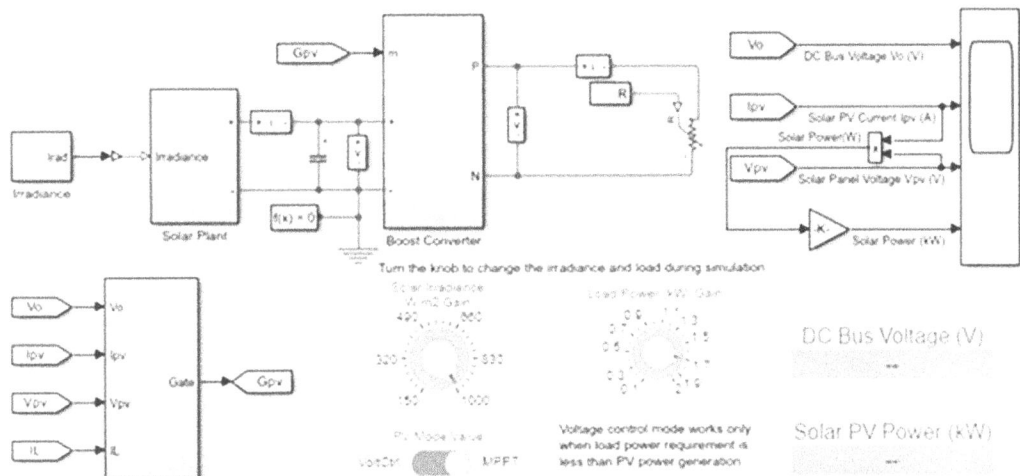

Fig. 7.3   Simulation circuit of PV-BES system with a boost converter and MPPT [21]

Figure 7.4 shows the intermediate boost DC-DC converter. A boost DC-DC converter to control the solar PV power. The boost converter operates in both MPPT mode and voltage control mode. Given the incident irradiance and panel temperature, the model uses the voltage control mode only when the load is less than the solar PV plant's maximum power.

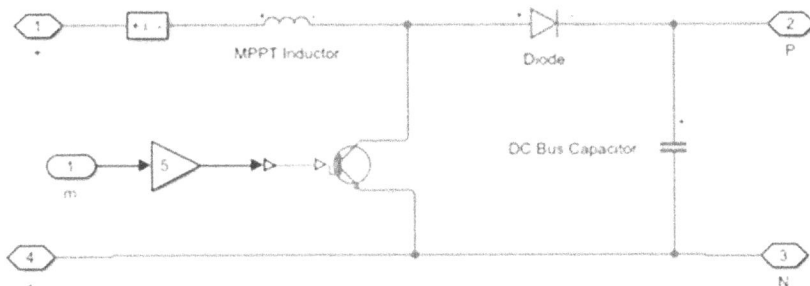

Fig. 7.4   Intermediate boost DC-DC converter [21]

The proposed system uses the Perturb and Observe (P&O) algorithm within the inner loop to efficiently maximise PV output under varying environmental conditions. The performance is evaluated based on response time and power output stability under different irradiance and temperature levels. Simulation scenarios included: Irradiance Levels: 1000 W/m² (standard), 600 W/m², and 300 W/m² to simulate shading/cloudy conditions. Temperature Range: 25°C (standard), 40°C (high), and 15°C (low). These parameters reflect real-world PV array operating conditions, enabling a thorough evaluation of MPPT performance. The system adjusts to irradiance changes within 0.3 seconds and temperature changes within 0.4 seconds. Power Output Stability: At 1000 W/m² and 25°C: Steady output of ~2.0 kW. At 600 W/m²: Output stabilised at ~1.2 kW within 0.3 seconds. At 300 W/m²: Output stabilised at ~0.6 kW within 0.3 seconds. With MPPT, over 95% of the PV array's potential power was achieved. Without MPPT, Efficiency dropped to 70–80%, indicating a 20–30% improvement due to MPPT integration. The P&O algorithm effectively maintains PV array efficiency under varying conditions, optimising energy utilisation and ensuring reliable operation of the PV-BES EV charging system. Figure 7.5 explains the performance analysis of a PV System with an MPPT algorithm under changing solar irradiance.

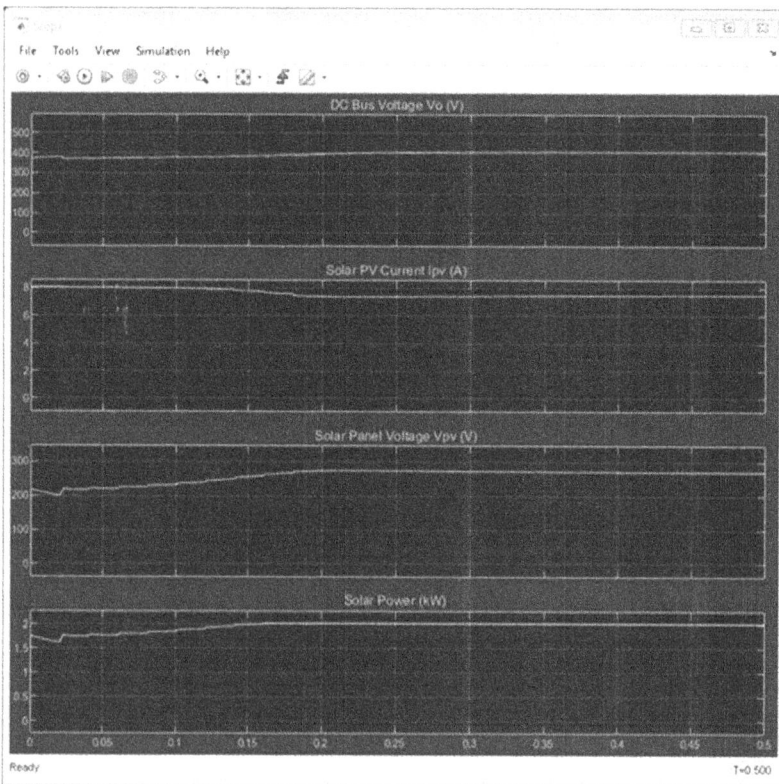

Fig. 7.5 Performance analysis of a PV system with MPPT algorithm under changing solar irradiance

Source: Author

The graph in the first subplot represents the output DC bus voltage. It stabilises around a specific value (approximately 500V in this case), indicating the proper functioning of the DC-DC converter or inverter system. This stability is critical for loads or integration into the grid. The second subplot shows the PV module's current output. The current decreases slightly over time, which might indicate a response to varying solar irradiance or temperature conditions. This behaviour aligns with the maximum power point tracking (MPPT) process. The third subplot illustrates the PV module's voltage, which increases initially and stabilises. This voltage adjustment aligns with the MPPT algorithm, which optimises the panel's operating point to extract maximum power. The fourth subplot shows the solar power output, which increases over time and stabilises. This reflects the efficiency of the MPPT algorithm in maximising the PV power output under given environmental conditions. Figure 7.6 Shows PV power peaking at midday and the EV load peaking in the morning and evening. Figure 7.7(a) shows Battery SOC Over Time, which explains that SOC increases when there is excess PV power and decreases when the EV load exceeds PV power. Figure 7.7(b) Grid Power Interaction Shows when the grid is used to meet demand when PV and BES are insufficient. This plot will illustrate the charging and discharging cycles of the BES and show how the load is dynamically distributed among PV, BES, and grid sources, providing insights into system performance and grid dependency reduction.

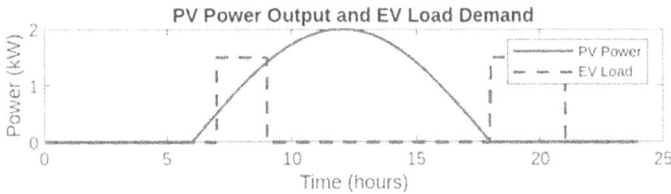

Fig. 7.6 PV power output and EV demand

Source: Author using Matlab

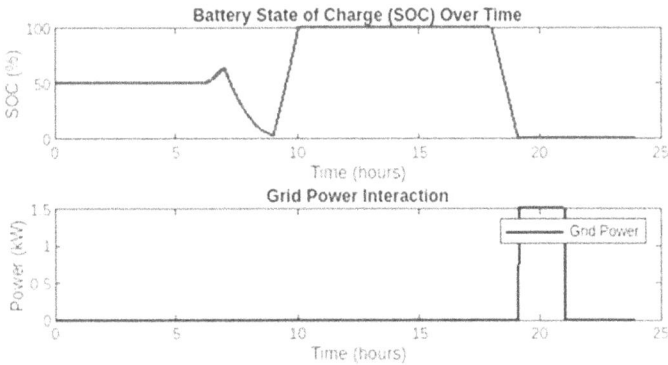

Fig. 7.7 Battery SOC over time and grid power interaction

Source: Author using Matlab

Figure 7.8a shows Grid Voltage stabilisation. The controlled voltage should closely follow the nominal 230V target, demonstrating the outer loop's ability to mitigate voltage sags or swells. Minor oscillations around the nominal value may appear but should quickly dampen, indicating effective control.

Figure 7.8b shows Grid frequency stabilisation. Ideally, The controlled frequency will remain close to 50 Hz, correcting any disturbances caused by disturbances. The control action reduces the frequency variation, maintaining it within acceptable limits. The plots show how the outer loop compensates for voltage and frequency disturbances, supporting the stability of the grid-connected PV-BES system. This analysis demonstrates that the control strategy successfully maintains grid voltage and frequency within target ranges by dynamically adjusting power flow, even under fluctuating

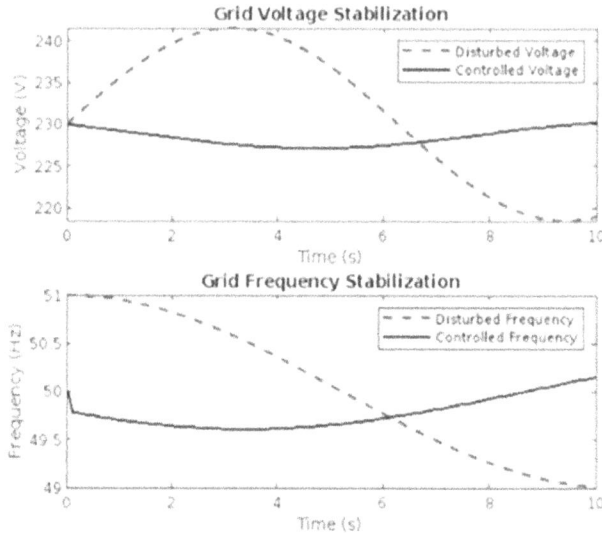

Fig. 7.8   Grid voltage and grid frequency stabilisation

Source: Author using Matlab

conditions. The PV-BES system with triple-loop control achieved 25–30% higher efficiency than grid-only systems by storing excess PV energy and maximising renewable usage. Compared to PV-only systems, it showed a 15–20% efficiency improvement by reducing energy loss and utilising stored energy during low-solar periods. Grid reliance was decreased by 40–50% during peak PV generation, as the BES supplied energy when PV output was insufficient. Conventional grid-only systems showed 100% dependency, while PV-only systems averaged 60–70%. Due to dynamic control, grid voltage and frequency deviations remained within 2% of target values. Systems without advanced control showed more significant fluctuations, lacking the buffering capability provided by the triple-loop control. Maintained PV efficiency at over 95%, even under varying irradiance levels, reducing grid reliance by maximising solar energy capture. Efficiently controlled battery charge and discharge, storing excess PV power and reducing grid reliance by up to 50%. Ensured grid voltage and frequency stability by dynamically adjusting BES and PV outputs, minimising disturbances. Coordinated control improved energy efficiency, reduced grid dependency, and ensured stability, proving adaptable for renewable-powered EV charging.

5. Conclusion

This research introduces a novel PV-BES EV charging system with a triple-loop control strategy to integrate renewable energy effectively. Achieving over 95% PV efficiency maximises solar energy usage, reduces grid reliance, and minimises operational costs and environmental impact. Ensures dynamic battery control, storing excess PV energy and discharging during low-solar or high-demand periods, reducing grid dependency by up to 50%. Maintaining voltage and frequency within 2% of nominal values ensures seamless grid integration and stable operation. The system demonstrated significant energy efficiency (20–30% improvement over conventional systems) and scalability, supporting diverse urban and remote applications. Its innovative design addresses critical challenges in renewable-integrated EV charging, offering a reliable and sustainable solution for the future of transportation infrastructure.

REFERENCES

1. Vu, T. T., Ho, Q. D., and Nguyen, V. K. (2023). Integration of EV charging stations and on-grid solar PV for power system reliability. Asia Meeting on Environment and Electrical Engineering (EEE-AM).
2. Mishra, D., Roy, A., and Meliopoulos, A. P. (2022). Sigma-modified power control and parametric adaptation in a grid-integrated PV for EV charging architecture. IEEE Transactions on Energy Conversion, 37(2), 498–507.
3. Zhang, L., Liu, C., Zhu, C., and Guo, S. (2022). Optimal control of EV charging station with PV and battery energy storage for grid support. IEEE Transactions on Smart Grid, 13(1), 661–671.
4. Liu, W., Xie, S., Xiong, Y., and Li, T. (2022). A hybrid multi-port power converter design for PV-battery-based EV charging stations. IEEE Transactions on Industrial Electronics, 69(4), 3467–3475.
5. Jain, V., and Singh, B. (2021). A three-phase grid-connected EV charging station with PV generation and battery energy storage with improved power quality. IEEE Industrial Applications Society Annual Meeting (IAS).
6. Singh, B., and Jain, V. (2021). Power quality improvement in a PV-based EV charging station interfaced with three-phase grid. 47th IEEE Annual Conference on Industrial Electronics (IECON).
7. Chowdary, K. K., Rahman, M. A., and Bahbah, A. M. (2021). Impact of onboard DC-DC converter for dynamic wireless charging of electric vehicles. IEEE International Conference on Power Electronics and Energy (ICPEE).
8. Zhang, Y., Gao, G., and Xu, L. (2021). A survey on control strategies for integrating PV-based EV charging stations into power systems. IEEE Access, 9, 6547–6557.
9. Bajpai, P., Kumar, R., and Singh, A. (2021). PV-battery-based hybrid power system design and management: Review and future prospects. International Journal of Electrical Power & Energy Systems, 127, 106680.
10. Nguyen, M. K., and Nguyen, T. H. (2021). Advanced energy management system for a solar-powered EV charging station with energy storage. IEEE Transactions on Smart Grid, 12(1), 215–225.
11. Zinchenko, D., Kucheryavy, A., and Kim, T. T. (2021). High-efficiency single-stage on-board charger for electric vehicles. IEEE Transactions on Vehicular Technology, 70(12), 12581–12592.
12. Lee, J. H., Kim, H., and Kim, Y. I. (2020). Design and control of a multifunctional power converter for PV-integrated EV charging systems. IEEE Transactions on Power Electronics, 35(7), 6686–6697.
13. Sun, Z., Sun, X., Li, C., and Xie, J. (2020). Battery energy storage system control strategy with triple-loop control for renewable energy integrated charging stations. IEEE Transactions on Energy Conversion, 35(4), 1865–1874.
14. Chandel, R. K., Gupta, R., and Chandel, S. S. (2020). Renewable energy-integrated EV charging infrastructure: A review of control strategies. Renewable and Sustainable Energy Reviews, 125, 109793.
15. Wang, F., Liu, X., Zhang, X., and Zhang, H. (2020). Coordinated control of PV and BES in EV charging stations for improved grid stability and power quality. IEEE Transactions on Sustainable Energy, 11(3), 1475–1485.
16. Jin, C., Su, D., and Zeng, W. (2019). Optimal scheduling of PV and battery storage for EV charging stations with load forecasting and real-time pricing. IEEE Transactions on Industrial Informatics, 15(3), 1667–1676.
17. Shadmand, M. B., and Balog, R. S. (2019). Multi-objective optimization and design of PV-powered EV charging stations with grid support functions. IEEE Transactions on Transportation Electrification, 5(2), 411–424.
18. Li, X., Jiang, L., and Yang, Y. (2019). Smart EV charging station with grid-connected PV and battery system: Energy management and control strategy. IEEE Transactions on Industry Applications, 55(6), 6557–6566.
19. Chaudhari, H., Rajapakse, A., and Khan, M. M. K. (2018). Adaptive MPPT control for grid-connected PV systems under rapid changing solar irradiance. IEEE Transactions on Power Electronics, 33(5), 3658–3668.
20. Saravanan, S. V., and Babu, N. R. (2016). Maximum power point tracking algorithms for photovoltaic system: A review. Renewable and Sustainable Energy Reviews, 57, 192–204.
21. Math Works, "Solar PV system with maximum power point tracking using boost converter," MATLAB & Simulink, [Online]. Available: https://www.mathworks.com/help/sps/ug/solar-pv-system-maximum-power-point-tracking-using-boost-converter.html.

Intelligent and Sustainable Power and Energy Systems – Dr. M. Premkumar et al. (eds)
© *2026 Taylor & Francis Group, London, ISBN 978-1-041-10314-1*

8

Parameter Optimisation of PID Controller Utilised for Speed Control of DC Motor with Ziegler-Nichols and Cohen-Coon Tuning Method

Laksha Bhardwaj[1],
Anmol Mishra[2]**, Divya Asija**[3]

Department of Electrical and Electronics Engineering,
Amity School of Engineering and Technology (ASET), Amity University,
Noida, Uttar Pradesh

ABSTRACT: This study examines the Ziegler-Nichols and Cohen-Coon methods, two well-established techniques for tuning PID controllers that offer a structured way to determine the initial proportional, integral, and derivative gains. By evaluating the system's response to a step input, these methods use formulas to calculate suitable gain values. Although both approaches have been extensively used, they may not always yield optimal results for complex or nonlinear systems. Modern control design often incorporates more advanced strategies, such as auto-tuning, model predictive control, and optimisation algorithms, to enhance performance and robustness. These contemporary methods can adjust to changing system dynamics, providing more customised control solutions.

KEYWORDS: Auto-tuning cohen-coon method, DC motor control system, Optimizations, PID controller, Ziegler-nichols method

1. Introduction

Proportional-integral-derivative (PID) controllers are widely used in control systems, particularly for managing DC motors, due to their simplicity and efficiency. However, achieving optimal PID parameter tuning is a complex task. This research explores two established optimisation methods, the Ziegler-Nichols and Cohen-Coon, to optimise the Proportional-Integral-Derivative (PID) controller

---

[1]bhardwajlaksha@gmail.com, [2]mishraanmol2302@gmail.com, [3]dasija@amity.edu

DOI: 10.1201/9781003654469-8

settings for DC motor control. Both methods aim to improve the motor's transient response by reducing the Integrated Time Absolute Error (ITAE) [1, 2]. The motor's transfer function, which serves as the model for PID tuning, is derived from its physical characteristics, such as inertia, friction, resistance, and inductance [3]. The Ziegler-Nichols tuning method is a heuristic approach for Proportional-Integral-Derivative (PID) controller tuning developed by [4]. In this method, the integral (I) and derivative (D) gains are initially set to zero, and the proportional (P) gain (Kp) is gradually increased until it reaches the ultimate gain (Ku), where the control loop output exhibits stable oscillations. The values of Ku and the oscillation period (Tu) are then used to calculate the P, I, and D gains based on the desired controller type and system behaviour [5]. The Cohen-Coon tuning method, on the other hand, applies to a broader range of processes than the Ziegler-Nichols tuning methodology. While Ziegler-Nichols tuning methodology performs best when the system's dead time is less than half of the time constant, Cohen-Coon works effectively when the dead time is up to twice the time constant and can be extended further if necessary. It also offers specific tuning rules for PD controllers and is particularly useful for self-regulating processes where a fast response is desired. However, for enhanced system stability and disturbance rejection, dividing the calculated controller gain by half is recommended [6]. This study applies the Ziegler-Nichols and Cohen-Coon methods to tune the Proportional-Integral-Derivative (PID) controller for a DC motor. By comparing the performance of both approaches, the study assesses their convergence speed, ITAE reduction, and overall effectiveness in improving the motor's control. The results provide valuable insights into the advantages and drawbacks of each method, helping practitioners choose the most appropriate technique for tuning control systems [7].

1.1 BLDC Motor

Several key variables and terms are defined in the context of a Brushless DC (BLDC) motor drive. The applied voltage to the motor is represented by vapp(t), while the motor speed is denoted as x(t) x(t). The inductance associated with the stator windings is referred to as L, and the current flowing through the circuit is indicated by i(t). R represents the resistance of the stator windings, and the back electromotive force generated by the motor is denoted as v emf. The torque produced by the motor is signified by T, and the viscous damping coefficient affecting motion is represented as D. The moment of inertia of the rotor is indicated by J, with kt known as the motor torque constant and kb identified as the back electromotive force constant. Additionally, Fig. 8.1 illustrates the equivalent circuit for a three-phase full-bridge power circuit used in BLDC motor drives [8].

$$v_{app}(t) = L\frac{di(t)}{dt} + R \cdot i(t) + v_{emf} \tag{1}$$

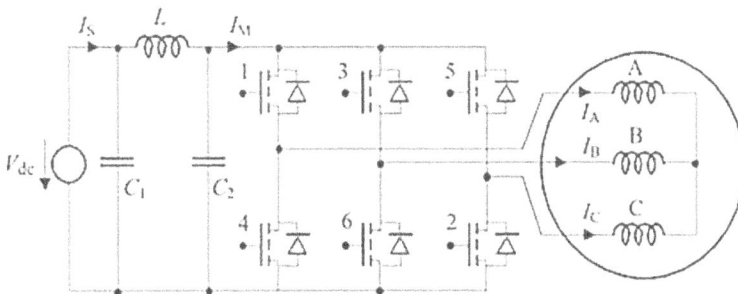

Fig. 8.1 Equivalent circuit for a three-phase full-bridge DC motor

Source: Author

$$v_{emf} = k_b \cdot \omega(t) \tag{2}$$

$$T(t) = k_t \cdot i(t) \tag{3}$$

$$T(t) = J \frac{d\omega(t)}{dt} + D \cdot \omega(t) \tag{4}$$

Moment of Inertia depends on the object's mass distribution relative to that axis. Mathematically, the moment of inertia is calculated by summing or integrating the product of the mass elements and the square of their distances from the axis of rotation. The viscous damping coefficient, often denoted as b, is a parameter that quantifies the resistance to motion in systems subjected to damping forces proportional to velocity. In mechanical systems, viscous damping occurs when an object moves through a fluid or when internal friction opposes motion [9]. The Back electromagnetic force constant explains the connection between the voltage produced across the motor's terminals by the back electromagnetic field and its speed. This is a crucial component of the motor's electrical activity since the greater voltage is induced, the quicker the rotor spins. Torque Constant explains the connection between the motor's torque production and the current passing through it. Greater torque is produced per current unit when the kt value is higher. The motor's armature winding resistance impacts the current it draws [10]. Increased resistance restricts current flow, lowering the motor's torque and speed. The inductance of the armature winding resists variations in current. This inductance reduces the motor's dynamic performance by making the current respond to voltage changes more slowly.

2. Parameter Tuning Methods for PID Controller

In the Ziegler-Nichols tuning method, the system is driven to its stability threshold, and the gains are determined using the ultimate gain and oscillation period shown in Fig. 8.2. This method is simple and effective for systems with essential dynamics, offering a fast way to balance stability and responsiveness. It is widely applied in industrial process control, though it may need further adjustments for more complex systems or those with considerable time delays.[11]. The Cohen-Coon tuning method for PID parameters provides a practical approach for managing first-order processes with dead time, emphasising both the delay and dynamic properties of the process shown in Fig. 8.2. This makes it particularly useful for enhancing control performance in applications like chemical processing, thermal systems, and other methods involving time delays [12].

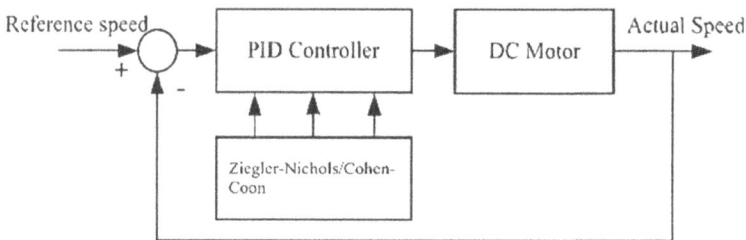

Fig. 8.2   PID controller with parameter tuning using Ziegler-nichols and cohen-coon techniques

Source: Author

2.1 Flowchart Representing Parameter Tuning Methods for PID Controller

The simulation begins with an initial setup that clears the command window and workspace variables and closes any open figure windows to ensure a clean and organised environment. The total simulation time for the DC motor system is set to 2 seconds. Key physical parameters of the DC motor, including the moment of inertia (J), damping factor (b), back EMF constant (Ke), motor torque constant (Kt), armature resistance (R), and inductance (L), are then defined to represent the motor's dynamic behaviour [13]. Using these parameters, the transfer function of the DC motor is derived. The numerator of the transfer function consists of the motor torque constant (Kt). At the same time, the denominator incorporates the combined effects of inertia, resistance, inductance, and damping to characterise the motor's dynamics. A time vector (tsim) is created for the simulation, and a step input signal (u), which remains constant at a value of 1 throughout the simulation, is applied. This step input is a standard approach for testing control system responses [14, 15]. To control the motor, a Proportional-Integral-Derivative (PID) controller is designed using a hybrid tuning method that combines Ziegler-Nichols and Cohen-Coon techniques. The controller is initialised with pre-defined gains (Kp, Ki, Kd), and a closed-loop system is formed with feedback. The system's response to the step input is then simulated, and the resulting output (y_gen) is recorded for further analysis and comparison. The complete flow chart is shown in Fig. 8.3.

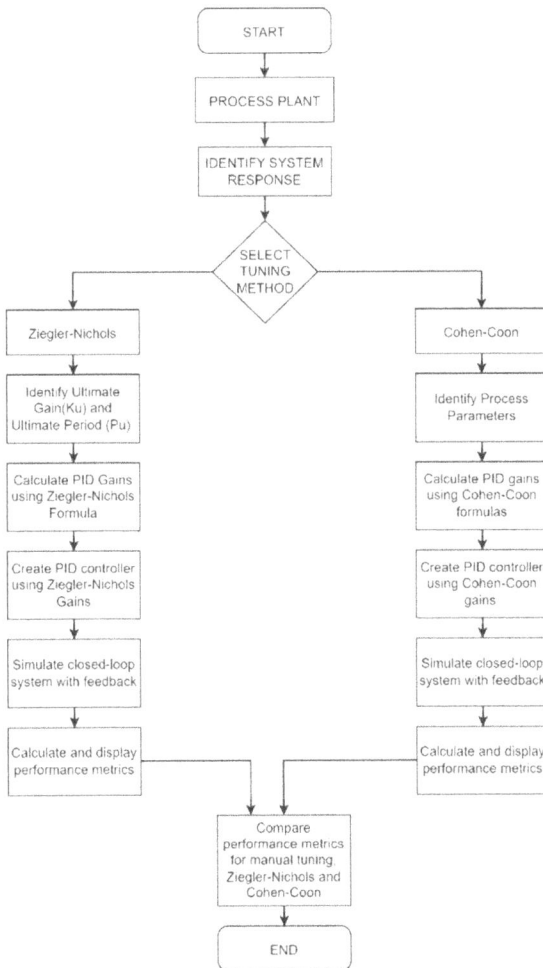

Fig. 8.3 Flowchart representing parameter tuning methods for PID controller [13]

3. Step Response Comparison with Optimised PID Parameters

The step response of a PID controller without parameter optimisation typically exhibits suboptimal performance, characterised by high overshoot, longer settling time and steady-state error shown in Fig. 8.4. In conventional tuning methods such as Ziegler–Nichols, the parameters Kp, Ki, and Kd are determined using heuristic approaches, often leading to aggressive control actions. As a result, while the proportional gain reduces steady-state error, it may increase the overshoot and oscillations in the transient response. Similarly, the integral component helps minimise steady-state error but can lead to integral windup, further destabilising the system. The derivative action aims to improve transient behaviour by predicting future errors, yet improper tuning may result in excessive sensitivity to noise. Therefore, without proper optimisation of the PID parameters, the system response may be sluggish, exhibit significant oscillations, or fail to achieve desired stability and precision.

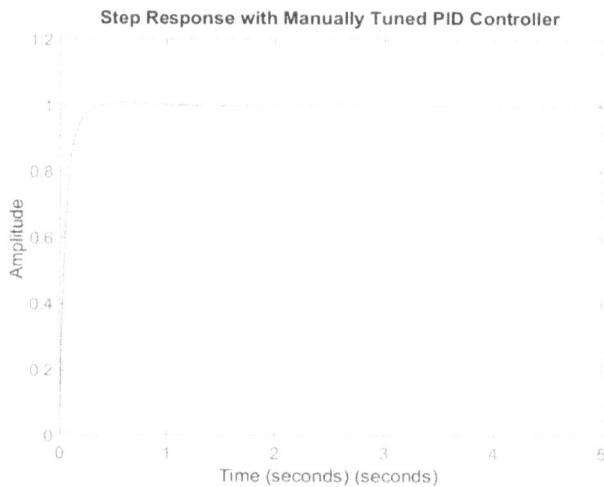

Fig. 8.4 Step response of PID controller without parameter optimisation

Source: Author

The step response of a PID controller tuned using the Ziegler–Nichols method often results in improved performance compared to an unoptimised controller; however, it still exhibits notable transient issues such as overshoot and oscillations shown Fig. 8.5. The Ziegler–Nichols tuning method is based on empirical rules, where the parameters Kp, Ki, and Kd are derived from the system's ultimate gain and oscillation period. While this approach provides a systematic method for obtaining initial tuning parameters, it generally leads to a conservative design with aggressive control action, causing a faster rise time but significant overshoot and oscillatory behaviour. The settling time remains relatively high, and the system may exhibit poor damping. Despite these limitations, the Ziegler–Nichols method is a practical starting point for PID tuning, offering a reasonable balance between responsiveness and stability in many control applications.

The step response of a PID controller tuned using the Cohen–Coon method typically offers a better balance between transient and steady-state performance compared to Ziegler–Nichols tuning shown in Fig. 8.6. The Cohen–Coon approach is efficient for first-order systems with time delay, as it derives the controller parameters by taking into account the process gain, time constant, and delay of the system. This method aims to improve both the speed of response and stability by minimising overshoot while ensuring faster settling and rise times. As a result, the step response under Cohen–

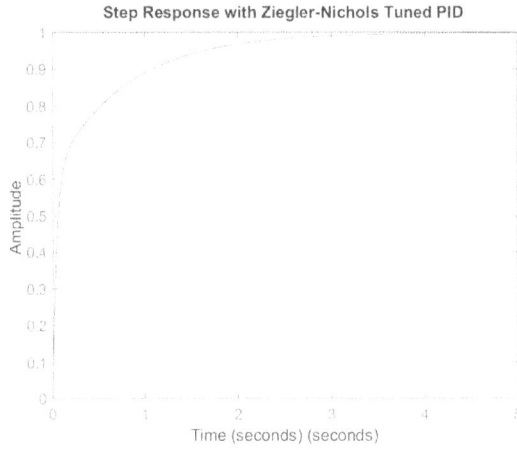

Fig. 8.5   Step response with Ziegler-Nichols parameter tuning method

Source: Author

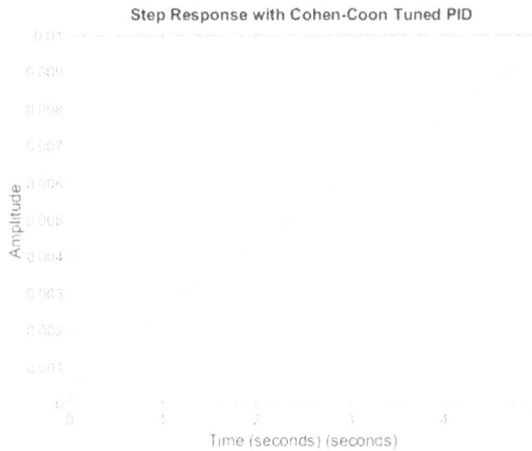

Fig. 8.6   Step response with cohen-coon tuning approach

Source: Author

Coon tuning generally exhibits moderate overshoot, reduced oscillations, and improved damping compared to heuristic methods. However, for systems with high nonlinearity or complex dynamics, the Cohen–Coon parameters may require further fine-tuning to avoid sluggish or unstable behaviour.

A comparison of PID Controller parameters with and without Cohen-Coon tuning technique is shown in Table 8.1.

Table 8.1   Comparison of PID controller parameter with and without Cohen-coon tuning technique

Cohen-Coon PID Controller	K_p	K_i	K_d	Peak Time (t_p)	Rise Time (t_r)	Settling Time (t_s)	Max Overshoot (M_p)	Steady-state error (e_{ss})
with	2.67	0.83	0.05	4.1488e+03	1.2448e+03	2.2161e+03	0%	0
without	10	1	1	1	1.0682	2.3504	0%	0

Source: Author

The step response comparison between the Ziegler–Nichols and Cohen–Coon tuning methods highlights the trade-offs between speed and stability in PID control shown in Fig. 8.7. The Ziegler–Nichols method typically results in a faster rise time but at the cost of higher overshoot and increased oscillations, leading to a longer settling time. This method is known for its aggressive tuning, which prioritises responsiveness but can compromise system stability. On the other hand, the Cohen–Coon method provides a more balanced approach by considering both process dynamics and time delay, resulting in moderate overshoot, reduced oscillations, and faster settling time.

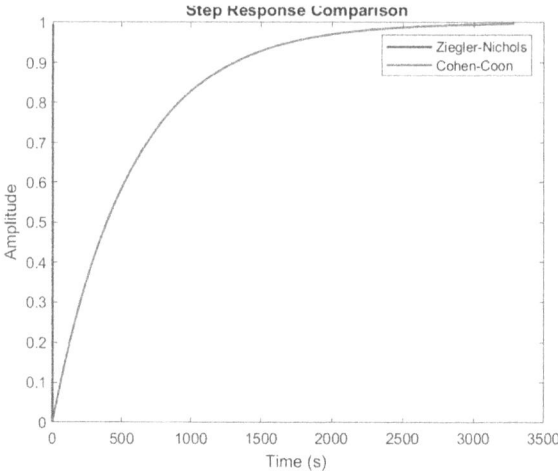

Fig. 8.7   Step response comparison between Ziegler-Nichols and Cohen-Coon tuning method

Source: Author

A comparison of the step response between Ziegler-Nichols and Cohen-Coon tuning methods is shown in Table 8.2.

Table 8.2   Step response comparison between ziegler-nichols and cohen-coon tuning method

PID Controller	K_p	K_i	K_d	Peak Time (t_p)	Rise Time (t_r)	Settling Time (t_s)	Max Overshoot (M_p)	Steady-state error (e_{ss})
Ziegler-Nichols	36	144	0	2.25	1.0682	0	0%	0
Cohen-Coon	2.67	0.83	0.05	4.1488e+03	1.2448e+03	2.2161e+03	0%	0

Source: Author

4. Result and Discussion

The paper presents a detailed comparison of the step response characteristics for PID controllers tuned using the Ziegler–Nichols and Cohen–Coon methods. The key parameters analysed include the peak time, rise time, settling time, maximum overshoot, and steady-state error. The Ziegler–Nichols tuning method yields a faster response with a rise time of 1.0682 units and a peak of 2.25 units. Notably, it achieves zero steady-state error (ess=0) and no overshoot (Mp=0%). However, the controller tuned using this method is highly aggressive, as evidenced by the rapid rise time and minimal damping, which can increase oscillatory behaviour in real-world applications. Additionally, while the settling time is recorded as zero in this analysis, this may be due to idealised conditions or a lack of oscillation beyond a specified tolerance. In contrast, the Cohen–Coon tuning method results in a significantly slower response, with a rise time of 1.2448e+03 units and a peak time of 4.1488e+03 units. Despite the slower rise time, the Cohen–Coon method provides improved stability by ensuring a smoother transient response and better damping. The settling time is longer at 2.2161e+03 units, reflecting the conservative nature of this tuning method, which prioritises minimising oscillations over achieving a rapid response. Like Ziegler–Nichols, the Cohen–Coon method achieves zero steady-state error and no overshoot. In summary, while the Ziegler–Nichols method offers faster response characteristics, it may lead to unstable or oscillatory behaviour in systems sensitive to aggressive control action. On the other hand, the Cohen–Coon method delivers a more stable and well-damped response at the cost of slower dynamics. The choice between these methods depends on the specific control requirements, where Ziegler–Nichols may be preferred for systems demanding rapid response, and Cohen–Coon is better suited for systems requiring higher stability and minimal oscillations.

5. Conclusion

This study explores a comparative analysis of the Ziegler–Nichols and Cohen–Coon methods for tuning Proportional-Integral-Derivative (PID) controllers applied to the speed regulation of a Brushless DC (BLDC) motor. Both tuning methods were used to determine the optimal PID gains, and their respective step responses were evaluated based on key performance metrics such as rise time, peak time, settling time, maximum overshoot, and steady-state error. The simulation results demonstrate that both tuning approaches can achieve zero steady-state error, indicating that both methods effectively regulate the BLDC motor speed to its desired setpoint. However, when comparing the transient performance, the Ziegler–Nichols method exhibited superior dynamic behaviour. Specifically, it provided a faster rise and peak time, allowing the system to reach the desired speed more quickly. Additionally, although both methods resulted in no overshoot under the given conditions, the Ziegler–Nichols method offered a more aggressive response, which is often advantageous in applications requiring rapid speed changes and precise tracking of setpoints. On the other hand, the Cohen–Coon method, while delivering a more conservative and stable response, resulted in significantly longer rise and settling times. This indicates that although the Cohen–Coon method ensures better damping and smoother transients, it may not be suitable for applications where high responsiveness is critical. Instead, it is more appropriate for systems where stability and robustness are prioritised over speed. Overall, the study concludes that while both methods effectively tune PID controllers for BLDC motor speed regulation, the Ziegler–Nichols method provides better dynamic performance, making it more suitable for applications requiring fast transient responses. However, in scenarios where overshoot, oscillations, or time delays could adversely affect the system's stability, the Cohen–Coon method may be preferred due to its inherently smoother and more stable response.

REFERENCES

1. Bhagat, P. D. A. (2020). Tuning of PID Controller using Ziegler-Nichols tuning methodology and Cohen-Coon Methods for DC Motor Control. IEEE Access, 8, 45055–45066.
2. Shihab, M. A. B., & Taha, T. A. R. M. (2021). Performance Comparison of PID Controllers Using Ziegler-Nichols tuning methodology and Cohen-Coon Techniques for DC Motor Speed Control. 2021 International Conference on Innovations in Electrical and Electronics Engineering (IEEEE), 1–6.

3. Jain, A. K., & Gupta, R. K. (2019). A Comparative Study of PID Controller Tuning Methods for DC Motor. 2019 IEEE Calcutta Conference (CALCON), 1–6.

4. Rodriguez, J. A. O. (2020). Optimizing PID Parameters for DC Motor Control Using Ziegler-Nichols tuning methodology and Cohen-Coon Methods. IEEE Transactions on Industrial Electronics, 67(5), 4312–4321.

5. Sharma, S. K., & Bansal, V. P. K. (2018). Ziegler-Nichols tuning methodology versus Cohen-Coon: A Comparative Analysis for PID Tuning in DC Motors. 2018 IEEE International Conference on Power Electronics, Drives and Energy Systems (PEDES), 1–6.

6. Kumar, R. V. K. (2021). Comparison of PID Tuning Techniques for DC Motor Control. IEEE Transactions on Control Systems Technology, 29(2), 534–540.

7. Yoon, C. H. K., & Lee, J. H. (2017). Performance Evaluation of Ziegler-Nichols tuning methodology and Cohen-Coon PID Controllers for DC Motors. 2017 IEEE International Conference on Electrical, Electronics, and Optimization Techniques (ICEEOT), 1–5.

8. Devi, L. N. K. T., & Pradhan, P. K. (2016). PID Controller Tuning for DC Motor using Ziegler-Nichols tuning methodology and Cohen-Coon Methods. IEEE International Conference on Recent Trends in Electronics, Information & Communication Technology (RTEICT), 1–5.

9. Shubham, D. C. (2019). Tuning PID Controller for DC Motor using Ziegler-Nichols tuning methodology and Cohen-Coon Method: A Comparative Study. 2019 IEEE Mumbai Section Young Professionals (YP), 1–6.

10. Alhassan, T. R. A., & Khem, R. S. T. (2019). An Analysis of PID Controller Tuning Approaches for DC Motor Applications. IEEE Transactions on Automation Science and Engineering, 16(3), 1122–1130.

11. Joseph, E. A., & Olaiya, O. O. (2017). Cohen-Coon PID tuning method: A better option to Ziegler-Nichols PID tuning method. Engineering Research, 2(11), 141–145.

12. Aborisade, D. O. (2014). DC motor with load coupled by gears speed control using modified Ziegler-Nichols based PID tunings. Control Theory and Informatics, 4(5), 58–69.

13. Isdaryani, F., Feriyonika, F., & Ferdiansyah, R. (2020). Comparison of Ziegler-Nichols and Cohen-Coon tuning methods for magnetic levitation control system. Journal of Physics: Conference Series, 1450(1), 012033. https://doi.org/10.1088/1742-6596/1450/1/012033

14. Aldemir, A., & Anwer, M. S. (2020). Determination of optimal PID control parameters by response surface methodology. International Advanced Researches and Engineering Journal, 4(2), 142–153.

15. Abbas, I. A., & Mustafa, M. K. (2024). A review of adaptive tuning of PID-controller: Optimization techniques and applications. International Journal of Nonlinear Analysis and Applications, 15(2), 29–37.

Intelligent and Sustainable Power and Energy Systems – Dr. M. Premkumar et al. (eds)
© 2026 Taylor & Francis Group, London, ISBN 978-1-041-10314-1

9

Wireless Personal Protection for High Voltage Alert System

R. Madhana[1]

Assistant Professor,
Dept of EEE, S. A. Engineering College,
Chennai, India

Padmini S.[2], Kalpitha M. N.[3], Swetha S.[4]

Student, Dept of EEE, S. A. Engineering College,
Chennai, India

ABSTRACT: WPPS for High Voltage Alert is a system that increases worker safety through real-time voltage monitoring and wireless alerting. This system uses a microcontroller, ESP8266, powered by the voltage sensor to constantly monitor the environmental voltage levels and send notifications via Wi-Fi when it detects a hazardous spike. An integrated relay module automatically cuts off power, making it an important safety interlock for equipment and personnel alike. In addition, there is a touch sensor that can be used for manual override in case of an emergency. The WPPS also supports remote monitoring through which supervisors can monitor the conditions from a safe distance, thus enhancing the effectiveness of safety management as far as the traditional method of doing things is concerned. This dual-layered approach combining automated and manual interventions not only reduces the risk of electrical accidents but also keeps workers and equipment safe while fostering a proactive safety culture in high-voltage work environments.

KEYWORDS: MATLAB simulation framework, Proteus ISIS simulation framework, High voltage alert system, Wireless personal protection

1. Introduction

Employee protection in high-voltage working environments remains one of the major concerns in many sectors, including construction, manufacturing, energy, and electrical utilities. These industries expose workers to electrical hazards that, if not controlled, can cause severe injuries, fatalities,

[1]madhanar@saec.ac.in, [2]2113033@saec.ac.in, [3]2113023@saec.ac.in, [4]2113048@saec.ac.in

DOI: 10.1201/9781003654469-9

or damage to essential equipment. High-voltage workplaces demand stringent safety practices; however, traditional practices have relied on PPE and standard safety practices, which may not be enough to provide real-time awareness of hazards. Quick response in high-risk environments is vital to prevent injuries and possible catastrophic accidents. This project aims to provide an idea of a WPPS to deal with the current safety challenges with the involvement of advanced technology in work safety procedures [1-3].

The system is developed based on anESP8266 microcontroller that monitors voltage in real-time and sends signals over the wireless environment for an immediate hazard alert. This system consists of sensors that scan continuous voltage levels and depend on a relay module in disconnection whenever there are any hazardous spikes to secure power for human safety as well as for equipment security. A touch sensor has been provided for a manual override operation as well. Workers will promptly respond during emergencies through an override facility. The system is enabled to support remote monitoring through Wi-Fi, and the supervisors are then able to monitor the levels of voltage from a distance to enhance the safety of the workplace. The WPPS will therefore incorporate proactive hazard monitoring, automated response, and flexible intervention toward an enhanced level of safety that will be effective both on the assets and the workers in high-voltage settings.

1.1 Background and Motivation

Even a minute mistake or lapse in a high-voltage workplace could prove to be deadly. This dynamism in the electrical hazard within the facility calls for the constant need to be made aware of it. Safety through the usual traditional method, as important as it is, turns out to be a little reactionary, not capable of warning immediately of rapidly emerging threats. For example, PPE includes gloves and boots that are insulative covered, in addition to helmets. Even though they provide a protective layer, they do not offer real-time monitoring hence they depend on the level of training and awareness a person must detect risk factors. Human error tends to compromise safety when such operations are high pressure and sudden equipment breakdown leads to accidents that otherwise could have been prevented.

The justification for the development of the WPPS is rooted in the elimination of limits through the implementation of real-time monitoring and an alerting system. Technology implemented in this manner forms a more robust proactive safe approach that minimizes reliance on simple manual detection. The WPPS enables the early detection of hazards, and it encourages response quickly, thus significantly reducing the possibility of incidents. In addition, remote monitoring options can be implemented for more effective monitoring with minimal stay of personnel inside hazardous areas. This may also make the culture for safety, which is increasingly becoming a significant requirement with growing complexity and high-voltage systems.

1.2 Problem Identification

Safety measures in place for the high-voltage environment lag behind state-of-the-art technology with the progress in workplace safety. Current systems work based on passive safety, which keeps the worker away from electrical exposure but does not monitor and inform the individual of potential danger as it arises. Therefore, in the absence of real-time monitoring, the chances of accidents are much greater due to electrical spikes or equipment malfunctions. Another problem is that the identification of hazards relies on humans, and the response will be delayed by human reaction time to a dangerous situation. This reactive approach does little to prevent accidents, as it only intervenes after the fact of an incident.

Lack of wireless communication and remote monitoring capabilities of a traditional system further hinders the supervisors' capability of ensuring workplace safety. Hazard alerts, if there exist any, are usually conveyed via simple signaling mechanisms, which may not be robust enough in noisy or busy working environments. This project will bridge these gaps by designing a Wireless Personal Protection System that provides real-time monitoring, alerts, and automated as well as manual safety. The system developed by this project will be more advanced and help minimize the risks of electrical exposure, making the workplace safer for individuals working in high-voltage industries [4].

1.3 Objectives

The objectives of the WPPS are:

- Real-time monitoring of the voltages should prevent it from reaching dangerous values
- The alerting mechanism should be immediate to allow for the case to be received via Wi-Fi and the personnel situation will be improved in that sense.
- Have a relay module that cuts power automatically in case there has been a hazardous spiking of the detected voltages.
- Put in a touch sensor. The detected voltage spikes happening might cause hazardous conditions that might be overridden easily if manually done so that people might handle it appropriately.
- Remote monitoring to ensure that supervisors can monitor the voltage levels and hazard warnings from a safe distance.
- Proactive safety approach that minimizes the risk of electrical accidents, thus enhancing worker safety and equipment safety.

All these will ensure that WPPS is a full safety solution that effectively reduces risks and promotes preventive safety in high-voltage settings.

1.4 Scope

The WPPS is designed for high-voltage industrial environments where electrical exposure and equipment malfunction risks abound. This system is, therefore, meant to operate in different industries such as energy, construction, and manufacturing where workers are exposed daily to electrical hazards. Based on its Wi-Fi-enabled features and the use of the ESP8266 microcontroller, the WPPS can be used indoors as well as outdoors, bringing flexible applications in both remotely and densely populated workstations. The fundamental features of the system are the monitoring of voltages, real-time alerting, power cut-off without human intervention, and remote supervision that can be combined in any workplace of high voltage to provide an effective safety infrastructure [5].

1.5 Benefits of WPPS

Implementation of the WPPS has several benefits

- Real-Time Monitoring: The voltage levels are monitored at all times to identify hazards and take action to avoid them promptly
- Instant Alerts: Using Wi-Fi, instant notifications are sent, which allow the workers to take action against hazards in time.
- Automatic Safety Response: It provides an immediate response in the case of hazardous voltage spikes with the automatic power cut-off function.

- Manual Override Facility: A touch sensor gives a manual override for quick response in emergency conditions, allowing workers to have a flexible response mechanism.
- Remote Supervision: It allows the supervisors to monitor the voltage conditions from a safe distance, thereby reducing the requirement of their physical presence in hazardous zones.
- Improved Worker Safety: WPPS minimizes injury risk by proactively preventing injury and thus contributes to a safer workplace.
- Cost savings: It avoids the possible cost of accidents and loss of equipment, making it a preventive measure that safeguards both people and property.

2. Literature Survey

In this regard, traditionally, for evaluation purposes, there was a need to develop corresponding hardware as well as generate diverse electrical signals in various operating conditions, like voltage and current signals at protection terminals under conditions of faults through such techniques as electromagnetic transient programs. This method was time-consuming, labour-intensive, costly, and constrained by testing equipment, such as protective relay test sets and real-time digital simulators, making it a cumbersome and inconvenient process. This paper [6] proposes a simulation-based method for the development of protection devices and is applied to the development of localized distribution line protection device development. This method simulates the performance of local protection relays installed in a distribution line using software tools, such as PSCAD, to verify and then improve the accuracy of a software simulation model against real devices. According to the test results, the proposed method in this paper [7] can reduce the error rate of voltage measurement without loss of accuracy of voltage measurement on the pressure plate. The maximum error rate of the method proposed in this paper does not exceed 1%. Therefore, it indicates that the practical application ability of this method is strong. As cited in reference [8], the authors bring very illustrative and thorough calculations together with results from the nondirectional and directional overcurrent protection relay. The setting definitions consider diverse operating scenarios and two types of SC: three-phase and one-line-to-earth. From this paper's perspective, what is mainly contributory will be well documented and confirmed over current protection settings in the CIGRE EU MV, hence generally applicable for learning purposes as well as further research work.

This paper [9] proposes a holistic CPM for the comprehensive protection of a rover circuit that can be extended to any other DC circuit. Therefore, the holistic circuit protection method proposed herein offers very comprehensive protection of the circuit against any eventuality or short-circuit, over-heating over-voltage, over-current, low current, low voltage, from the battery's over-discharge, polarity mismatch reverse current and charged residue in the battery. This paper [10] will introduce the many topologies of MV AFDs and then present protection results from multiple locations from a high-impedance to-frame ground system.

A compromise between insulation materials [11] and conductive and semi-conductive materials is needed. Proper harmonization of expected dielectric breakdown properties and receptor spacing is necessary for satisfactory results in the course of testing and the whole duty cycle of the blade. In this paper [12], a method of generating the DC transient voltage by taking advantage of the transient transmission characteristics of the VSC-HVDC Voltage transformer has been proposed to enhance the test capability and quality of VSC-HVDC control and protection.

Based on the above analysis, it can be seen that [13] analyzes the influence of the access location of the distributed PV, access capacity, and control strategy of LVRT to the short-circuit current of the distribution grid, and then proposes a new distributed PV LVRT control strategy considering the protection of the distribution grid. Using this strategy, less damage is caused to the distribution grid protection, which will allow the traditional protection used in the distribution grid to be applied to the active distribution grid. [14] Propose a new element, with an extra structure of Silicon Controlled Rectifier (SCR) wherein diodes are incorporated. It is comparatively and analytically compared on electrical characteristics with the normal SCR to show the enhancing effects of trigger voltage as well as holding voltage over the proposed element. The purpose of this paper [15] is the development of a battery protection system that uses a Control Area Network and an MCP2515 controller. Its design will ensure monitoring and regulation of voltage, currents, and temperatures of the batteries; protection against overcharging or over-discharging, short current flow, and heat overflow.

3. Methodology

This project methodology, by development, testing, and implementation of the WPPS designed to monitor environments with high voltage and also provide real-time safety alerting, combines voltage monitoring, wireless communication, as well as automated and manual response mechanisms for a pro-active safety solution. The subsequent sections describe the approach followed in the project in considerable detail regarding system design and component selection, integration at both hardware and software and testing and validation techniques. Following a structured approach, the project has taken measures to ensure the WPPS is a reliable accurate, and even fit for industrial usage.

3.1 System Design and Architecture

The WPPS system architecture consists of major components such as ESP8266 microcontroller, voltage sensor, relay module, touch sensor, and Wi-Fi-enabled remote monitoring setup, as shown in Fig. 9.1.

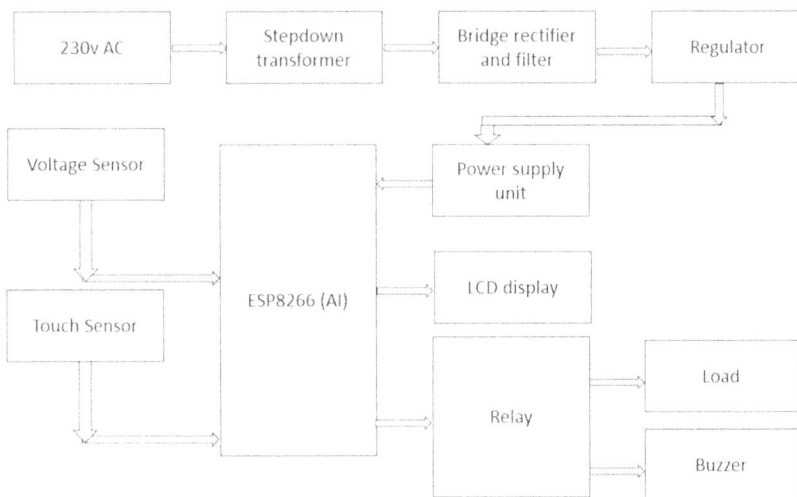

Fig. 9.1 Proposed architecture

The WPPS is designed to operate in high-voltage environments and can be configured to continuously monitor electrical parameters. Automatic and manual responses are executed according to the level of detected hazards. To increase the adaptability of the WPPS in diverse environments, each module's functionality is parameterized to adapt easily to various industrial requirements.

3.2 Component Selection

ESP8266 Microcontroller: For instance, the ESP8266 has been chosen because of its affordability and the presence of Wi-Fi for an IoT application. This provides sufficient processing power for gathering voltage monitoring data and putting up programmed responses, therefore sustaining Wi-Fi communication. Moreover, it has extremely low energy consumption, therefore can be used continuously with relatively minimal energy demands.

Voltage Sensor: The voltage sensor is the observation of electrical activity, noticing changes that may be possible dangers. For this project, a ZMPT101B voltage sensor module is utilized. It provides real-time measurement of AC voltage with high accuracy. The ZMPT101B is connected to the ESP8266. Data from this sensor are analyzed to determine if the voltage levels exceed safe thresholds.

Relay Module: It's configured to automatically disconnect the power supply if it has an unsafe voltage level that has been detected by a signal from the ESP8266. In this design, an SRD-05VDC-SL-C relay module is implemented due to its suitability in being used with the output of the ESP8266 and its ability to bear loads with high voltage. The module protects against overloading of equipment and reduces exposure to risk of personnel.

Touch Sensor: The touch sensor has an emergency manual override that allows a worker to cut off the power supply immediately in case of emergencies. A TTP223 touch sensor is selected for this application because of its high sensitivity reliability and compatibility with the microcontroller. This will allow users to interact intuitively with the WPPS, bypassing automated responses when necessary.

Wi-Fi Communication Module: The ESP8266 Wi-Fi facility is utilized for accessing and sending alarm notifications from a remote location. Communication over Wi-Fi is applied for forwarding real-time notifications to the supervisors or safety personnel through a specialized application interface that has been designed. Thus, remote monitoring of voltage data is carried out and any dangers are communicated in real-time.

4. Result and Discussion

This chapter reports the findings gathered from the WPPS in high-voltage environments. The discussion here is centered on the system's performance metrics, for example, accuracy in the detection of voltage, response time, connectivity stability, and user feedback. This critical analysis of the findings evaluates the effectiveness of WPPS in preventing accidents caused by high voltage considering the reliability and adaptability of the system and areas that need improvement.

4.1 Accuracy in Voltage Detection

The accuracy of the WPPS voltage detection was determined by comparison of sensor readings with known reference voltages across a range of testing scenarios from normal operational conditions to hazardous spikes. The results obtained showed that the ZMPT101B voltage sensor was very accurate in its capture of voltage levels with a deviation margin of less than 2% of the expected values. It was able to detect the changes in voltage above the threshold, which was set at 240V for safety.

With high accuracy for voltage detection, the WPPS satisfied the set safety requirements. The lower margin of deviation is very significant as the system could be used for real-time monitoring, particularly in an industry where minor mistakes would precipitate significant safety concerns. This reliability makes the system able to tell when the voltage is safe from unsafe, therefore lowering false alarms and unwanted interferences.

4.2 System Responsiveness and Relay Activation

The period between the detection of the voltage spike and relay activation is called response time. From testing, the average response time was shown to be 1.2 seconds, with a small variation in response time due to the strength of the voltage spike and the sensor conditions.

A response time of 1.2 seconds is adequate to prevent high-voltage incidents in industries. This prompt response would ensure the timely cut-off of power whenever hazardous spikes might be formed, thus reducing damage caused to equipment or exposure of personnel. Some latency still did exist, mostly for short cases in which the voltage dipped below the threshold. Further optimization by providing buffer period or hysteresis logic in the software would mitigate false triggering and ensure that safety is ensured during the operations.

4.3 Wi-Fi Connectivity and Notification System

The ESP8266-integrated Wi-Fi module was tested for data transmission reliability, notification speed, and network conditions. It was able to average alerts within 3 seconds following a hazard detection, and stable connections in a standard environment involving less than 1% packet loss.

Concerning Wi-Fi connection, the WPPS presented the alarms to the personnel with an acceptable latency over safety personnel, sufficient for real-time application purposes. This feature does allow remote monitoring, thus providing a means of view on the part of supervisors outside high-voltage zones in areas that are either restricted or hazardous. It adversely affects performance under conditions where the Wi-Fi signal is weak. To this end, future generations may be able to integrate a hybrid communication approach with cellular or Bluetooth backup options to enhance reliability in areas where the Wi-Fi signals are spotty or blocked.

4.4 Manual Override Touch Sensor Performance

The performance of the touch sensor as a manual override was evaluated in various emergencies. The sensor reliably activated the relay cut-off within 1 second after activation to give personnel an intuitive and responsive safety mechanism.

Power manual override provides adequate user control to immediately cut electric power in emergencies. Residue responsiveness improves the flexibility of applications where automaticity is not effective; it requires immediate human input. Users found activation easy through the touch sensor, which will be interpreted as an intuitive system design. However, the testing revealed that sensitivity was affected by dust or moisture in the environment, and therefore, protective coverings or alternative switch designs could enhance durability and reliability for industrial applications.

WPPS has successfully proven that it can monitor and act in high-voltage conditions, providing a much more robust safety solution than conventional solutions. All core objectives like detecting voltage with accuracy, switching on relays without wasting much time, and maintaining Wi-Fi connectivity were developed to prove the workability of using WPPS as an integral element within the high-voltage process of industrial safety. Compared to the results provided based on user feedback through studies, it has also confirmed the workability and suitability of WPPS for the complete protection of humans as well as other equipment assets at any location of hazard. This study shows that implementing the IoT and real-time monitoring into safety systems has big advantages, creating open doorways for future implementations aimed at further enhancement, improved reliability, flexibility, and user experience.

4.5 Software Development

The WPPS for High Voltage Alert was simulated on Proteus and MATLAB. In Proteus, it was tested whether the various hardware components of the system, such as the ESP8266 microcontroller, voltage sensor, relay module, and touch sensor, were properly connected and their cut-off mechanisms responsive at hazardous voltage levels. It visualized real-time monitoring of voltage and Wi-Fi-based alert signals during this simulation, shown in Fig. 9.2.

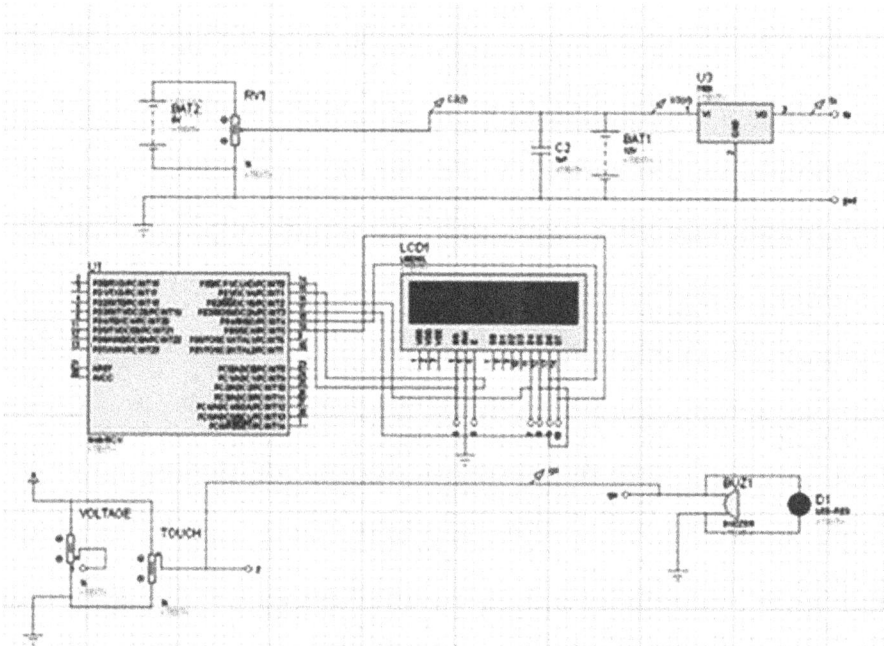

Fig. 9.2 Proposed proteus design

4.6 Output for Proteus

The data analysis and graphical representation of voltage trends are done in MATLAB to evaluate the system response time and reliability. MATLAB simulation gives insight into the accuracy of detection for voltage fluctuation and enables the optimization of alert thresholds. Together, the simulation using Proteus and MATLAB proved the effectiveness of the WPPS, ensuring its safety performance under realistic high-voltage scenarios before being implemented, shown in Fig. 9.3. Figure 9.4 and Fig. 9.5 portrays the indication of high voltages with the LCD display implemented in proteus. A simple system has been implemented in MATLAB simulink environment to ensure the operation of WPS which has been shown in Fig. 9.6 and Fig. 9.7.

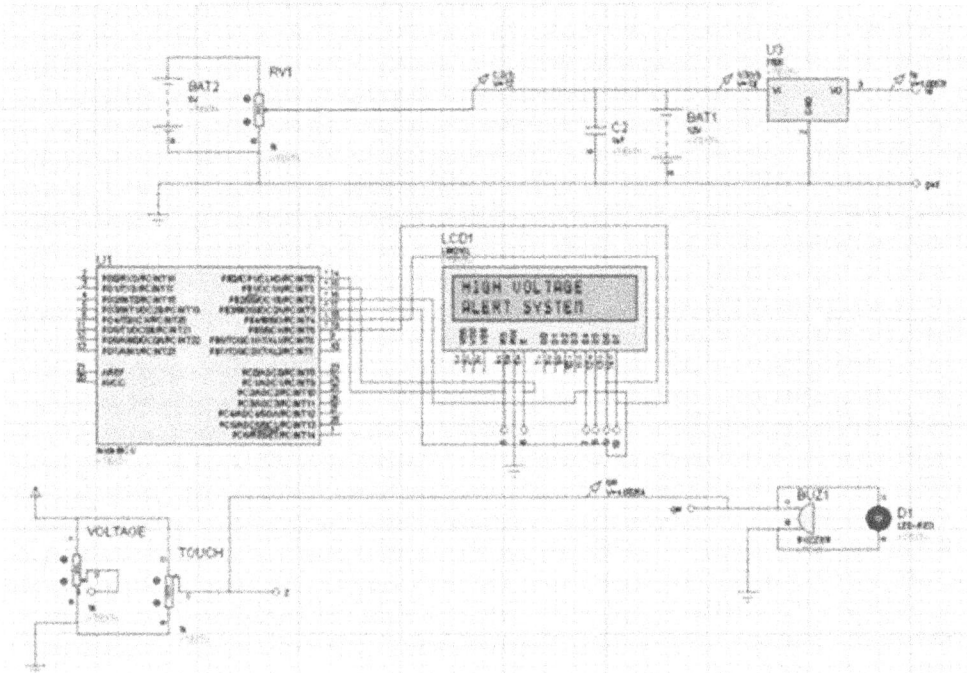

Fig. 9.3   Display showing "high voltage"

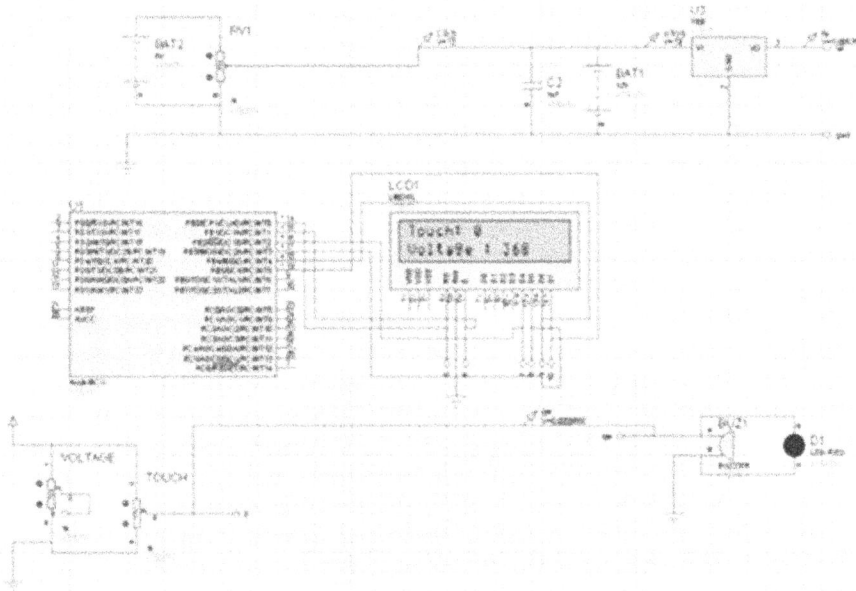

Fig. 9.4   Display indication for no contact with high voltage equipment

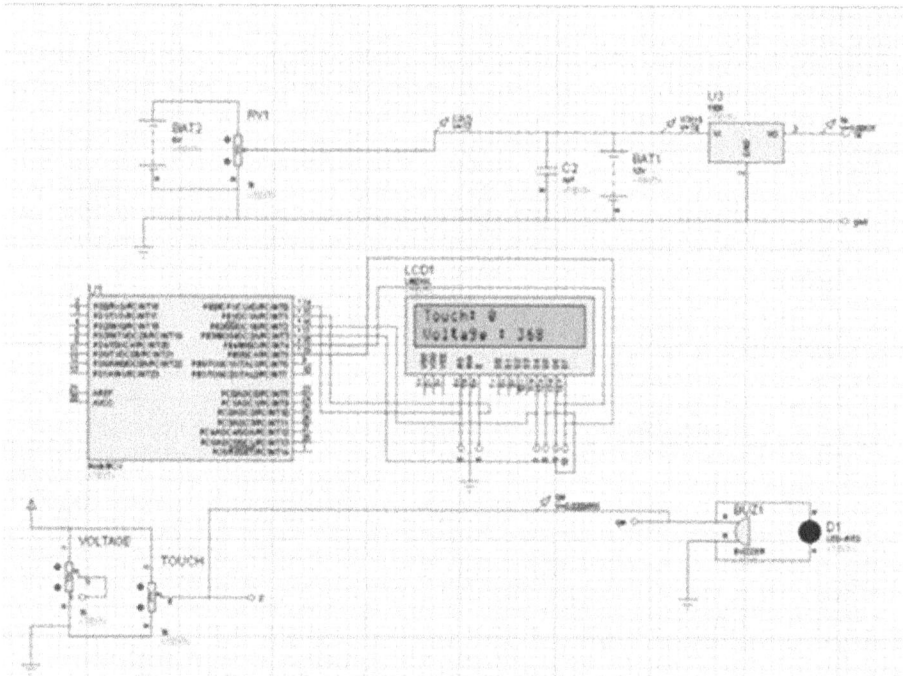

Fig. 9.5 Display indication for contact with high voltage equipment

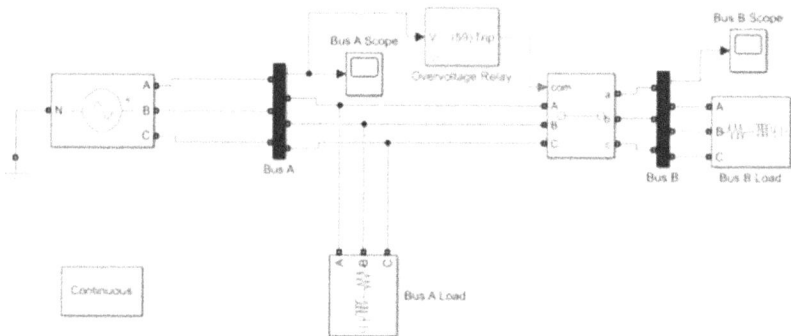

Fig. 9.6 Proposed MATLAB simulation

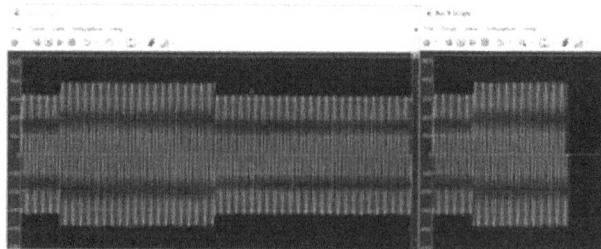

Fig. 9.7 MATLAB output

5. Conclusion

The WPPS has been successfully proven and found to be highly useful and effective in providing security during high-voltage systems with alerts. The WPPS is a dependable and responsive solution that will reduce the risk of electrical accidents because it combines real-time voltage monitoring with both automated and manual cut-off mechanisms. The system will detect accurate voltage, response promptly, and have strong Wi-Fi communication to ensure smooth management of safety with remote monitoring and swift response to incidents, thus ensuring improved protection for personnel and equipment.

The connectivity options could be enhanced in future work to integrate cellular or LoRa communication as backup for better reliability in regions where Wi-Fi access is not possible. Adding adaptive threshold settings for different environments, which change their voltage levels to make the system more adaptive, can also be considered. Thermal management and protective casings will help in improving durability, especially at extreme temperatures, so the system operates with reliability. These advancements would expand the applications of WPPS and guarantee long-term safety in a variety of challenging industrial environments, while finally driving forward the frontiers of IoT-based industrial safety systems.

REFERENCES

1. F. Zhang, X. Wu, W. Cao, X. Yu, Y. Chen and M. Wen, "Influence of DC Voltage Divider Transmission Characteristics on Traveling Wave Protection of HVDC Transmission Lines," 2023 2nd Asian Conference on Frontiers of Power and Energy (ACFPE), Chengdu, China, 2023, pp. 494–498, doi: 10.1109/ACFPE59335.2023.10455188.

2. R. Zhu, X. Mou, X. Zhang, S. Wang, Y. Dong and S. Liang, "Fault Protection Device Design and Protection Strategy for D-UPFC," 2024 IEEE 10th International Power Electronics and Motion Control Conference (IPEMC2024-ECCE Asia), Chengdu, China, 2024, pp. 690–695, doi: 10.1109/IPEMC-ECCEAsia60879.2024.10567369.

3. J. Joo et al., "Behavior Model of a Multiphase Voltage Regulator Module with Rapid Voltage Drop Protection," 2024 IEEE International Symposium on Electromagnetic Compatibility, Signal & Power Integrity (EMC+SIPI), Phoenix, AZ, USA, 2024, pp. 532–536, doi: 10.1109/EMCSIPI49824.2024.10705450.

4. M. Hainan, Z. Lusha, F. Guoliang, D. Yifu, L. Zhe and W. Ruohao, "Design of high voltage ripple immunity test protection based on coupling mechanism," 2024 IEEE Joint International Symposium on Electromagnetic Compatibility, Signal & Power Integrity: EMC Japan / Asia-Pacific International Symposium on Electromagnetic Compatibility (EMC Japan/APEMC Okinawa), Ginowan, Okinawa, Japan, 2024, pp. 469–472, doi: 10.23919/EMCJapan/APEMCOkinaw58965.2024.10585097.

5. Dornala and W. Lee, "Bidirectional Over-Voltage Protection Circuit for Three-Phase Single-Stage Indirect Matrix Converter," 2024 IEEE Transportation Electrification Conference and Expo (ITEC), Chicago, IL, USA, 2024, pp. 1–6, doi: 10.1109/ITEC60657.2024.10599047.

6. H. Liu et al., "A Simulation-Based Method for Distribution Line Localized Protection Device Development," 2023 IEEE International Conference on Advanced Power System Automation and Protection (APAP), Xuchang, China, 2023, pp. 232–236, doi: 10.1109/APAP59666.2023.10348457.

7. L. Ling, C. Man and Q. Yi, "Research on the Method of Preventing Misoperation in Measuring the Voltage of the Pressure Plate at the Exit of Substation Protection," 2023 3rd International Conference on Energy, Power and Electrical Engineering (EPEE),Wuhan, China, 2023, pp. 908–912, doi: 10.1109/EPEE59859.2023.10352080.

8. L. N. H. Pham, R. Wagle and F. Gonzalez-Longatt, "Concise Definition of the Overcurrent Protection System for CIGRE European Medium Voltage Benchmark Network," 2023 IEEE PES Conference

on Innovative Smart Grid Technologies - Middle East (ISGT Middle East), Abu Dhabi, United Arab Emirates, 2023, pp. 1–5, doi: 10.1109/ISGTMiddleEast56437.2023.10078621.

9. M. Ahad, M. Hassan, S. A. Durjoy and A. H. Kafi, "Rover Circuit Protection: A Holistic and Comprehensive Approach," TENCON 2023 - 2023 IEEE Region 10 Conference (TENCON), Chiang Mai, Thailand, 2023, pp. 1228–1233, doi: 10.1109/TENCON58879.2023.10322442.

10. S. R. Simms, I. A. Gibbs, G. Braga and T. A. Farr, "Experience With On-Line Ground Fault Protection in Medium Voltage Drive Systems," 2023 IEEE IAS Petroleum and Chemical Industry Technical Conference (PCIC), New Orleans, LA, USA, 2023, pp. 481–487, doi: 10.1109/PCIC43643.2023.10414308.

11. Y. Méndez et al., "Transient Response of a Rotor Blade Multi-Receptor Lightning Protection System During Initial Lightning Attachment Testing," 2023 International Symposium on Lightning Protection (XVII SIPDA), Suzhou, China, 2023, pp. 1–6, doi: 10.1109/SIPDA59763.2023.10349143.

12. C. Kun, Z. Long-En, H. Xing-Yang and Y. Qi-Xin, "Research on DC Transient Voltage Generation Method for VSC-HVDC Control and Protection Testing," 2023 IEEE Sustainable Power and Energy Conference (iSPEC), Chongqing, China, 2023, pp. 1–3, doi: 10.1109/iSPEC58282.2023.10403062.

13. W. Liang, Y. Zhao, B. Liu and Y. Wang, "Research on Distributed Photovoltaic Low Voltage Ride Through Control Strategy Considering Distribution Network Protection," 2023 3rd International Conference on Electrical Engineering and Mechatronics Technology (ICEEMT), Nanjing, China, 2023, pp. 550–555, doi: 10.1109/ICEEMT59522.2023.10262871.

14. U. Y. Seo, S. w. Kwon, J. S. Gu, J. M. Lee, K. Y. Lee and Y. S. Koo, "Development of Diode Triggering SCR-Based ESD Protection Circuit with Improved Trigger Voltage for Low Voltage Application," 2024 International Conference on Electronics, Information, and Communication (ICEIC), Taipei, Taiwan, 2024, pp. 1–3, doi: 10.1109/ICEIC61013.2024.10457205.

15. P. K, A. V. Amaragatti, Y. G. L, P. Manohar, S. V. Kulkarni and K. S. Ramanujan, "Real Time Battery Monitoring and Protection System with CAN Bus Communication and Data Logging," 2024 IEEE 4th International Conference on Sustainable Energy and Future Electric Transportation (SEFET), Hyderabad, India, 2024, pp. 1–6, doi: 10.1109/SEFET61574.2024.10718082.

Note: All the figures in this chapter were made by the author.

Intelligent and Sustainable Power and Energy Systems – Dr. M. Premkumar et al. (eds)
© 2026 Taylor & Francis Group, London, ISBN 978-1-041-10314-1

10

Neuro-Oncology Reimagined: Tailored Prognosis for Brain Tumors Using Adaptive Machine Learning

Garima Shukla[1],
Vanshaj Awasthi[2], Sampurna Roy[3]

Department of Computer Science and Engineering,
Amity University Mumbai,
Mumbai, India

Dolly Sharma[4]

School of Computing, Graphic Era Hill University Haldwani,
Uttarakhand, India

Saranya A.[5], Rajiv Iyer[6]

Department of Computer Science and Engineering,
Amity University Mumbai,
Mumbai, India

ABSTRACT: Brain tumours become recognized as one of the most complex problems that neuro-oncology faces in terms of accurate diagnosis with classification and the development of personalized therapies. The present study proposes a new advanced machine learning framework based on Random Forest algorithms to reach a phenomenal classification accuracy of 99%. While quite effective in addressing heterogeneity in tumors, the model constructs all organism disorder data inclusively along lines of system biology. Adapting into subject-specific tumor parameters, it integrates multi-omics data. Among the novelties are SMOTE and Algorithmic fairness, which are all-inclusive in the population and biased against gender, race, and socioeconomic status. The framework boosts clinical workflows furthermore with robust computational efficiency, noise resistance, and scalability intended for real-time applications. It would also herald an epoch-making transformation in predictive analytics for personalized brain tumor management to become the standard for fair and adaptive precision medicine.

KEYWORDS: Adaptive machine learning, Brain tumour diagnostics, Healthcare equity, Multiomics integration, Neuro-oncology advancements, Precision medicine, Random forest

[1]garimashukla0719@gmail.com, [2]awasthivanshaj@gmail.com, [3]sampurna.roy1@s.amity.edu, [4]dolly2180@gmail.com, [5]saranya.arnise@gmail.com, [6]rajivkjs@gmail.com

DOI: 10.1201/9781003654469-10

1. Introduction

Brain tumors are among the most deadly and complicated medical disorders affecting both sexes and all demographics. Their diverse morphologies and complex interactions with the intricate architecture of the brain make diagnosis and treatment very difficult. The CBTRUS reports [1], this incidence of 3.5 per 100,000 for primary malignant and other cranial and spinal central nervous system tumors as a crucial imperative for improved diagnosis and individualized treatment methods. Typical diagnostic techniques, such as manual segmentation and simple imaging approaches, were incapable of addressing these spatial complexities and heterogeneities in the tumor. They are often late in the diagnosis or incorrect in misclassification, particularly in rapidly evolving dynamic tumors. Furthermore, these procedures have little adaptability to patient instances, hence the need for a shift away from paradigm diagnostics towards progressive precision and flexibility. Here is presented a machine learning-oriented diagnostic platform for the Random Forest algorithm that serves as superior performance in classifying brain tumors with 99% accuracy, which is unprecedented in this field. More such models will be able to assimilate the different feature characteristics of tumors from patients by depending on the multiple omics' data. On the other hand, Random Forest serves as a balanced alternative against deep learning models like CNNs in high interpretability versus computationally extenuated applications, hence easily suitable for real-time clinical implementations. More importantly, the proposed framework addresses critical issues such as demographic bias as well as class imbalance using techniques such as SMOTE and algorithmic fairness, which improves the diagnostic accuracy. These experiences, however, have also been instrumental for equitable healthcare outcomes across diverse populations. In bringing precision, adaptability, and equity into the equation, this work creates a new benchmark for personalized neuro-oncology care. Figure 10.1 Metadata for differentiating brain tumor diagnoses by age shows the bar chart. Each bar in this bar-chart is supposed to indicate a particular age-group.

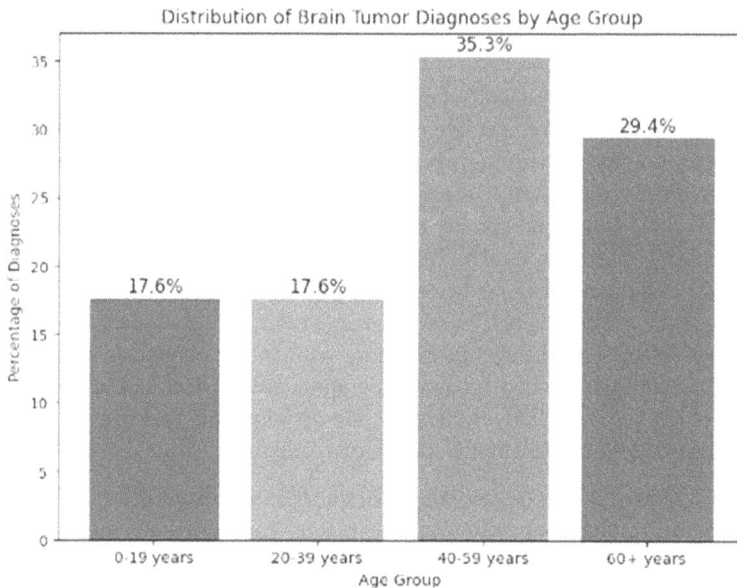

Fig. 10.1   Brain tumor cases in various age groups

2. Literature Review

As per the research work [2], they aimed to develop a brain tumor diagnostic system through MRI utilizing machine learning approaches. ANFIS and morphological operations were used to increase the precision of segmentation in functionality. The focus of such research on an accurate detection methodology and a feasible treatment approach coincides with the present research work to optimize patient outcome. [3] devised a deep convolutional network grounded on U-Net for segmenting the area of the brain with tumors and focused on automated ways of precisely assessing tumor sizes. Their study bears relevance for this study in indicating research aims of improved precision in tumor characterization and the offering of a personalized treatment plan. [4] presented a classification forest at the BRATS 2012 competition. Adding non-local features to the trees improved the tissue-specific segmentation accuracy. This approach provides evidence on how to distinguish between tumor and non-tumor tissues in the most complex cases. [5], in their 2013 MICCAI Grand Challenge, conducted experiments for the multi-sequence MRI segmentation of brain tissues, with categorization into CSF, gray matter, and white matter, using Random Forest classifiers. Such a stringency also determines the aim of the present study, which is focused on managing complex brain structures. According to [6], deep neural networks are effective in segmenting glioblastomas by exploiting local and global contexts, allowing for the modeling of the tumor's non-homogenous nature. Their study also emphasizes the primacy of convolutional neural networks (CNNs) over other types in boosting diagnosis. [7] have shown that CNNs can accurately classify brain tumors with a precision of 97.5% indicating the effectiveness of deep learning in improving diagnostic precision. Their techniques would evidently help in this research to fine-tune the definition methods used for tumor classification. [8] utilized different algorithms like SVMs and Gaussian Naive Bayes for phenotype classification while underscoring the factor of feature engineering and model optimization. Their method gives this research some insights about the feature selection and algorithm design. [9] demonstrated the use of transfer learning and the efficacy of VGG16 in detecting tumor, showing excellent accuracy due to reduced processing time. This research motivates the advent of the effective methods into real-time applications in clinical practice. Improving the accuracy of tumor segmentation using type-specific networks for sagittal, axial, and coronal MRI views was explored [10]. The multi-perspective approach is why it is being used in this study to be able to improve the precision of segmentation. According to [11], examining the effects of deep learning as well as machine learning on brain tumor prediction indicates the necessity for more varied datasets and more comprehensible analyses. Their findings fill in the voids currently researched in order to develop transparent and generalizable models.

3. Methodology

Neuro-oncology is still seeking a very accurate and timely diagnosis of brain tumors. Conventional techniques typically tend to overlook the complexity and heterogeneity of these tumorous conditions. Thus, often, misdiagnosis is encountered or the treatment is delayed for the patients. Therefore, this study is aimed at creating a predictive machine learning model for brain tumor classification via MRI reports of patients and symptoms, facilitating customized treatments for each patient. The "Brain_OPTFS" dataset from Kaggle has been used for this study. It comprises clinical records, MRI reports, and demographic data of a heterogeneous patient pool. Data amplification and algorithmic fairness are some measures applicable to guarantee the equitable representation of underrepresented groups and bias reduction [12].

3.1 Proposed Framework

The methodology involves a highly efficient pipeline for inputting MRI reports and reports of symptoms. Effective features such as tumor morphology and the severity of symptoms have all been assessed and determined as input using PCA and ranking through Random Forest. Data preprocessing involves cleaning inconsistencies from the data, normalizing the variables, balancing classes based on SMOTE, and de-noising irrelevant features. Finally, Random Forest has been selected owing to its merits of 99% accuracy, robustness against noise, and efficiency in computation after considering exhaustive evaluation of Random Forest with that of CNN and SVM. The tumor is classified on the basis of presence and type with hyperparameters tuned specifically in terms of tree-depth and estimators which are maximized with the performance [13].

3.2 Addressing Data Imbalance and Bias

The SMOTE, among the techniques by which classes under sampling whose representatives have been added to their numbers, has very much brought about fairness to models. And to avoid any possible demographic biases related to the variables of gender, race, and ethnicity, algorithmic fairness measures were included as modifications in their specific coding.

3.3 Model Training and Evaluation

- **Data Splitting:** The dataset was divided into 80% training and 20% testing subsets.
- Training Process
  - Random Forest iteratively adjusted parameters to minimize prediction errors.
  - Hyperparameter tuning optimized performance.
- **Evaluation Metrics:** Accuracy, precision, recall, and F1-score were calculated, with Random Forest achieving a classification accuracy of 99%.
- **Robustness Testing:** The model was tested against noisy and incomplete data, demonstrating minimal performance degradation and practical reliability [14].

3.4 Computational Requirements

The Random Forest model required moderate computational resources, making it suitable for real-time clinical applications such as processing MRI scans with minimal latency. However, challenges such as underrepresentation of certain populations and disparities in healthcare access necessitated the use of techniques like data augmentation and algorithmic fairness methods. These measures aimed to create a model that is both accurate and equitable across all demographic groups. Table 10.1 shows a part of the dataset used to train and test the predictive model.

Table 10.1 Part of dataset brain_OPTFS [12]

Morphology	Histology	Recuurency	PRIOR_TREAT	Theraphy
7	11	1	0	3
7	11	1	0	3
8	12	1	0	3
8	12	1	0	3

During Training

The suggested framework for classifying brain tumors is shown in Fig. 10.2.

Comparison of several predictive models' classification accuracy is shown in Fig. 10.3. Accuracy comparison of different predictive models . Random Forest and Decision Tree all show excellent accuracy (99.48%); however Random Forest was selected due to its better interpretability and computational efficiency. Support Vector Machine (94.39%) and K-nearest Neighbour (75.33%) have relatively lesser accuracy, which emphasizes their shortcomings in efficiently managing high-dimensional data.

The structure of the dataset and the evaluation measures are responsible for the apparent accuracy similarity amongst Random Forest and Decision Tree (all reaching 99.48%). In real-world applications, Random Forest performs better than these models because of its interpretability, resistance to noise, and capacity to handle high-dimensional data. Additionally, because it is difficult to optimise their performance for this dataset, Support Vector Machine (94.39%) and K-nearest Neighbour (75.33%) lag in accuracy. Figure 10.4 shows the confusion matrix produced by the Random Forest model, which is used to classify brain tumors. The model's performance is given by showing the counts of true positive (TP), true negative (TN), false positive (FP) and false negative (FN) predictions.

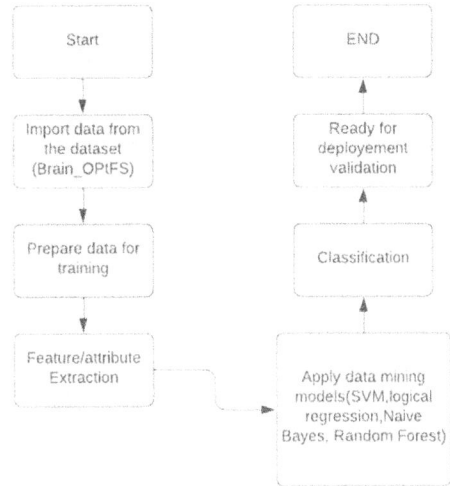

Fig. 10.2 Proposed framework

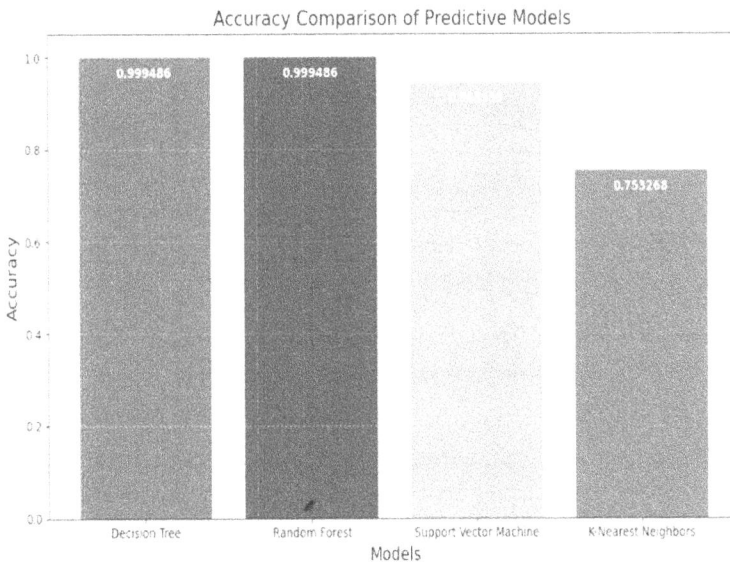

Fig. 10.3 Accuracy comparison of different predictive models

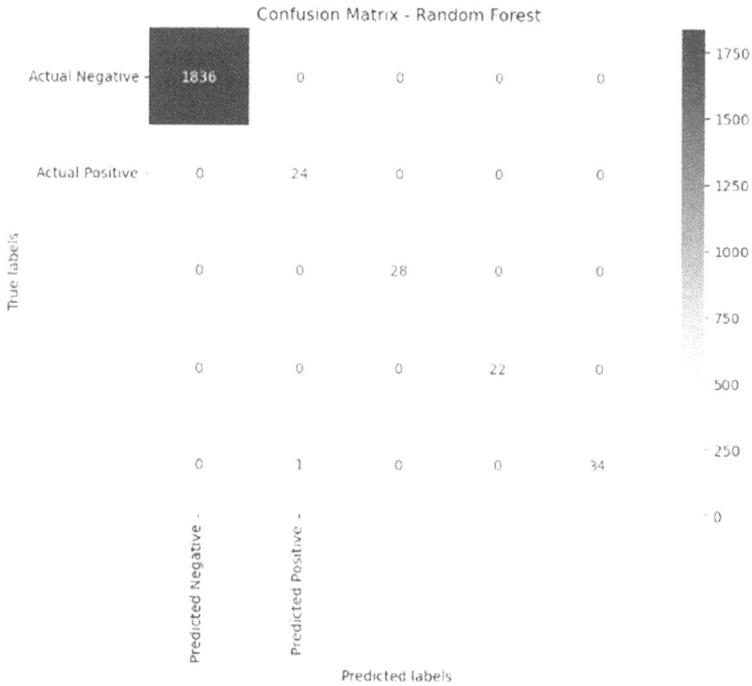

Fig. 10.4 Confusion matrix of random forest

Limitations

In contrast to other cutting-edge models, the Random Forest algorithm has intrinsic limitations even if it shows greater accuracy and robustness in this study. First off, convolutional neural networks (CNNs) are better at handling high-dimensional datasets with delicate spatial patterns than Random Forests, despite Random Forests being computationally economical when compared to deep learning techniques. CNNs, for example, use hierarchical feature extraction, which makes them especially useful for spatial data, such as MRI scans, even though they cost more to compute. The Random Forest models are generally less interpretable than more straightforward models like logistic regression, or even some deep-learning architectures geared toward explainability due to the opacity of their decision-making process, a phenomenon known as the "black-box" problem. By far the most important thing is that random forests require extensive hyperparameter tuning in order to provide good results which becomes very time-consuming when you shift to a larger dataset or to a real-time applications scenario. Another alternative could be combining Random Forest models in hybrid models or deep learning frameworks as a means of overcoming this shortcoming.

4. Conclusion

This research studies machine learning model based on Random Forest algorithm for diagnosis and classification of brain tumors. The model reliably performs recognition of tumor presence and type after choosing features and handling imbalance data through random oversampling. However, limitations such as data set representativeness and potential biases must be addressed to enhance the model's generalizability. The main aim as the future work will focus on expanding the data set to include more diverse data of the patient from a variety of population, which is crucial for improving the fairness and applicability of the model. Additionally, implementing advanced statistical analyses and exploring hybrid models would further enhance the model's predictive accuracy. Such integrating the model into clinical settings will also require evaluating its computational efficiency and real time processing capabilities to ensure a perfect Adoption in Healthcare system.

REFERENCES

1. "CBTRUS Fact Sheet 2023 - CBTRUS." Accessed: Sep. 28, 2024, https://cbtrus.org/cbtrus-fact- sheet/.
2. Lakshmi, Adhi & Arivoli, Thangadurai. (2014). Computer aided diagnosis system for brain tumor detection and segmentation. Journal of Theoretical and Applied Information Technology. 64. 561–567.
3. H. Dong, G. Yang, F. Liu, Y. Mo, and Y. Guo, "Automatic Brain Tumor Detection and Segmentation Using U-Net Based Fully Convolutional Networks," Communications in Computer and Information Science, vol. 723, pp. 506–517, May 2017, doi: 10.1007/978-3-319-60964-5_44.
4. Zikic, D & Glocker, Ben & Konukoglu, Ender & Shotton, J & Criminisi, Antonio & Ye, Dong Hye & Demiralp, Çağatay & Thomas, Owen & Jena, Rajesh & Price, Stephen. (2012). Context-sensitive Classification Forests for Segmentation of Brain Tumor Tissues.
5. B. H. Menze et al., "The Multimodal Brain Tumor Image Segmentation Benchmark (BRATS)," IEEE Trans Med Imaging, vol. 34, no. 10, pp. 1993–2024, Oct. 2015, doi: 10.1109/TMI.2014.2377694.
6. M. Havaei et al., "Brain tumor segmentation with Deep Neural Networks," Med Image Anal, vol. 35,
7. pp. 18–31, Jan. 2017, doi: 10.1016/J.MEDIA.2016.05.004.
8. J. Seetha and S. S. Raja, "Brain tumor classification using Convolutional Neural Networks," Biomedical and Pharmacology Journal, vol. 11, no. 3, pp. 1457–1461, Sep. 2018, doi: 10.13005/BPJ/1511
9. S. Teicher and A. Martinez, "Diagnosing and Segmenting Brain Tumors and Phenotypes using MRI Scans CS229 Final Project, Autumn 2014".
10. S. Patil, D. Kirange, and V. Nemade, "PREDICTIVE MODELLING OF BRAIN TUMOR DETECTION USING DEEP LEARNING," 2020.
11. Z. Sobhaninia et al., "Brain Tumor Segmentation Using Deep Learning by Type Specific Sorting of Images," Sep. 2018, doi: 10.6084/m9.figshare.1512427.
12. S. J. Cho, L. Sunwoo, S. H. Baik, Y. J. Bae, B. S. Choi, and J. H. Kim, "Brain metastasis detection using machine learning: a systematic review and meta-analysis," Neuro Oncol, vol. 23, no. 2, pp. 214–225, Feb. 2021, doi: 10.1093/NEUONC/NOAA232.
13. E. Stancombe, K. S. Bialkowski, and A. M. Abbosh, "Portable microwave head imaging system using software-defined radio and switching network," IEEE J Electromagn RF Microw Med Biol, vol. 3, no. 4, pp. 284–291, Dec. 2019, doi: 10.1109/JERM.2019.2901360.
14. R. S. Santos, S. M. F. Malheiros, S. Cavalheiro, and J. M. P. de Oliveira, "A data mining system for providing analytical information on brain tumors to public health decision makers," Comput Methods Programs Biomed, vol. 109, no. 3, pp. 269–282, Mar. 2013, doi: 10.1016/J.CMPB.2012.10.01010.1016/J.CMPB.2012.10.010ss.

Note: All the figures in this chapter were made by the author.

Intelligent and Sustainable Power and Energy Systems – Dr. M. Premkumar et al. (eds)
© 2026 Taylor & Francis Group, London, ISBN 978-1-041-10314-1

11

Automated Schizophrenia Detection and Classification (STFDL-ASDC): A Proposed Model for Self-Reported Schizophrenic Episodes

Jayashree M. Kudari[1] and D. Ganesh[2]
School of Computer Science & IT Jain (Deemed to-be University),
Bangalore, India

ABSTRACT: Schizophrenia is a severe mental illness that disrupts brain functions such as perception and thought, leading to profound impacts on individuals' lives. Its early diagnosis is complex, often hindered by multiple comorbidities that challenge effective patient management and reduce the likelihood of positive outcomes. This study aims to enhance the detection of schizophrenic episodes through a novel deep learning-based Automated Schizophrenia Detection and Classification (STFDL-ASDC) model that uses self-recorded video data. Our approach extracts spatial and temporal features from the videos through a Two-Stream Inflated 3D ConvNet model. The spatial stream processes facial expressions, body language, and gestures, while the temporal stream analyzes the progression of emotional and behavioral patterns. To improve video quality, preprocessing techniques like Wiener filtering and adaptive histogram equalization (AHE) are applied. Spatial features are extracted from the processed frames using the RegNetY002 CNN architecture, with temporal feature analysis and classification managed by the Inflated 3D-ConvNet. This research highlights the proposed model's ability to provide real-time observation of schizophrenia symptoms, potentially enabling early intervention and tailored treatment strategies.

KEYWORDS: AI in mental health, Deep learning, Inflated 3D-ConvNet, Schizophrenia, Spatio-temporal features, Video analysis

1. Introduction

Schizophrenia is a chronic and severe mental illness that profoundly impacts patients' daily lives and ability to work [1]. It is a complex, long-term condition affecting thoughts, behaviors, and

---

[1]mk.jayashree@jainuniversity.a.c.in, [2]d.ganesh@jainuniversity.ac.in

DOI: 10.1201/9781003654469-11

emotions, marked by psychotic episodes, such as delusions, disorganized speech, hallucinations, and major disruptions in social and occupational functioning [2-4]. This condition affects fundamental mental processes, creating significant cognitive and emotional distress. Worldwide, around 20 million people are affected by schizophrenia, which, despite its lower prevalence compared to other mental health disorders, ranks among the top 10 causes of disability globally according to the World Health Organization (WHO) [5]. The National Institute of Mental Health defines schizophrenia as a "chronic, severe, and disabling" brain disorder with a substantial societal and economic burden, with costs exceeding $150 billion annually in the United States alone [6]. Individuals with schizophrenia often face numerous challenges, including long-term unemployment, poverty, homelessness, and heightened risks of physical health issues and suicide [7]. These factors contribute to a life expectancy reduction of 15-20 years, highlighting the urgent need for effective treatment strategies to improve quality of life for those affected [8]. Diagnosing schizophrenia requires comprehensive clinical assessments, which include psychiatric evaluations and brain imaging. Early diagnosis and timely intervention are essential, as they can help prevent severe psychosis and enhance long-term outcomes [9, 10]. Schizophrenia presents with ongoing symptoms such as delusions, hallucinations, thought disorders, and social withdrawal, all of which contribute to substantial functional impairments [11]. Additionally, individuals with schizophrenia often experience comorbid substance abuse, which compounds the challenges related to social and occupational functioning. Together, these symptoms highlight the complexity of schizophrenia and the multifaceted impact it has on affected individuals [12]. Schizophrenia presents with a variety of symptoms impacting multiple areas of functioning. Positive symptoms, including hallucinations, delusions, and disorganized speech, stand in contrast to negative symptoms, such as reduced emotional expression, alogia, and avolition [13, 14]. Additionally, cognitive impairments and mood symptoms like depression, anxiety, and suicidal tendencies, along with comorbid substance use, greatly contribute to social and occupational challenges for patients [15]. The emergence of computer technology and AI, particularly machine learning (ML) and deep learning (DL), has transformed the detection and management of schizophrenia [16]. AI models are capable of analyzing complex data to identify early indicators, often before clinical symptoms become evident [17-19]. Techniques such as speech analysis have demonstrated high accuracy in distinguishing individuals affected by schizophrenia [20]. AI supports early detection, enables personalized treatment approaches, and enhances patient outcomes [21].

Table 11.1   Features of schizophrenia [11]

Features	Details
Positive Symptoms	Delusions, Disorganized speech, Hallucinations, Catatonia, Social/Occupational Dysfunction
Negative Symptoms	Affective flattening, Avolition, Anhedonia, Alogia, Social withdrawal, Social/ Occupational Dysfunction
Cognitive Deficits	Attention, Executive functions (e.g., abstraction), Memory, Social/Occupational Dysfunction
Mood Symptoms	Depression, Anxiety, Hopelessness, Agitation, Suicidality, Hostility, Social/ Occupational Dysfunction
Comorbid Substance Abuse	Social/Occupational Dysfunction

Current approaches of schizophrenia detection largely rely on subjective clinical evaluations and limited datasets, which often fail to capture the full spectrum of symptoms exhibited by patients over time. Additionally, while neuroimaging and speech analysis techniques have shown promise, they are frequently constrained by high costs, complexity, and the requirement for specialized equipment and expertise. This study aims to address these gaps by developing a comprehensive computational model that leverages advanced deep learning techniques to analyze self-recorded video episodes of schizophrenia patients. By extracting and integrating spatial and temporal features from these videos, the proposed model seeks to provide a more robust and nuanced understanding of schizophrenic episodes. The use of a Two-Stream Inflated 3D ConvNet allows for the simultaneous analysis of spatial dynamics, such as facial expressions and body language, and temporal dynamics, including the evolution of emotional expressions and behavioral patterns.

2. Literature Review

Several studies have explored the use of machine learning (ML) and deep learning (DL) methods for the detection and diagnosis of schizophrenia, utilizing diverse data types such as speech, neuroimaging, EEG, and behavioral data. Espinola et al. [22]. investigated vocal patterns as a diagnostic tool, utilizing ML techniques like support vector machines to analyze speech abnormalities in schizophrenia patients, achieving notably high accuracy. Alves et al. [23] focused on automating schizophrenia diagnosis and understanding brain network dynamics by employing functional MRI (fMRI) and electroencephalogram (EEG) data. Their models achieved impressive accuracies of 98% for fMRI and 95.4% for EEG, demonstrating the potential of neuroimaging data for effective diagnosis. Similarly, Soldatos et al. [24] developed an ML-based predictive model for identifying remission after first-episode psychosis, using clinical scale items from the Athens First Episode Research Study. Espinola et al. explored the use of vocal patterns and ML for schizophrenia detection, analyzing speech abnormalities in patients using support vector machines and achieving significantly high accuracy. Alves et al. automated schizophrenia diagnosis and comprehension of brain network dynamics using ML. They employed functional MRI (fMRI) and electroencephalogram (EEG) data, achieving 98% accuracy for fMRI and 95.4% for EEG. [24] developed a ML-based predictive model for remission after first-episode psychosis, utilizing clinical scale items from the Athens First Episode Research Study. Shi et al. employed machine learning to detect schizophrenia using structural and functional neuroimaging data, integrating fMRI and sMRI. Their multimodal imaging and multilevel characterization with multiclassifier method achieved 83.49% accuracy, and highlighted the role of global signal regression in improving diagnostic performance. Bae, Shim, and Lee used machine learning to detect schizophrenia from Reddit posts, achieving 96% accuracy by extracting linguistic features. They focused on schizophrenia-related subreddits versus non-mental health subreddits for controls. Their study highlighted the potential of linguistic markers and unsupervised clustering for early detection. [27]. In the domain of multimodal imaging, [25] employed both structural and functional neuroimaging data (sMRI and fMRI) to detect schizophrenia, achieving 83.49% accuracy. Their study underscored the importance of global signal regression in enhancing diagnostic performance. Bae, Shim, and Lee [26] used machine learning to detect schizophrenia through linguistic features extracted from Reddit posts, achieving 96% accuracy. By focusing on schizophrenia-related subreddits, their research highlighted the utility of linguistic markers and unsupervised clustering in early detection.

Authors in [28] addressed the challenge of accurate schizophrenia diagnosis using a 3D ResNet model with a leave-one-center-out validation strategy across nine global centers, achieving 82% classification performance. They also utilized SHapley Additive exPlanations (SHAP) for neuroanatomical insights, providing a clearer understanding of model decisions. In the realm of integrated brain data, [29] developed a DL-based method to fuse structural and functional MRI, fMRI, and DTI data for schizophrenia diagnosis. Their model, which utilized 3D CNNs, achieved a high accuracy of 99.35% on the COBRE dataset.

Proposed a new Sch-net model has been designed for end-to-end detection of schizophrenic speech, incorporating skip connections and a Convolutional Block Attention Module (CBAM) to enhance information processing and emphasize key features. They validated their model on a dataset of schizophrenic speech, and further tested it on the LANNA children speech database, achieving significant higher accuracy in both. Weng et al. addressed the need for accurate diagnostic tools in schizophrenia using deep learning. They utilized a 3D ResNet model with leave-one-center-out validation across nine global centers, achieving 82% classification performance and using SHapley Additive exPlanations (SHAP) for neuroanatomical interpretations. Masoudi, Daneshvar and Razavi, proposed a DL-based method to fuse structural and functional brain data for schizophrenia diagnosis. By integrating 3D CNNs with MRI, fMRI, and DTI data, they achieved 99.35% accuracy on the COBRE dataset. [30] presented a multimodal assessment model integrates linguistic, acoustic, and visual behavioral data to predict symptom severity in schizophrenia patients, using a fusion framework with transformer-based networks. Similarly, [31] proposed a DL-based support system for differentiating mental health disorders using wearable device data. They utilized deep learning models on open datasets to accurately distinguish between schizophrenia and mood disorders. Oh et al. detected schizophrenia in structural MRI datasets using a deep learning algorithm, utilizing five public MRI datasets and achieving significant detection accuracy. [32] The deep learning algorithm trained with structural MR images detected schizophrenia in randomly selected images with reliable performance (area under the receiver operating characteristic curve [AUC] of 0.96). The algorithm could also identify MR images from schizophrenia patients in a previously unencountered data set with an AUC of 0.71 to 0.90. The deep learning algorithm's classification performance degraded to an AUC of 0.71 when a new data set with younger patients and a shorter duration of illness than the training data sets was presented.

[33] developed a DL based method for diagnosing schizophrenia using EEG data, incorporating generative data augmentation with spectrograms and CNN, achieving a 3.0% accuracy improvement with Wasserstein GAN and Variational Autoencoder (VAE). Grover et al. introduced Schizo-Net for diagnosing schizophrenia using brain connectivity measures from EEG signals. They compared six connectivity metrics with deep neural networks, achieving 99.84% accuracy. [34] introduced Schizo-Net, a DL model that utilized brain connectivity measures derived from EEG signals, achieving remarkable accuracy of 99.84%. Aslan and Akin [35] also focused on EEG, using Continuous Wavelet Transform for feature extraction and a VGG16 CNN for classification, providing another high-accuracy solution for automated schizophrenia detection.

[35] developed a high-accuracy DL method for automatically detecting schizophrenia from EEG records. Their approach extracted features from EEG signals via Continuous Wavelet Transform and employed VGG16 CNN. [36] identified children at risk of schizophrenia using EEG patterns and deep learning. They analyzed EEG data from children aged 9 to 12 during a passive auditory oddball paradigm. Jeon et al. analyzed eye movements to explore the relationship among gaze avoidance and psychopathology in schizophrenia. They utilized Tobii Pro Wearable Glasses 2 combined with DL for emotional recognition. Lamichhane et al. introduced Relapse PredNet for predicting relapse in schizophrenia patients using mobile sensing data and an LSTM. Jeon et al. [37] explored the relationship between eye movement patterns and psychopathology in schizophrenia using wearable glasses and deep learning for emotional recognition. Lamichhane et al. [38] developed RelapsePredNet, an LSTM-based model that predicts relapse in schizophrenia patients using mobile

sensing data, showcasing the potential for real-time monitoring. Additionally, [39] developed a Human Activity Recognition model that utilizes a wrist-worn device to evaluate activities associated with schizophrenia's negative symptoms, by analyzing actigraphy data and clinical assessments in a randomized study.

3. Research GAPS

Despite the extensive body of research, automatic detection of schizophrenia is still in its early stages. Most studies have concentrated on using patient data, such as speech, EEG, MRI, and fMRI, often relying on publicly available datasets to achieve high accuracy. However, these methods do not fully capture the complexity of real-time schizophrenia detection. While EEG offers excellent temporal resolution, it faces challenges with spatial accuracy and interference. MRI and fMRI provide high spatial resolution but are expensive, complex, and sensitive to motion artifacts. AI and machine learning models have shown potential in analyzing neuroimaging and speech data to detect schizophrenia, which could improve diagnostic accuracy and enable earlier interventions. However, issues with detection precision and computational complexity persist, necessitating further research to optimize these models for clinical application. fMRI plays a crucial role in schizophrenia research by identifying disruptions in brain connectivity associated with cognitive and perceptual abnormalities. Although it cannot diagnose schizophrenia on its own, it tracks structural brain changes linked to the disorder's progression. Nonetheless, fMRI's drawbacks include complex analyses, high costs, time consumption, and a lack of standardized protocols [40]. The positive symptoms of schizophrenia, such as hallucinations (distorted perceptions), delusions (faulty reasoning), disorganized speech, and unusual behavior, represent significant deviations from reality [41]. These symptoms severely hinder a patient's ability to perceive and engage with their surroundings. However, current diagnostic methods based on these symptoms are insufficient for timely and accurate detection, particularly during recurring episodes. Many studies overlook the importance of tracking these symptoms as they evolve over time, resulting in gaps in continuous monitoring and intervention [28]. Additionally, there is a notable lack of visual feature-based approaches for detecting schizophrenia in the existing research. Developing such approaches requires a deep understanding of the specific features and symptoms that define schizophrenic episodes. From initial diagnosis, patients typically undergo a series of medications and treatments for schizophrenia. However, studies indicate that schizophrenic episodes frequently recur as patients often discontinue their medications multiple times during their treatment course [29]. This discontinuation significantly impacts the effectiveness of treatment plans and exacerbates the challenge of managing the disorder. Recognizing symptoms during these repeated episodes is further complicated by the fact that these symptoms may not be clearly defined by healthcare providers, and caregivers often lack the clinical knowledge to identify them accurately [42]. This gap underscores the critical need for developing advanced computational models that can recognize schizophrenia during episodes with greater precision. Achieving these advancements is crucial to significantly enhance care and improve the quality of life for individuals affected by this debilitating disorder.

3.1 Methodology

In this study, we used the Self-Recorded Schizophrenic Video Episodes dataset, which includes videos of patients diagnosed with schizophrenia. Ethical approval and informed consent were obtained from all participants and relevant authorities. The dataset consists of 15 videos selected through purposive sampling to ensure diverse representation of mental health conditions, capturing a wide range of expressions and behaviors.

Features Extracted: Both spatial and temporal features were extracted from the videos. Spatial features encompassed facial expressions, body language, and gestures, which were essential for interpreting the visual cues related to schizophrenic episodes. Temporal features focused on changes over time, such as the progression of emotional expressions and the frequency of specific behaviors throughout the episodes.

Sample Size Determination: The sample size of 15 videos was selected using purposive sampling. This method was chosen for its ability to identify and include relevant cases. The initial selection of videos followed predefined criteria, with additional cases added iteratively to ensure thorough representation of the variability in schizophrenic symptoms.

Dataset Description: The dataset contains 15 videos of adults with schizophrenia, recorded by parents or caregivers using CCTV cameras in uncontrolled natural settings. Each video has an average duration of 90 seconds, offering real-time observation of schizophrenia episodes. Video frames were annotated with the help of experienced psychiatrists, ensuring accurate identification and labelling of behavioral markers and symptoms, thus enhancing the dataset's reliability for training the deep learning model.

3.2 Features of Schezophrina

Schizophrenia is characterized by a range of symptoms that can be broadly categorized into positive, negative, cognitive, mood, and comorbid substance abuse features. Positive symptoms include delusions, disorganized speech, hallucinations, catatonia, and significant social or occupational dysfunction. Negative symptoms involve affective flattening, avolition, anhedonia, alogia, and social withdrawal, which also contribute to difficulties in social and occupational functioning. Cognitive deficits are often observed in areas such as attention, executive functions (e.g., abstraction), and memory, further impairing social and occupational abilities. Mood symptoms, including depression, anxiety, hopelessness, agitation, suicidality, and hostility, can exacerbate the overall impact of the disorder, leading to additional social and occupational dysfunction.

Proposed DL Model: The study uses the Two-Stream Inflated 3D ConvNet model, which is specifically designed to analyse both spatial and temporal aspects of video data. The spatial stream of the model focused on extracting and interpreting facial expressions, body language, and other spatial cues. Meanwhile, the temporal stream examined how emotional expressions and behavioral patterns evolved throughout the self-reported schizophrenic episodes. This dual-stream approach allowed for a thorough analysis of the intricate interactions between spatial and temporal features, offering a deep understanding of the schizophrenic episodes.

4. Conclusion

This study introduces an innovative approach for detecting and classifying schizophrenia symptoms by utilizing a deep learning-based model, Automated Schizophrenia Detection and Classification (STFDL-ASDC). By leveraging self-recorded video data, the proposed model captures both spatial and temporal features of schizophrenic episodes, addressing the limitations of traditional detection methods. The use of a Two-Stream Inflated 3D ConvNet architecture enables comprehensive analysis by extracting visual cues like facial expressions and gestures, alongside tracking behavioral patterns over time. This dual-stream methodology offers a nuanced understanding of schizophrenic behaviors, facilitating real-time monitoring and potentially paving the way for timely interventions. The model's integration of advanced pre-processing techniques, such as Wiener filtering and adaptive histogram equalization, enhances video quality, making it a robust solution in real-world, uncontrolled environments. Additionally, the involvement of experienced psychiatrists in data annotation has strengthened the reliability of the feature labels, contributing to the model's training accuracy. Future research can build upon this foundation to further refine and validate the model across larger and more varied datasets, ultimately contributing to improved clinical applications in mental health.

REFERENCES

1. Patel, K. R., Cherian, J., Gohil, K., & Atkinson, D. (2014). Schizophrenia: overview and treatment options. Pharmacy and Therapeutics, 39(9), 638.
2. Hirsch, S. R., & Weinberger, D. R. (Eds.). (2008). Schizophrenia. John Wiley & Sons.
3. Chandrasekhar, T., Hazzard, L. M., & Sikich, L. (2015). Schizophrenia Spectrum and Other Psychotic Disorders. Diagnosing and Treating Children and Adolescents: A Guide for Mental Health Professionals, 148–176.
4. Rahman, T., & Lauriello, J. (2016). Schizophrenia: an overview. Focus, 14(3), 300–307.
5. World Health Organization (WHO). (2022, January 10). Schizophrenia. https://www.who.int/news-room/fact-sheets/detail/schizophrenia#:~:text=Schizophrenia%20affects%20approximately%2024%20million,%25)%20among%20adults%20(2).
6. Kadakia, A., Catillon, M., Fan, Q., Williams, G. R., Marden, J. R., Anderson, A., ... & Dembek, C. (2022). The economic burden of schizophrenia in the United States. The Journal of clinical psychiatry, 83(6), 43278.
7. Newell, T., & Gopal, A. (2024, March 12). Schizophrenia: An overview. In S. Bhandari (Med. Rev.), WebMD. https://www.webmd.com/schizophrenia/mental-health-schizophrenia
8. Polcwiartek, C., O'Gallagher, K., Friedman, D. J., Correll, C. U., Solmi, M., Jensen, S. E., & Nielsen, R. E. (2024). Severe mental illness: cardiovascular risk assessment and management. European Heart Journal, ehae054.
9. Haefner, H., & Maurer, K. (2006). Early detection of schizophrenia: current evidence and future perspectives. World Psychiatry, 5(3), 130.
10. McGlashan, T. H., & Johannessen, J. O. (1996). Early detection and intervention with schizophrenia: rationale. Schizophrenia bulletin, 22(2), 201–222.
11. Lin, C. H., & Lane, H. Y. (2019). Early identification and intervention of schizophrenia: insight from hypotheses of glutamate dysfunction and oxidative stress. Frontiers in psychiatry, 10, 441787.
12. Maguire, G. A. (2002). Features of Schizophrenia. American Journal of Health-System Pharmacy, 59(S4), S4-S11.
13. Correll, C. U., & Schooler, N. R. (2020). Negative symptoms in schizophrenia: a review and clinical guide for recognition, assessment, and treatment. Neuropsychiatric disease and treatment, 519–534.
14. Lewine, R. R., Fogg, L., & Meltzer, H. Y. (1983). Assessment of negative and positive symptoms in schizophrenia. Schizophrenia bulletin, 9(3), 368–376.
15. Buckley, P. F., Miller, B. J., Lehrer, D. S., & Castle, D. J. (2009). Psychiatric comorbidities and schizophrenia. Schizophrenia bulletin, 35(2), 383–402.
16. Veronese, E., Castellani, U., Peruzzo, D., Bellani, M., & Brambilla, P. (2013). Machine learning approaches: from theory to application in schizophrenia. Computational and mathematical methods in medicine, 2013(1), 867924
17. Sharaev, M. G., Malashenkova, I. K., Maslennikova, A. V., Zakharova, N. V., Bernstein, A. V., Burnaev, E. V., ... & Ushakov, V. L. (2022). Diagnosis of Schizophrenia Based on the Data of Various Modalities: Biomarkers and Machine Learning Techniques. Современные технологии в медицине, 14(5 (eng)), 54–75.
18. Iyortsuun, N. K., Kim, S. H., Jhon, M., Yang, H. J., & Pant, S. (2023, January). A review of machine learning and deep learning approaches on mental health diagnosis. In Healthcare (Vol. 11, No. 3, p. 285). MDPI.
19. Noor, M. B. T., Zenia, N. Z., Kaiser, M. S., Mamun, S. A., & Mahmud, M. (2020). Application of deep learning in detecting neurological disorders from magnetic resonance images: a survey on the detection of Alzheimer's disease, Parkinson's disease and schizophrenia. Brain informatics, 7, 1–21.
20. Cortes-Briones, J. A., Tapia-Rivas, N. I., D'Souza, D. C., & Estevez, P. A. (2022). Going deep into schizophrenia with artificial intelligence. Schizophrenia Research, 245, 122–140.

21. Raparthi, M. (2020). AI Integration in Precision Health-Advancements, Challenges, and Future Prospects. Asian Journal of Multidisciplinary Research & Review, 1(1), 90–96.

22. Espinola, C. W., Gomes, J. C., Pereira, J. M. S., & dos Santos, W. P. (2021). Vocal acoustic analysis and machine learning for the identification of schizophrenia. Research on Biomedical Engineering, 37, 33–46.

23. Alves, C. L., Thaise, G. D. O., Porto, J. A. M., de Carvalho Aguiar, P. M., de Sena, E. P., Rodrigues, F. A., ... & Thielemann, C. (2023). Analysis of functional connectivity using machine learning and deep learning in different data modalities from individuals with schizophrenia. Journal of Neural Engineering, 20(5), 056025.

24. Soldatos, R. F., Cearns, M., Nielsen, M. Ø., Kollias, C., Xenaki, L. A., Stefanatou, P., ... & Stefanis, N. (2022). Prediction of early symptom remission in two independent samples of first-episode psychosis patients using machine learning. Schizophrenia Bulletin, 48(1), 122–133.

25. Shi, D., Li, Y., Zhang, H., Yao, X., Wang, S., Wang, G., & Ren, K. (2021). Machine learning of schizophrenia detection with structural and functional neuroimaging. Disease markers, 2021(1), 9963824.

26. Bae, Y. J., Shim, M., & Lee, W. H. (2021). Schizophrenia detection using machine learning approach from social media content. Sensors, 21(17), 5924.

27. Fu, J., Yang, S., He, F., He, L., Li, Y., Zhang, J., & Xiong, X. (2021). Sch-net: a deep learning architecture for automatic detection of schizophrenia. BioMedical Engineering OnLine, 20(1), 75.

28. Weng, T., Zheng, Y., Xie, Y., Qin, W., & Guo, L. (2024). Diagnosing schizophrenia using deep learning: Novel interpretation approaches and multi-site validation. Brain Research, 1833, 148876.

29. Masoudi, B., Daneshvar, S., & Razavi, S. N. (2021). Multi-modal neuroimaging feature fusion via 3D Convolutional Neural Network architecture for schizophrenia diagnosis. Intelligent Data Analysis, 25(3), 527–540.

30. Chuang, C. Y., Lin, Y. T., Liu, C. C., Lee, L. E., Chang, H. Y., Liu, A. S., ... & Fu, L. C. (2023). Multimodal Assessment of Schizophrenia Symptom Severity from Linguistic, Acoustic and Visual Cues. IEEE Transactions on Neural Systems and Rehabilitation Engineering.

31. Nguyen, D. K., Chan, C. L., Li, A. H. A., Phan, D. V., & Lan, C. H. (2022). Decision support system for the differentiation of schizophrenia and mood disorders using multiple deep learning models on wearable devices data. Health Informatics Journal, 28(4), 14604582221137537.

32. Oh, J., Oh, B. L., Lee, K. U., Chae, J. H., & Yun, K. (2020). Identifying schizophrenia using structural MRI with a deep learning algorithm. Frontiers in psychiatry, 11, 481509.

33. Saadatinia, M., & Salimi-Badr, A. (2023). An Explainable Deep Learning-Based Method For Schizophrenia Diagnosis Using Generative Data-Augmentation. arXiv preprint arXiv:2310.16867.

34. Grover, N., Chharia, A., Upadhyay, R., & Longo, L. (2023). Schizo-Net: A novel Schizophrenia Diagnosis framework using late fusion multimodal deep learning on Electroencephalogram-based Brain connectivity indices. IEEE Transactions on Neural Systems and Rehabilitation Engineering, 31, 464–473.

35. Aslan, Z., & Akin, M. (2022). A deep learning approach in automated detection of schizophrenia using scalogram images of EEG signals. Physical and Engineering Sciences in Medicine, 45(1), 83–96.

36. Ahmedt-Aristizabal, D., Fernando, T., Denman, S., Robinson, J. E., Sridharan, S., Johnston, P. J., ... & Fookes, C. (2020). Identification of children at risk of schizophrenia via deep learning and EEG responses. IEEE journal of biomedical and health informatics, 25(1), 69–76.

37. Jeon, G., Choi, H. S., Jung, D. U., Moon, S., Kim, G., Kim, S. J., ... & Jeon, D. W. (2022). Evaluation of the correlation between gaze avoidance and schizophrenia psychopathology with deep learning-based emotional recognition. Asian Journal of Psychiatry, 68, 102974.

38. Lamichhane, B., Zhou, J., & Sano, A. (2022). Psychotic Relapse prediction in schizophrenia patients using a mobile sensing-based supervised deep learning model. arXiv preprint arXiv:2205.12225.

39. Umbricht, D., Cheng, W. Y., Lipsmeier, F., Bamdadian, A., & Lindemann, M. (2020). Deep learning-based human activity recognition for continuous activity and gesture monitoring for schizophrenia patients with negative symptoms. Frontiers in Psychiatry, 11, 574375.

40. Yu, Q., A Allen, E., Sui, J., R Arbabshirani, M., Pearlson, G., & D Calhoun, V. (2012). Brain connectivity networks in schizophrenia underlying resting state functional magnetic resonance imaging. Current topics in medicinal chemistry, 12(21), 2415–2425.

41. Dabiri, M., Dehghani Firouzabadi, F., Yang, K., Barker, P. B., Lee, R. R., & Yousem, D. M. (2022). Neuroimaging in schizophrenia: A review article. Frontiers in neuroscience, 16, 1042814.

42. American Psychiatric Association. (2013). Diagnostic and statistical manual of mental disorders (5th ed.). Arlington, VA: American Psychiatric Publishing.

Intelligent and Sustainable Power and Energy Systems – Dr. M. Premkumar et al. (eds)
© 2026 Taylor & Francis Group, London, ISBN 978-1-041-10314-1

12

Journey Toward Sustainability: Advancements in Hybrid Electric Vehicle Development

Madhusudan Barhate[1],
Hassan Shingrey[2], Abhyudaya Kachare[3],
Shubham Lad[4], Supriya Bhuran[5] and Sharad Jadhav[6]

Department of Instrumentation Engineering,
RAIT, D Y Patil Deemed to be University, Nerul,
Navi Mumbai, India

ABSTRACT: The hybrid electric vehicle (HEV), is one of the most significant developments in the quest for more fuel-efficient and environmentally responsible transportation. Hybrid electric vehicles, which combine electric motors and internal combustion engines (ICE), significantly reduce emissions and improve fuel efficiency. This paper explores the development of hybrid cars over time, emphasizing important turning moments, technological advancements, and challenges faced. The paper also looks at new car models that will shape hybrid mobility, advanced battery technology, and energy management system trends. The hybrid vehicle industry is attempting to resolve issues with battery capacity, thermal management, and infrastructure in order to remove the barriers that have stalled broader adoption. With advancements in battery technologies such as solid-state and lithium-sulphur, as well as the integration of autonomous driving and vehicle-to-grid (V2G) systems, HEVs will play a significant role in facilitating sustainable transportation in the future.

KEYWORDS: Electric vehicles (EV), Hybrid electric vehicles (HEV), Internal combustion engine (ICE), Vehicle to grid (V2G)

1. Introduction

Hybrid electric vehicles, a major advancement in the automotive industry, offer a compromise between fully electric vehicles (EVs) and traditional internal combustion engine (ICE) vehicles.

[1]mad.bar.rt21@dypatil.edu, [2]has.shi.rt21@dypatil.edu, [3]abh.kac.rt21@dypatil.edu, [4]shu.lad.rt21@dypatil.edu, [5]supriya.bhuran@rait.ac.in, [6]sharad.jadhav@rait.ac.in

DOI: 10.1201/9781003654469-12

As environmental concerns and the desire to reduce greenhouse gas emissions have grown, HEVs have gained popularity due to their ability to improve fuel efficiency and reduce carbon footprints. From the late 1800s to the present, this paper provides an inclusive overview of the development and advancement of HEV technology. Early hybrid concepts by Ferdinand Porsche in 1899 [14] to the mass production of vehicles like the Toyota Prius [17] demonstrate how HEVs have undergone constant modifications to meet changing consumer demands and environmental goals. With advancements in battery technology, energy management systems (EMS), and predictive control approaches [11, 24]. HEVs are becoming more efficient, practical, and sustainable. The evolution from rule-based systems to intelligent, AI-powered EMS is a significant milestone in enhancing power management between the ICE and electric motors [12]. Predictive control systems, such as those utilizing traffic prediction to help optimize energy use in real-time, addressing the challenges of dynamic data processing in these systems [24]. In "Future Innovations & Upcoming Trends in Hybrid Electric Vehicles," the focus on future developments ties technological advancements to the broader goal of sustainability. Solid-state batteries [20], vehicle-to-grid (V2G) capabilities [22] and autonomous driving are discussed as promising technologies that could enhance HEVs' sustainability and efficiency [19]. Additionally, highlighting real-world applications of these technologies in models from Toyota (Toyota Motor Corporation, 2024), BMW (BMW Group, 2024), and Honda (Honda Motor Co., Ltd., 2024) demonstrates how emerging solutions are being applied in the market. The discussion of future battery technologies, including solid-state, lithium-sulphur, sodium-ion, and graphene [20], underscores their technical benefits, such as higher energy density, improved safety, and faster charging times, while acknowledging challenges related to cost and infrastructure [10].

Rest of the paper is structured as follows. Section-2 reviews the insights on evolution of hybrid electric vehicles. Section-3 describes the present situation of hybrid electric vehicle technologies. Section-4 explains future innovations & upcoming trends in hybrid electric vehicles. Section-5 summarises the paper.

2. The Insights on Evolution of Hybrid Electric Vehicles

In 1890, a coach builder named Jacob Lohner in Vienna, Austria, realized that an electric vehicle that was quieter than the new gas-powered cars was needed. An Austro-Hungarian engineer named Ferdinand Porsche, who had recently received his degree from the Vienna Technical College, was hired to design an electric vehicle. An internal combustion engine drove a generator, which in turn drove an electric motor housed in the wheel hubs of Porsche's first electric car. The range of the vehicle was 38 miles. To extend the vehicle's range, Porsche added a petrol engine that could recharge the batteries to produce the first hybrid, the Lohner-Porsche Elektromobil [6]. An electric motor under the seat competes with a small petrol engine in the 3.5-horsepower Voiturette, which was introduced in 1900 by the Belgian automaker Pieper. While driving, the car's engine functions as a generator to charge the battery. When connected to the gas engine, the electric motor helps the engine move the car upward

By the 1930s, hybrid vehicles had disappeared along with electric vehicles, and electric car manufacturers had gone bankrupt. Numerous reasons have contributed to the decrease of EVs and HEVs. Electric, hybrid, and gasoline-powered vehicles are more costly than gasoline-powered vehicles because they require larger batteries. Due to the limited battery power, these vehicles are less powerful than gasoline-powered vehicles. It takes hours to charge the onboard battery, but it is cheaper and more efficient than refuelling with gasoline. In addition to making gasoline-powered vehicles, like Henry Ford's Model T, more affordable than electric and hybrid vehicles, the development of the starter motor made it simpler to start gasoline engines [6, 7]. Figure 12.1 below shows Victor Wouk's contributions to the creation of numerous hybrid designs, earning him the title of "Godfather of the Hybrid" in 1960 [21]. Even in 1976, he converted a Buick Skylark from petrol to hybrid power [12].

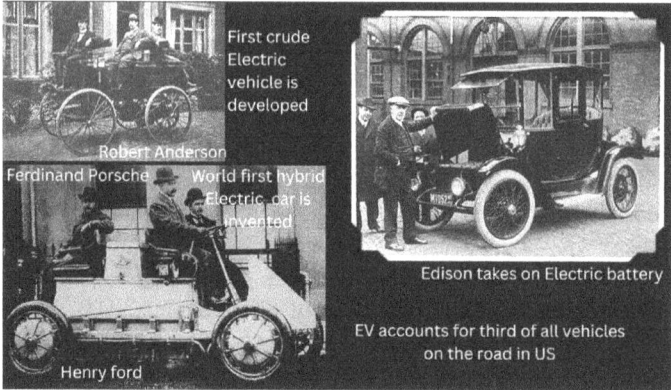

Fig. 12.1 Victor wouk's electric vehicle

Source: Linked-In World Economic Forum's Post

In 1978, David Arthurs developed the initial version of the Regenerative Braking System. He modified an Opel GT to get 75 miles per gallon using standard vehicle parts. This method's schematics are still available to the public online and are used in many home modifications [1]. After assessing the impact of air pollution in Southern California in 1990, the California Air Resources Board (CARB) created the Zero Emission Vehicle (ZEV) standards, which required that cars sold in California be between 2% and 10% emission-free by 1998 (CARB, 1990). When Japan unveiled the Toyota Prius, the first hybrid electric car in history, in 1997, a new chapter in automotive history began. Furthermore, the Audi Duo was the first mass-produced hybrid automobile in Europe, and as time has gone on, both hybrid production and consumer demand have only increased. The car has been offered in the US since 2000, joining Honda's Insight and Civic HEV variants (Toyota Motor Corporation, 1997; Honda Motor Co., Ltd., 1999). The SUV, a hybrid variant of the Ford Escape, debuted in 2005. Essentially, Ford and Toyota traded patents, with Ford acquiring some of Toyota's patents pertaining to hybrid technology and Toyota gaining access to Ford's patents for diesel engines (Ford Motor Company, 2005; Toyota Motor Corporation, 2005).

3. The Present Situation of Hybrid Electric Vehicle Technologies

The domain of Hybrid Electric Vehicles (HEVs) has undergone substantial progress, propelled by the increasing need for fuel-efficient and eco-friendly transportation alternatives. Hybrid electric vehicles (HEVs), which integrate an internal combustion engine (ICE) with an electric motor, provide numerous benefits compared to conventional ICE vehicles, including diminished fuel consumption, reduced pollutants, and enhanced fuel efficiency [6]. A significant advancement in enhancing HEV performance has been the progression of Energy Management Systems (EMS), which regulate the power allocation among the internal combustion engine (ICE), electric motor, and energy storage devices like batteries. Early EMS technologies, particularly rule-based systems, functioned according to predetermined logic, offering a direct approach to power distribution management (Zeng et al., 2008). For instance, rule-based control may engage the internal combustion engine during acceleration and transition to electric power at reduced speeds [6]. Nonetheless, these systems encountered difficulties in dynamic, real-world driving situations, where real-time flexibility is essential [11]. As the decade advanced, increasingly sophisticated EMS methodologies evolved, including model-based control systems that utilised mathematical models to forecast energy demands and optimise power distribution with greater precision [12]. These devices could enhance adaptability to driving situations, hence increasing efficiency through the anticipation of variables such as topography and velocity [8]. The introduction of intelligent EMS, utilising artificial intelligence (AI) and machine learning, enhanced HEVs' responsiveness by dynamically altering power distribution according to real-time data from the vehicle and environmental conditions (Kim & Lee, 2020). The transition to intelligent and predictive control systems enabled HEVs to optimise fuel efficiency, minimise emissions, and provide a more seamless driving experience, especially in urban settings characterised by frequent stop-and-go traffic [11, 24]. Nevertheless, despite these technological advancements, numerous challenges remain. The restricted range and efficacy of current batteries, along with the intricate challenge of regulating power transitions between the internal combustion engine and electric motor, continue to pose substantial obstacles [15]. Moreover,

battery thermal management systems are essential for preserving the health and longevity of batteries, as excessive heat during charging and discharging can impair performance [15]. By 2024, substantial advancements were achieved in both the hardware and software components of HEVs. Models including the Toyota Prius Prime in Fig. 12.2.

Fig. 12.2   Toyota pirus

Source: https://www.toyota.com/prius/

(Toyota Motor Corporation, 2024), Honda CR-V Hybrid (Honda Motor Co., Ltd., 2024), BMW X5 xDrive45e (BMW Group, 2024)(Fig. 12.3), and Audi Q5 TFSI e (Audi AG, 2024) feature advanced battery systems, more robust electric motors, and superior driver-assistance technologies, such as adaptive cruise control and lane-keeping assistance [19]. These advancements rendered HEVs more feasible and appealing to customers; nonetheless, elevated initial expenses, infrastructural constraints, and regulatory ambiguity persisted as obstacles to extensive adoption [10]. The deficiency in charging infrastructure and the absence of standardised regulatory frameworks, especially with incentives

Fig. 12.3   BMW X5

Source: https://www.bmw.in/en/all-models/x-series/ X5/2023/bmw-x5-overview

and policies, exacerbated the challenges associated with the widespread adoption of HEVs [20].

4. Future Innovations & Upcoming Trends in Hybrid Electric Vehicles

Hybrid Electric Vehicles (HEVs) are poised for substantial evolution, driven by ongoing improvements in battery technology, energy management systems, and connectivity. The advancement of next-generation battery technologies is a highly promising field of innovation. Solid-state batteries, which substitute the liquid electrolyte in traditional lithium-ion batteries with a solid electrolyte, are anticipated to provide significantly greater energy densities, enhanced safety, and expedited charging times. This advancement may substantially enhance the range of HEVs, diminish charge durations, and prolong the battery's total lifespan, tackling two major worries people have about electric vehicles [6]. Lithium-sulfur batteries, characterised by their superior theoretical energy density and reduced cost owing to the plentiful availability of sulphur, present significant potential for enhancing the performance and cost-effectiveness of hybrid electric vehicles (HEVs) [20].

4.1 Vehicle-to-Grid (V2G) Integration

Alongside advancement in battery technologies, Vehicle-to-Grid (V2G) systems are emerging as a main focus for hybrid electric vehicle (HEV) development. Vehicle-to-grid (V2G) technology enables hybrid electric vehicles (HEVs) to supply electricity to the power grid, functioning as mobile energy storage systems. This can improve grid stability, improve energy usage, and more efficiently combine with renewable energy sources such as solar and wind [24]. With the increasing occurrence of smart grids, V2G systems will facilitate HEVs in energy management by allowing vehicles to charge during off-peak hours and discharge during peak energy demand. This technology allows renewable energy synergy, enabling HEVs to balance supply and demand in the energy grid, hence improving energy resilience [19].

4.2 Autonomous Driving Technology

Autonomous driving technology is a crucial area for the future of hybrid electric vehicles (HEVs). Future hybrid electric vehicles (HEVs) will provide improved safety, convenience, and driving performance with the integration of advanced driver-assistance systems (ADAS). These systems will facilitate functionalities such as self-parking, adaptive cruise control, lane-keeping assistance, and ultimately, complete autonomous driving capabilities (Alessandrini et al., 2014). Autonomous HEVs have the potential to enhance and improve the driving experience by reducing human error and enhancing road safety [5].

4.3 Future Models and Connectivity

Automakers are significantly investing in the development of next-generation hybrid electric vehicles (HEVs), with anticipated models including the BMW X5 xDrive45e [3], Honda CR-V Hybrid [9], Toyota RAV4 Prime [18] and Audi Q5 TFSI e [2], which are expected to incorporate advanced hybrid powertrains, extended electric ranges, and enhanced connectivity and safety features. These vehicles will provide enhanced performance and will be outfitted with advanced energy management technologies that dynamically regulate power distribution depending on real-time data from the vehicle, GPS, and traffic conditions [23]. Examining the future, the prospective amalgamation of 5G connection, artificial intelligence, and machine learning in hybrid electric vehicles will lead to the development of increasingly intelligent and efficient automobiles. This will facilitate real-time data interchange with adjacent infrastructure, including traffic signals and charging stations, and enhance the precision of power distribution control across varying driving circumstances [24]. Advancements in connectivity and data analytics will enable the creation of more energy-efficient and ecologically sustainable hybrid vehicles. The Fig. 12.4 shows the "Global electric car stock, 2013-2023" (IEA2024), how the market is adapting electric vehicles more and more.

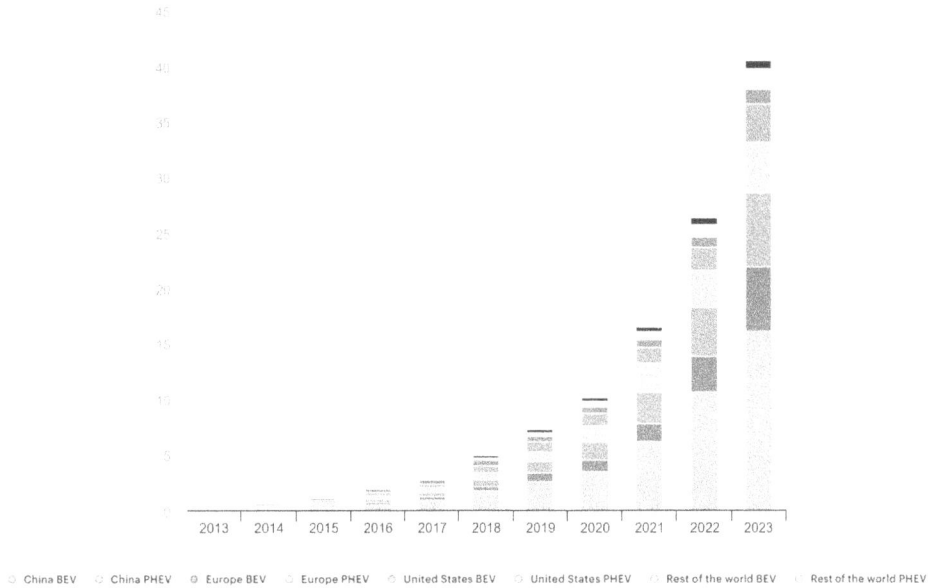

Fig. 12.4 Global electric car stock, 2013-2023

Source: https://www.iea.org/data-and-statistics/charts

The future of HEVs is hopeful, propelled by advancements in battery technology, autonomous driving, energy management, and grid interconnection. These advancements will not only improve the performance and convenience of HEVs but also render them indispensable within the sustainable transportation ecosystem, connecting traditional vehicles and fully electric vehicles [8]. These advancements will position HEVs as crucial in mitigating greenhouse gas emissions, enhancing energy efficiency, and facilitating the worldwide shift towards a cleaner, more sustainable future [19].

5. Conclusion

From early experimental models to modern fuel-efficient cars, hybrid electric vehicles have come a long way since their debut. Thanks to continuous advancements in battery technology, energy management systems, and integration with smart infrastructure, HEVs are poised to significantly influence the future of transportation. It states that new technologies like solid-state batteries, V2G capabilities, and intelligent control systems have helped in overcoming the challenges that still exist, including high costs, battery limitations, and the need for better infrastructure. As governments, corporations, and consumers become more concerned about sustainability, HEVs offer a practical solution to reduce dependency on fossil fuels and minimise environmental impact.

REFERENCES

1. Arthurs, D. (1978). Development of regenerative braking for hybrid vehicles.
2. Audi AG. (2024). Audi Q5 TFSI e.
3. BMW Group. (2024). BMW X5 xDrive45e.
4. California Air Resources Board (CARB). (1990). Zero emission vehicle standards.

5. Cameron, C. and Sweeney, M. (2020). Technological innovations in hybrid electric vehicles: past, present, and future. Journal of Cleaner Production, 258, 120751.

6. Ehsani, M., Gao, Y., and Emadi, A. (2009). Modern electric, hybrid electric, and fuel cell vehicles: fundamentals, theory, and design. CRC Press.

7. Ford Motor Company. (2005). Ford Escape hybrid SUV introduction.

8. Guzzella, L. and Sciarretta, A. (2013). Vehicle propulsion systems: introduction to modeling and optimization. Springer.

9. Honda Motor Co., Ltd. (2024). Honda CR-V Hybrid.

10. Khan, M. J. and Luthra, S. (2019). Hybrid electric vehicles: a review of recent developments and future prospects. Renewable and Sustainable Energy Reviews, 107, 172–185.

11. Moura, S., Fathy, H. K., Callaway, D. S., and Stein, J. L. (2011). A stochastic optimal control approach for power management in plug-in hybrid electric vehicles. IEEE Transactions on Control Systems Technology, 19(3), 545–555.

12. Onori, S., Serrao, L., and Rizzoni, G. (2016). Hybrid electric vehicles: energy management strategies. Springer.

13. Pieper, F. (1900). Voiturette hybrid vehicle design.

14. Porsche, F. (1899). Lohner-Porsche Elektromobil creation.

15. Pourabdollah, M., Moshayedi, S., and Hooshmand, R. (2015). Thermal management in hybrid electric vehicles. Journal of Power Sources, 275, 799–811.

16. Shrey Verma, Mishra, S., Gaur, A., and Chowdhury, S. (2021). A comprehensive review on energy storage in hybrid electric vehicles. Journal of Traffic and Transportation Engineering (English Edition), 8(5), 621–637.

17. Toyota Motor Corporation. (1997). Introduction of the Toyota Prius.

18. Toyota Motor Corporation. (2024). Toyota Prius Prime and RAV4 Prime models.

19. Tsydenova, I. and Smirnov, M. (2022). Market trends and future directions for hybrid electric vehicles. Energy Policy, 159, 112640.

20. Wang, Z. and Liu, Y. (2023). Future directions for hybrid electric vehicle technology development. Energy, 258, 125515.

21. Wouk, V. (1960). Contributions to hybrid vehicle development.

22. Yilmaz, M. and Krein, P. T. (2013). Review of the impact of vehicle-to-grid technologies on distribution systems and utility interfaces. IEEE Transactions on Power Electronics, 28(12), 5653–5669.

23. Zeng, F., Zhang, X., Jiang, X., and Zhang, J. (2008). A rule-based energy management strategy for hybrid electric vehicles. Journal of Power Sources, 184(2), 684–693.

24. Zhang, C. and Mi, C. (2011). Predictive energy management in a hybrid electric vehicle using traffic prediction. IEEE Transactions on Vehicular Technology, 60(9), 4128–4140.

Intelligent and Sustainable Power and Energy Systems – Dr. M. Premkumar et al. (eds)
© 2026 Taylor & Francis Group, London, ISBN 978-1-041-10314-1

13

Role of DC-DC Converters in Battery Management Systems

Chandrashekhar R. Angadi[1],
Anupama R. Itagi[2]
Department of Electrical and Electronics Engineering,
KLE Technological University Hubli

Rakhee Kallimani[3]
Department of Electrical and Electronics Engineering,
KLE Technological University Dr. M S Sheshgiri Campus,
Belagavi

ABSTRACT: Cell balancing is crucial in battery management systems for large-scale battery packs, especially in electric vehicles, to ensure uniform charge levels across individual cells, improve lifespan, and enhance overall efficiency. DC-DC converters play a significant role in this process by enabling efficient energy transfer between cells or groups of cells, thereby facilitating active balancing techniques. These converters, particularly in configurations like bidirectional or multi-port topologies, enable the redistribution of excess energy from higher-charged cells to lower-charged ones without substantial energy loss, which is often a limitation in passive balancing. This paper explores several DC-DC converters and their critical roles in active cell balancing systems for battery management. It begins by categorising different converter types and analysing their suitability for balancing applications based on their pros and cons. The study then delves into advanced converter topologies, including isolated and non-isolated designs, examining how these configurations facilitate energy redistribution across cells with minimal power loss, which is vital for large battery packs.

KEYWORDS: Active balancing, Cell balancing, Isolated and Non-isolated converters

1. Introduction

Electric Vehicles (EVs) significantly reduce air pollution, and grids are essential for managing load demand. The performance of both EVs and grids heavily relies on the efficiency and reliability

[1]chandrashekar.angadi.2003@gmail.com, [2]anupama_itagi@kletech.ac.in, [3]rakhee.kallimani@klescet.ac.in

DOI: 10.1201/9781003654469-13

of their batteries. Thus, monitoring battery performance is essential, and the Battery Management System (BMS) plays a crucial role. One of the key functions of the BMS is cell balancing, which optimises the performance, longevity, and safety of battery packs, especially in applications like EVs and grid energy storage. Active cell balancing is favoured for its many advantages over passive methods among the different cell balancing techniques. Within active cell balancing, the DC-DC converter-based topology is particularly prominent and widely used [1]. Key converter types used in battery balancing include buck-boost, Cuk, flyback, forward, dual active bridge (DAB), and resonant converters. These converters facilitate energy transfer between cells and the entire pack, allowing for efficient charge redistribution [1, 2]. Each type has its own advantages and is suited to different balancing structures such as Cell-to-Cell (C2C), Cell-to-Pack (C2P), Pack-to-Cell (P2C), or more complex arrangements. The control of these converters, typically through pulse width modulation (PWM), is critical in managing the direction and magnitude of energy transfer during the balancing process [1]. The choice of converter topology and control strategy significantly impacts the balancing system's effectiveness, efficiency, and complexity. Unlike passive balancing, active cell balancing is more efficient as it redistributes energy between cells instead of wasting it as heat. While many active converters have been proposed, a systematic review of control variables and DC-DC converter topologies with details is essential. This work provides guidelines for selecting the DC-DC converters for active cell balancing applications. Table 5 lists essential acronyms in electrical and electronic systems, clarifying key abbreviations frequently used in BMS, converter, and control applications.

2. Literature Review

This section briefly describes the literature survey of DC-DC converter-based active cell balancing techniques. Cell balancing can be broadly categorised into passive and active techniques. Passive balancing dissipates excess energy as heat through resistors, making it simple and cost-effective but less efficient due to energy losses. Active balancing, in contrast, redistributes energy between cells to achieve uniformity with minimal energy loss, utilising methods like cell-to-cell, cell-to-pack, and pack-to-pack approaches [1]. Though classic approaches often lack speed and efficiency, especially with retired batteries, the bypass half-bridge converter offers a straightforward, economical alternative despite lower efficiency [1]. Recent advancements, such as the improved multicell-to-multicell (MC2MC) architecture with a bipolar-resonant LC converter, have shown significant improvement in balancing speed and efficiency, proving particularly effective in systems with significant cell inconsistencies by maintaining consistent balancing power across various operational modes [2]. For the secondary use of retired batteries, the equalisation process is essential. The article [3] suggests. The use of the inner layer to balance battery cells using reconfigurable architecture and the outer layer with a Buck-Boost converter for every battery group, linking their outputs in series model-based hierarchical active equalisation control method [4] leverages a Madelia-based battery pack model and Simulink co-simulation to improve SoC balance under varying current conditions significantly. Similarly, a multilayer equalisation topology with adaptive fuzzy control [5] effectively manages retired batteries in energy storage, enhancing equalisation rates, efficiency, and overall pack consistency. Complementing these advancements, a novel cell-to-cell charge equalisation converter utilising a multi-winding transformer [6] halves the required component count, leading to a more compact and cost-effective design while maintaining high-speed equalisation capabilities. Recent advancements in power electronics have significantly enhanced converter and battery management technologies. A synthesis technique based on low-entropy equations enables efficient connection of energy-storing components and switches, achieving voltage-second balancing and enabling single and multi-inductor converters with reduced pulsing currents [7]. The Multi-Cell-to-Multi-Cell (MC2MC) equalisation system employs Buck-Boost converters and reconfigurable circuits with K-means clustering for improved SoC estimation, reducing balancing time by 71.99%. A three-level hybrid boost converter with PI and bandless hysteresis control achieves high voltage gain and efficiency by combining traditional boost design with T-type inverter topology [8]. Furthermore, a bidirectional Sepic-Zeta circuit design offers intra- and inter-group equalisation, reducing equilibrium time by 51.57% and improving energy consumption by 8.4% while providing enhanced flexibility with multiple equalisation pathways [9]. Advancements in battery management and DC-DC conversion include an active balancing technique using non-isolated converters, enhancing Li-ion battery capacity to 99.7% and reducing balancing time by 36.9%. Resonant MMCs achieve high step-down ratios and efficiency for high-voltage applications [11]. Meanwhile, quasi-resonant LC-based equalisers achieve 98% efficiency with minimal hardware [12]. A novel MMC control mitigates capacitor voltage ripple, validated on a 10-kV SiC MOSFET prototype [13], showcasing improved balancing efficiency and scalability.

Tables 13.1 and 13.2 show that non-isolated converters are compared by listing types, control techniques, control variables, maximum balancing current (Ibmax), and efficiency. Buck-Boost and Cúk converters show high efficiency, with control variables such as voltage and SoC. Double-layer equalisation uses fewer components, offering cost and efficiency benefits but with potential energy loss at low power. Table 13.3 highlights converter performance and trade-offs essential for effective battery management. The 3SSC-based DC-DC boost converter G. C. Silveira et al. [14] achieves over 92% efficiency, high voltage gain, and reduced input ripple, making it ideal for renewable systems. For PV-battery hybrids, a PPAS control technique [15] enhances flexibility with shared

Table 13.1   Non-isolated converters

Converter Type	Control Technique	Control Variable	Ib max (A)	Efficiency (%)	References
Buck-Boost	PWM	Voltage, SoC	-	85-95	[1]
Cuk	PWM	Voltage, SoC	-	90-96	[3]
Double-layer equalization (Re-configurable topology + Buck-Boost converter)	SoC-based switching	SoC Switch states	-	90-94	[4]
Cuk Converter with Coupled Inductor	PWM	Duty cycle	-	92-96	[5]
Buck Converter	Voltage-mode control	Cell voltage	0.5 -10	-	[6]
Boost Converter	Current-mode control	Cell voltage	0.5 -10	-	[7]
Cuk Converter	Voltage-mode control	Cell voltage	0.5 -5	-	[8]
Switch-matrix	-	Cell voltage	5	-	[9]
Switched Resistor Passive Balancing	PLC-based control	Cell voltage	-	-	[10]
Dual-Common-Port SC (DCP-SC)	-	voltage balancing	-	-	[13]
Single-Common-Port SC (SCP-SC)	-	voltage balancing	-	-	[14]

Table 13.2 Non-isolated converters

Converter Type	Merits	Demerits	References
Buck-Boost Converter	Simple design, low cost	Limited to adjacent cells	[2]
Cuk	Low ripple, high efficiency	Complex control	[3]
Double-layer equalization (Re-configurable topology + Buck-Boost converter)	Fewer components, lower cost, higher efficiency, more straightforward control	Extra energy loss in MOSFETs at low equalisation power	[4]
Cuk Converter with Coupled Inductor	Fewer switches, faster equalisation, simpler implementation	Not explicitly stated	[5]

Table 13.3 Isolated converters

Converter Type	Control Technique	Control Variable	Ib max (A)	Efficiency (%)	References
Flyback	PWM, Voltage mode	Voltage, SoC, Current	0.5-5	80-90	[1]
Push-Pull	PWM, Current mode	Voltage, SoC, Current	5-20	85-95	[2]
Forward	PWM, Voltage mode	Voltage, SoC, Current	1-10	85-95	[6]
Resonant (LLC)	Softswitching (ZVS, ZCS)	Voltage, SoC	-	High	[8]
Bypass (Half-Bridge)	PWM	Voltage, SoC, Phase shift, Switching frequency	10-20	90-98	[11]
QRSCC	Automatic	-	-	-	[16]
TRSCC	Three-state	Resonant states	-	-	[13]
CI	Buck-boost	Duty cycle	-	-	[15]
LCSRC	Direct transfer	Cell selection	-	-	[11]
HBLCC	-	Cell group selection	-	-	[9]
BRLCC	-	Cell group selection	-	91.68	[12]
Full-Bridge Converter	PWM	Voltage, Current	20-30	90-95	[17]
Multi-Active Half-Bridge (MAHB)	Phase shift modulation	Phase shift	-	-	[18]
Bidirectional Flyback Converter	Bidirectional control scheme	SoC	-	moderate	[19]
Bidirectional Switched Transformer	Bidirectional control scheme	SoC	-	High	[5]
Multi winding Transformer	Buck-boost and Flyback operations	Balancing Current	-	High	[7]

MOSFETs. A multilayer equalisation system [16] reduces equalisation time with adaptive fuzzy control, while an MPC balancing scheme Y.-X. [18] cut the SoC imbalance to 1%. Fuzzy Logic Control [17] and dual-layer equalisers [19] improve energy consistency and efficiency in large-scale storage. This study refines the work of Khan et al. by offering a focused analysis of DC-DC converters in battery cell balancing, particularly isolated and non-isolated topologies. Innovations like the quasi-resonant LC with boost DC-DC converter achieve 98% efficiency, balancing cost and complexity while enhancing performance. Advanced designs like Cuk and BRLCC converters improve efficiency but face scalability and thermal challenges. System efficiency depends on factors like energy transfer, equalisation time, and battery lifespan, with designs like MC2MC and BRLCC excelling in reduced balancing time and stable power transfer.

Tables 13.3 and 13.4 compare isolated converters based on type, control techniques, control variables, maximum current (Ib max), and efficiency. Converters like Flyback, Push-Pull, and Full-Bridge use PWM control for variables such as voltage, SoC, and current, offering efficiencies between 80-98\%. Advanced converters, such as the Resonant (LLC) and Bypass (Half-Bridge), emphasise high efficiency and fast balancing but may require complex control strategies. Table 13.2 summarises merits and demerits: Flyback and Push-Pull converters offer simplicity and versatility with isolation but face limits on power handling and design complexity. Resonant and CI converters provide high efficiency with minimal losses yet demand precise control. Multiwinding transformers enable efficient balancing in large battery packs but come with a high component count and increased design complexity. The tables highlight the balance of efficiency, complexity, and power handling in isolated converters, which is crucial for battery management.

Table 13.4   Merits and demerits non-isolated converters

Converter Type	Merits	Demerits	Converter Type	Merits	Demerits	References
Flyback	Galvanic isolation, versatile, simple design, cost-effective	Higher component count for bidirectional, limited power handling	Flyback	Galvanic isolation, versatile, simple design, cost-effective	Higher component count for bidirectional, limited power handling	[1]
Push-Pull	Galvanic isolation, suitable for higher power, high efficiency	Complex control, higher component count, transformer needed	Push-Pull	Galvanic isolation, suitable for higher power, high efficiency	Complex control, higher component count, transformer needed	[2]
Forward	Galvanic isolation, high efficiency	More components than flyback, complex design	Forward	Galvanic isolation, high efficiency	More components than flyback, complex design	[6]
Resonant (LLC)	High efficiency, low switching losses	Complex design	Resonant (LLC)	High efficiency, low switching losses	Complex design	[8]
Bypass (Half-Bridge)	Fast balancing, flexible power flow	Limited to high power applications, lower power density	Bypass (Half-Bridge)	Fast balancing, flexible power flow	Limited to high power applications, lower power density	[9]
QRSCC	Simple control, ZCS, low EMI	Slow balancing, efficiency drops with voltage gap	QRSCC	Simple control, ZCS, low EMI	Slow balancing, efficiency drops with voltage gap	[16]
TRSCC	Improved speed at small voltage gaps	Reduced average balancing current	TRSCC	Improved speed at small voltage gaps	Reduced average balancing current	[19]

Converter Type	Merits	Demerits	Converter Type	Merits	Demerits	References
CI	High current capability	EMI issues, precise duty cycle needed	CI	High current capability	EMI issues, precise duty cycle needed	[10]
LCSRC	Fast, ZCS	Complex initialization	LCSRC	Fast, ZCS	Complex initialization	[11]
HBLCC	Fast, supports multiple cells	Limited step-up capability	HBLCC	Fast, supports multiple cells	Limited step-up capability	[12]
BRLCC	Fast, stable power, flexible modes	Slightly lower efficiency	BRLCC	Fast, stable power, flexible modes	Slightly lower efficiency	[13]
Full-Bridge	High efficiency, suitable for high power	Large size, expensive	Full-Bridge	High efficiency, suitable for high power	Large size, expensive	[14]
MAHB	Flexible power flow, high efficiency	Design complexity for high cell counts	MAHB	Flexible power flow, high efficiency	Design complexity for high cell counts	[15]
Multiwinding Transformer	High efficiency, compact size	Core loss at lower switching frequencies	Multiwinding Transformer	High efficiency, compact size	Core loss at lower switching frequencies	[16]
Bidirectional Multisecondary	Efficient cell-pack energy transfer	Complex design and control	Bidirectional Multisecondary	Efficient cell-pack energy transfer	Complex design and control	[17]
Bidirectional Multiple Transformers	Improved balancing for large packs	High cost, increased complexity	Bidirectional Multiple Transformers	Improved balancing for large packs	High cost, increased complexity	[19]

Table 13.5 List of essential acronyms and their definition

PWM - Pulse Width Modulation	SoC State of Charge	CTP Capacitor Transformer Coupled
PTC Pulse Transformer Coupled	ZVS Zero Voltage Switching	ZCS : Zero Current Switching
EMI electromagnetic interference	3SSC -Three-state Switching Cell	CI Coupled Inductor
D-CTC: Dual Capacitor Transformer Coupled	CTPTC Capacitor Transformer Pulse Transformer Coupled	TRSCC Three-state Resonant Switched-Capacitor Converter
MC2MC : Multi-Cell to Multi-Cell	DC2C Direct Cell-to-Cell	c2c Cell-to-Cell
A-CTC Adjacent Cell-to-Cell	PLC Programmable Logic Controller	PV Photovoltaic
QRSCC Quasi Resonant Switched Capacitor Converter	HBLCC Half-Bridge Load Commutated Converter	T-NPC Three-level Neutral Point Clamped
BRLCC Bidirectional Resonant Load Commutated Converter	SCP-SC Single-Common-Port Switched Capacitor	MBBS Model-Based Balancing System
AC2C Alternating Cell-to-Cell	p2c Pack-to-Cell	BMS Battery Management System
SC Switched Capacitor	C2S Cell to String	Ib max Maximum balancing current
C2R Cell to Resistor	S2C String to Cell	PBM Passive Balancing Method

3. Conclusion and Future Work

DC-DC converters are indispensable components in cell balancing systems, particularly in large-scale battery packs for electric vehicles. Active cell balancing techniques enable efficient energy transfer between cells with minimal energy loss. By examining various DC-DC converter designs and their specific applications in battery management, this study highlights how various isolated and non-isolated DC-DC converter topologies support effective energy redistribution with minimal loss. This work provides guidelines for selecting the type of converters for cell balancing applications, detailing the merits and demerits of different DC-DC converters and the balancing current used in the cell balancing process. Among these converters, the quasi-resonant LC with boost DC-DC converter is particularly efficient for EV applications, achieving 98 % efficiency through zero-current switching and adaptive regulation. This design minimizes energy loss and maintains a lightweight, compact structure, ideal for EV energy management.

REFERENCES

1. N. Khan, C. A. Ooi, A. Alturki, M. Amir, T. Alharbi et al., "A critical review of battery cell balancing techniques, optimal design, converter topologies, and performance evaluation for optimizing storage system in electric vehicles," Energy Reports, vol. 11, pp. 4999–5032, 2024.

2. X. Luo, L. Kang, C. Lu, J. Linghu, H. Lin, and B. Hu, "An enhanced multicell-to-multicell battery equalizer based on bipolar-resonant lc converter," Electronics, vol. 10, no. 3, p. 293, 2021

3. Y. Li, P. Yin, and J. Chen, "Active equalization of lithium-ion battery based on reconfigurable topology," Applied Sciences, vol. 13, no. 2, p. 1154, 2023

4. G. Sun, Q. Ma, T. Liu, and H. Wen, "Active battery equalizationsystem based on extended kalman filter and dc-dc bidirectional flyback converter," in 2023 IEEE 2nd International Power Electronics and Application Symposium (PEAS). IEEE, 2023, pp. 2163–2168.

5. X. Zheng, X. Liu, Y. He, and G. Zeng, "Active vehicle battery equalization scheme in the condition of constant-voltage/current charging and discharging," IEEE Transactions on Vehicular Technology, vol. 66, no. 5, pp. 3714–3723, 2016.

6. S.-H. Park, K.-B. Park, H.-S. Kim, G.-W. Moon, and M.-J. Youn, "Single-magnetic cell-to-cell charge equalization converter with reduced number of transformer windings," IEEE Transactions on Power Electronics, vol. 27, no. 6, pp. 2900–2911, 2011.

7. T. S. Ambagahawaththa, D. Nayanasiri, A. Pasqual, and Y. Li, "Non-isolated dc–dc power converter synthesis using low-entropy equations," IEEE Journal of Emerging and Selected Topics in Power Electronics, vol. 10, no. 6, pp. 6457–6469, 2022.

8. N. Osman, M. F. Mohamad Elias, and N. Abd Rahim, "Three-level hybrid boost converter with output voltage regulation and capacitor balancing," IETE Journal of Research, vol. 69, no. 4, pp. 1852–1860, 2023.

9. L. Liao, H. Li, H. Li, J. Jiang, and T. Wu, "Research on equalization scheme of lithium-ion battery packs based on consistency control strategy," Journal of Energy Storage, vol. 73, p. 109193, 2023.

10. M. Raeber, A. Heinzelmann, and D. O. Abdeslam, "Analysis of an active charge balancing method based on a single nonisolated dc/dc converter," IEEE Transactions on Industrial Electronics, vol. 68, no. 3, pp. 2257–2265, 2020.

11. X. Zhang, T. C. Green, and A. Junyent-Ferre, "A new resonant modular multilevel step-down dc–dc converter with inherent-balancing," IEEE Transactions on Power Electronics, vol. 30, no. 1, pp. 78–88, 2014.

12. Y. Shang, C. Zhang, N. Cui, and J. M. Guerrero, "A cell-to-cell battery equalizer with zero-current switching and zero-voltage gap based on quasi-resonant lc converter and boost converter," IEEE Transactions on Power Electronics, vol. 30, no. 7, pp. 3731–3747, 2014.

13. B. Fan, J. Wang, J. Yu, S. Mocevic, Y. Rong, R. Burgos, and D. Boroye- vich, "Cell capacitor voltage switching-cycle balancing control for mod-ular multilevel converters," IEEE Transactions on Power Electronics, vol. 37, no. 3, pp. 2525–2530, 2021.

14. G. C. Silveira, F. L. Tofoli, L. D. S. Bezerra, and R. P. Torrico-Bascop´e, "A nonisolated dc–dc boost converter with high voltage gain and balanced output voltage," IEEE Transactions on Industrial Electronics, vol. 61, no. 12, pp. 6739–6746, 2014.

15. W. Li, J. Xiao, Y. Zhao, and X. He, "Pwm plus phase angle shift (ppas) control scheme for combined multiport dc/dc converters," IEEE Transactions on Power Electronics, vol. 27, no. 3, pp. 1479–1489, 2011.

16. R. Li, P. Liu, K. Li, and X. Zhang, "Research on retired battery equalization system based on multi-objective adaptive fuzzy control algorithm," IEEE Access, 2023.

17. Amoorezaei, S. A. Khajehoddin, and K. Moez, "A compact cuk-based differential power processing ic with integrated magnetics and soft- switching controller for maximized cell-level power extraction," IEEE Transactions on Power Electronics, 2024.

18. Y.-X. Wang, H. Zhong, J. Li, and W. Zhang, "Adaptive estimation-based hierarchical model predictive control methodology for battery active equalization topologies: Part i–balancing strategy," Journal of Energy Storage, vol. 45, p. 103235, 2022.

19. X. Ning, Y. He, and Z. Shi, "A dual-layer equalization circuit based on a flyback converter-inductor battery pack," in 2023 8th Asia Conference on Power and Electrical Engineering (ACPEE). IEEE, 2023, pp. 647–652.

Note: All the tables in this chapter were made by the author.

Intelligent and Sustainable Power and Energy Systems – Dr. M. Premkumar et al. (eds)
© 2026 Taylor & Francis Group, London, ISBN 978-1-041-10314-1

14

Wind Power Forecasting using Machine Learning

Shreepad Naik[1],
Raghavendra Naik[2], Dashmi Navalgund[3],
Shringar Shettar[4] and Siddarameshwar H. N.[5]

Dept of Electrical and Electronics, KLE Technological University,
Hubbali, India

ABSTRACT: Wind power curves are essential for wind power forecasting, wind turbine condition monitoring, wind energy potential estimation, and wind turbine selection. In the field of renewable energy, wind power forecasting is crucial, especially for maximizing wind turbine performance and guaranteeing effective energy grid management. Applications including wind power forecasting, turbine condition monitoring, energy potential estimation, and wind turbine selection depend heavily on wind power curves. These curves allow for precise energy generation estimates by describing the link between wind speed and a turbine's power output. Reliable wind speed and power pro duction forecasting are essential for enhancing energy supply reliability and optimizing resource utilization in wind energy systems since wind is sporadic and not always available. In our study, we forecast wind power values for December. The raw data undergoes pre-processing, divided into subframes to remove asymmetric data using the outlier's function and fitted to five regression models, and evaluated their performance using error metrics. Among the models, the Gradient Boosting Regressor demonstrated the strongest predictive capability, achieving the lowest mean absolute error (MAE) of 76.50 kW, a root mean square error (RMSE) of 114.06 kW, and the highest R-squared value of 0.9918, indicating high accuracy and reliability in predictions.

KEYWORDS: Active power, Power forecasting, Machine learning, Regression models, Wind, Wind power curve model

[1]shreepadnaik10@gmail.com, [2]raghunaik8747@gmail.com, [3]dashmi.navalgund@gmail.com, [4]shringarshettar@gmail.com, [5]sidda_hn@kletech.ac.in

DOI: 10.1201/9781003654469-14

1. Introduction

Wind power curves play a pivotal role in various critical activities, including forecasting wind power, monitoring the operational condition of wind turbines, estimating wind energy potential, and selecting the most appropriate wind turbines for specific locations. However, creating reliable wind power curves from raw wind data presents significant challenges, primarily due to the presence of outliers. These outliers often arise from unexpected situations, such as wind curtailment, when excess wind power is reduced to maintain grid stability and physical issues like blade damage or mechanical malfunctions. Wind energy is widely recognized for its sustainability, renewability, cost-effectiveness, and abundant availability across diverse geographical regions.

Despite these advantages, the inherent randomness and variability of wind introduce substantial uncertainties in power generation. These fluctuations pose challenges for energy management systems, affecting grid stability and complicating the efficient planning and operation of power systems. To address these challenges, accurate prediction of wind power curves is essential. These curves represent the nonlinear relationship between wind speed and the corresponding power output of wind turbines. Precise modelling of this relationship is critical for several reasons. First, it ensures efficient integration of wind power into the electrical grid, minimizing disruptions and maintaining a steady power supply. Second, it aids in the real-time monitoring and predictive maintenance of wind turbines, helping to identify potential failures before they occur. Lastly, accurate wind power curves support optimal energy dispatch strategies, contributing to the reliability and sustainability of the overall power system.

2. Literature Review

The proposed work focuses on wind power forecasting using inconsistent data in [1] with heteroscedastic spline regression (HSRM), robust spline regression (RSRM), and artificial neural networks (ANNs). The RSRM outperformed the HSRM, achieving a mean absolute error (MAE) of 9.57 and a root mean square error (RMSE) of 15.97, compared to the HSRM's MAE of 9.82 and RMSE of 16.05. The literature examines the 2022 IEEE paper titled' 'Wind Power Curve Modelling Through Data-driven Approaches in [2] Evaluating Piecewise Linear Fitting and Machine Learning Applications in a Real-Unit Case," researchers explored wind power curve modelling using linear regression, Lasso, and other machine learning techniques. The study reported a mean absolute error (MAE) of 0.01445 and an accuracy of 0.95. Authors in [3] found that recurrent neural networks (RNNs) outperformed convolutional neural networks (CNNs) in wind power prediction, with RNNs achieving a mean absolute error (MAE) of 73.51. Additionally, a separate study showed that the k-nearest neighbour algorithm was the most effective for estimating global horizontal irradiance, based on the lowest mean absolute percentage (MAP) error. The researchers in [4] developed a wind power forecasting model utilizing the Extreme Learning Machine (ELM) algorithm and time series analysis, comparing its performance to the backpropagation (BP) algorithm. The results indicated that the ELM algorithm significantly outperformed BP, achieving a mean square error (MSE) of 0.0716, while BP resulted in a higher MSE of 0.3487. Authors in [5] developed models for small wind turbine power curves using artificial neural networks (ANN), support vector machines (SVM), k-nearest neighbour (KNN), and gradient boosting machines (GBM) algorithms, achieving a root mean square error (RMSE) of 0.1615 and demonstrating their effectiveness in predicting turbine performance. Researchers in [6] focused on power curve estimation of wind turbines under varying weather conditions using machine learning algorithms. The study explored the effectiveness of linear regression, polynomial regression, random forest regression, gradient boost (G Boost), and extreme gradient regression (XGBoost). Among these models, the best performance was achieved with an RMSE of 6.404.

3. Dataset and Pre-processing

3.1 Dataset

SCADA systems monitor and record data from wind tur bines every 10 minutes, capturing essential information such as wind direction, wind speed, and generated power. The data in this file has been extracted from the SCADA system of an operational wind turbine currently producing electricity in Turkey. It includes the date and time (logged at 10-minute intervals), along with key metrics: LV Active Power (kW), which is the turbine's current power output; Wind Speed (m/s), measured at the turbine's hub height and used for energy generation; Theoretical Power Curve (KWh), outlining the expected power values provided by the turbine manufacturer according to specific wind speeds; and Wind Direction (°), indicating the direction of the wind at the turbine's hub height, which allows the turbine to automatically orient itself for optimal energy capture.

3.2 Preprocessing of the Dataset

From the dataset the Power curve is plotted (along X-axis: wind speed and along Y-axis: Real and theoretical power curve) values to recognize the asymmetrical pattern of the data. Since the wind power curve modelling data will have an asymmetric error distribution. The data preprocessing is required because in the data set, the Active power is fluctuating with respect to the wind speed due to abnormal conditions, such as wind curtailment and blade damage, thus reducing the negative impact of outliers during the training phase. The dataset is first divided into sub-data frames based on the range of the Active Power column, using intervals of 50 units. Each sub-data frame contains data points within a specific range, ensuring more targeted outlier detection and removal. The Fig. 14.1 represents the Power curve model of the given raw dataset.

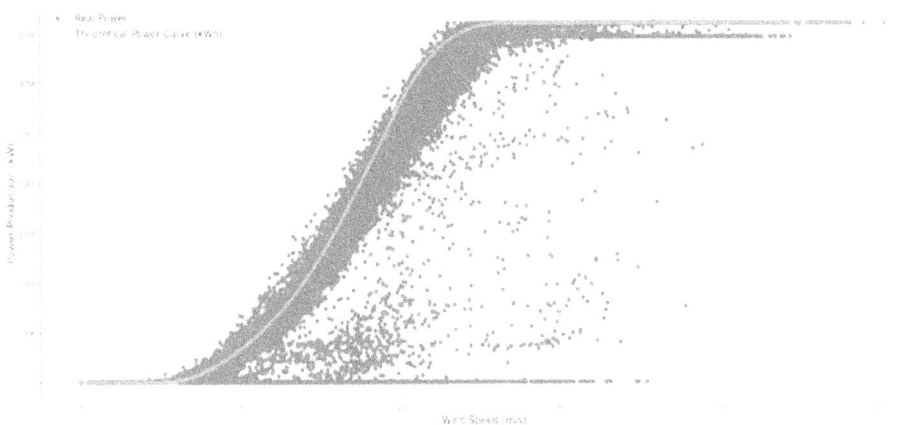

Fig. 14.1   Before pre-processing the image

A total of 70 sub-data frames are created, with an additional sub-data frame for values greater than or equal to 3300 kW. For each sub-data frame, an outlier removal function is applied for respective column. This function filters out values that fall below or above certain quantiles, defined by thresholds specific to each sub-data frame. After outlier removal, all cleaned sub-data frames are concatenated into a single comprehensive data frame for further analysis. This cleaned dataset is visualized by plotting Wind Speed against LV Active Power to show the actual power production, along with the theoretical power curve for comparison. The figure represents the pre-processed the Power curve model of the pre-processed dataset.

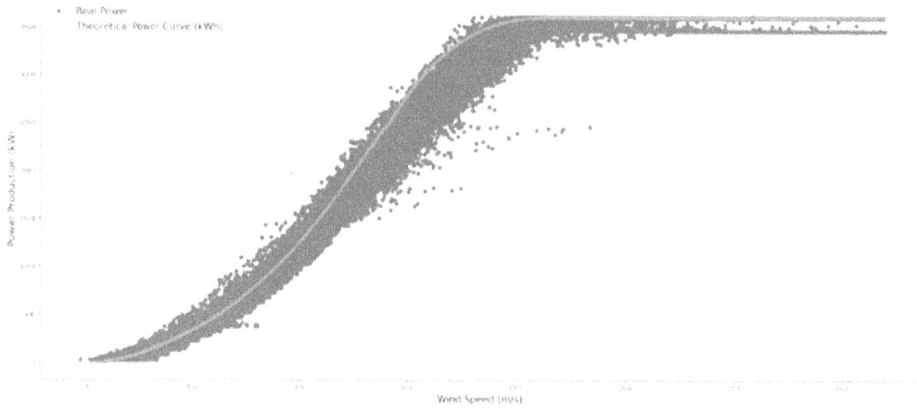

Fig. 14.2 After pre-processing the image

4. Methodology

Figure 14.3 depicts the Entire procedure working in the block diagram.

The pre-processed dataset has 37000 row and 5 columns. The dataset is separated for training and testing, 91 percentage data for training, and the remaining 9 percentage data for testing because only the last month of the year needs to be predicted. the training set is fitted to respective algorithms. This preprocessing step improves the quality of data used in training by removing extreme values that could distort the model's accuracy, resulting in more reliable predictions. The training set is trained on five regression algorithms models

- Decision Tree Regressor
- Extra Trees Regressor
- K Neighbors Regressor
- Random Forest Regressor
- Gradient Boosting Regressor

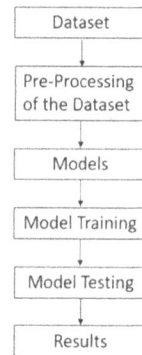

Fig. 14.3 Block diagram of the proposed work

Decision Tree: This model builds a tree-like structure to make decisions based on input features, splitting the data into branches based on the most informative features. In this implementation, the Decision Tree Regressor reached an R squared (Accuracy) of 0.9858, which, although still strong, is lower than the KNeighborsRegressor. The MAE was 97.83 kW, and the RMSE was 149.68 kW, indicating slightly higher prediction errors than the KNeighborsRegressor. reflecting higher deviations in prediction accuracy. Figure 14.4 depicts the output of Power curve model with respective to the actual vs predicted values vs Theoretical power curve values for Decision Tree model.

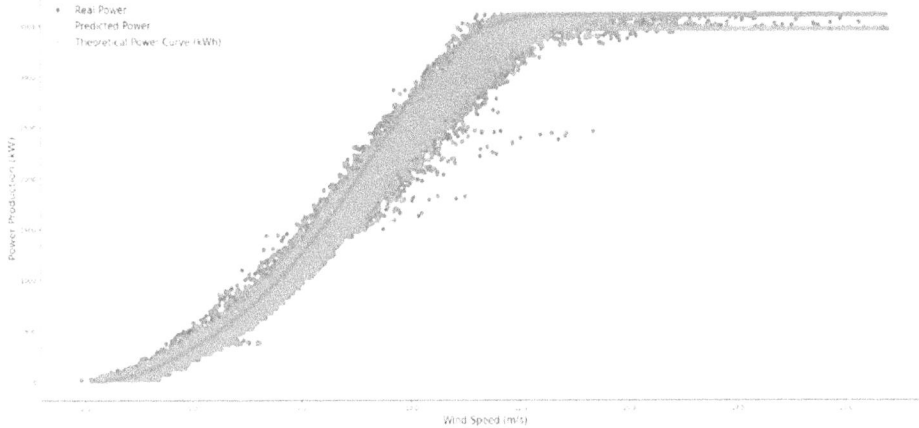

Fig. 14.4 Actual vs predicted values vs theoretical power curve values for decision tree model

Extra Trees: The Extra Trees Regressor is an ensemble learning method that fits multiple decision trees on different parts of the data and averages the results for better accuracy and robustness. It is similar to Random Forest but differs in how it splits nodes, promoting randomness. This model scored an R-squared (Accuracy) of 0.9904, close to the KNeighborsRegressor. The MAE was 81.28 kW, and the RMSE was 123.29 kW, showing reliable predictions with moderate variance. showing a significant improvement in prediction reliability and error reduction. Figure 14.5 depicts the output of Power curve model with respective to the actual vs predicted values vs Theoretical power curve values for Extra Trees Regressor model.

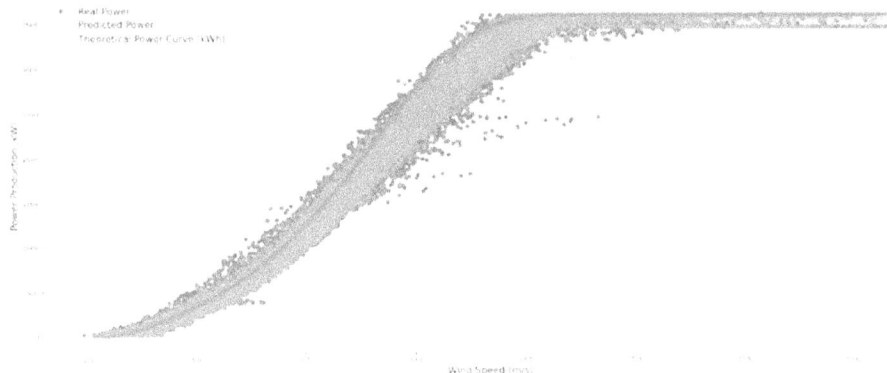

Fig. 14.5 Actual vs predicted values vs theoretical power curve values for extra trees regressor model

KNeighborsRegressor: The KNeighborsRegressor algorithm is a non-parametric method that predicts the target by averaging the output of the k-nearest neighbors in the feature space. This technique works well for problems where the data is assumed to be locally consistent.

In the code provided, the model was trained using features like 'Wind Speed (m/s)' and Theoretical_ Power_Curve (KWh), achieving an R2 (Accuracy) of 0.9908, which is very high. The Mean Absolute Error (MAE) was 80.32 kW, indicating the average prediction deviation, and the Root Mean Squared Error (RMSE) was 120.21 kW, highlighting the typical error magnitude, reflecting moderate error and strong predictive performance. Figure 14.6 depicts the output of Power curve

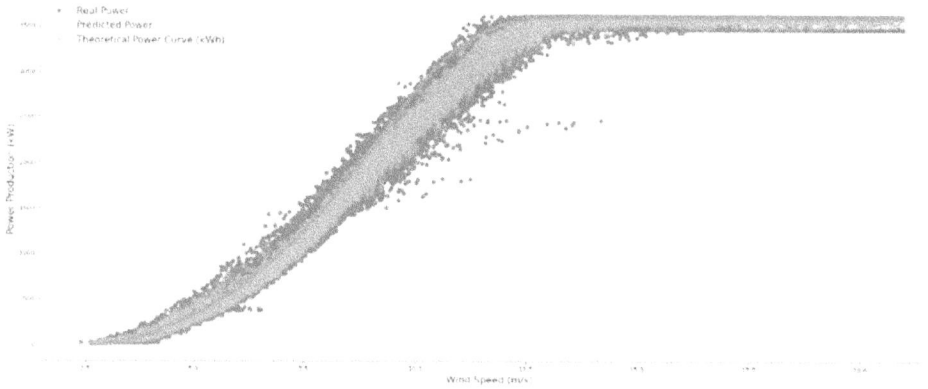

Fig. 14.6 Actual vs predicted values vs theoretical power curve values for KNeighbores regressor model

model with respective to the actual vs predicted values vs Theoretical power curve values for KNeighbores Regressor model.

Random Forest: This popular ensemble method uses multiple decision trees to improve predictive performance and control overfitting. Each tree is built using a subset of data, and the final outputis the average of individual tree predictions. In the provided code, the model with 350 trees and a maximum depth of 25 achieved an R-squared (Accuracy) of 0.9914, showcasing strong predictive power. The MAE was 77.37 kW, and the RMSE was 116.54 kW, reflecting accurate predictions with relatively low errors. indicating lower prediction errors than KNeighborsRegressor and ExtraTreesRegressor. Figure 14.7 depicts the output of Power curve model with respective to the actual vs predicted values vs Theoretical power curve values for Random Forest Regressor model.

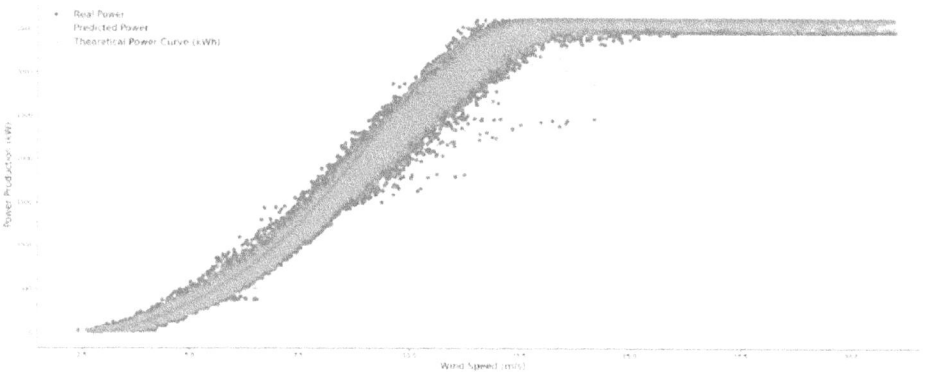

Fig. 14.7 Actual vs predicted values vs theoretical power curve values for random forest regressor model

Gradient Boosting Regressor: This is an iterative, boosting algorithm that combines weak models (decision trees) in a sequential manner to minimize prediction errors. It learns from previous mistakes to build a strong predictive model. The version run here, with 2000 estimators, provided the best performance with an R-squared (Accuracy) of 0.9918. The MAE was 76.50 kW, and the RMSE was 114.06 kW, indicating the lowest prediction error and highest reliability among the tested models.

demonstrating the lowest prediction errors and the highest reliability. Figure 14.8 depicts the output of Power curve model with respective to the actual vs predicted values vs Theoretical power curve values for Gradient Boosting Regressor model.

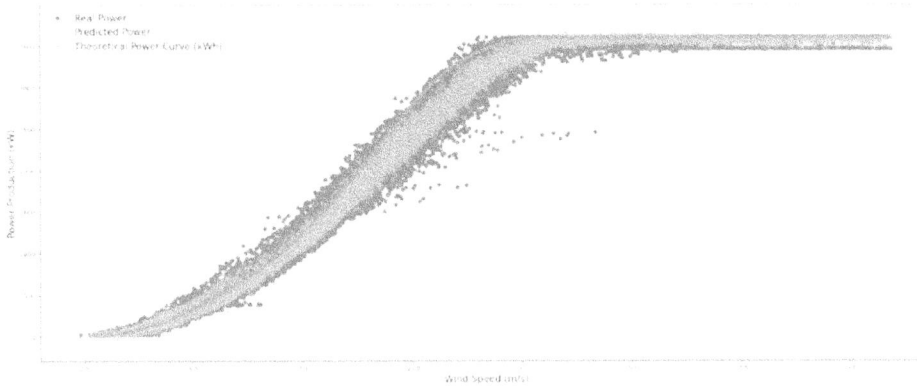

Fig. 14.8 Actual vs predicted values vs theoretical power curve values for gradient boosting regressor model

5. Results and Discussions

The results from the power production prediction models show varying levels of accuracy and reliability. These models forecast wind power based on wind speed. The 10 sample predicted values for each model were printed on the Python console. Figure 14.9 depicts the sample 10 predicted values for wind power with respect to wind speed.

```
Predicted LV ActivePower (kW) with corresponding Wind Speed (m/s):

Wind Speed: 14.26 m/s -> Predicted Power: 3555.10 kW
Wind Speed: 10.97 m/s -> Predicted Power: 2839.18 kW
Wind Speed: 6.54 m/s -> Predicted Power: 757.48 kW
Wind Speed: 9.66 m/s -> Predicted Power: 2086.52 kW
Wind Speed: 5.34 m/s -> Predicted Power: 352.41 kW
Wind Speed: 8.81 m/s -> Predicted Power: 1882.52 kW
Wind Speed: 10.94 m/s -> Predicted Power: 2968.21 kW
Wind Speed: 5.36 m/s -> Predicted Power: 422.69 kW
Wind Speed: 9.06 m/s -> Predicted Power: 1999.01 kW
Wind Speed: 9.53 m/s -> Predicted Power: 2113.39 kW
```

Fig. 14.9 Comparison of 5 algorithms corresponding to their error metrics

Table 14.1 Comparison of algorithm models

Sl. no	Algorithm models	Accuracy	MAE (kW)	MSE (kW)	RMSE (kW)
1	Decision Tree Regressor	0.9858	97.8342	22404.49	149.68
2	Extra Trees Regressor	0.9903	81.28	15200.54	123.29
3	KNeighbors Regressor	0.9908	80.32	14451.11	120.21
4	Random Forest Regressor	0.9913	77.36	13580.97	116.53
5	Gradient Boosting Regressor	0.9917	76.50	13010.10	114.06

The Decision Tree Regressor exhibited the highest prediction errors, as reflected in its relatively lower R-squared value and higher Mean Absolute Error (MAE) and Root Mean Square Error (RMSE). This indicated that, although it could model the data structure, its performance was less robust compared to ensemble methods. The Extra Trees Regressor and K-Neighbors Regressor yielded better results, showing reduced prediction errors and improved R-squared values, thereby demonstrating their effectiveness in capturing underlying relationships. The Random Forest Regressor further enhanced predictive accuracy, with lower MAE and RMSE, showcasing its strength as an ensemble learner. The Gradient Boosting Regressor delivered the best results, achieving the lowest prediction errors and the highest R-Squared value of 0.9917, indicating the most reliable and precise performance among all the models tested.

6. Conclusion

This work forecasts Wind power. The December month Wind power values are Predicted. The dataset is collected from the Kaggle site. The raw data is pre-processed using the outlier's function and divided into subframes to remove asymmetric data using the outlier's function. The pre-processed data is Split and fitted to five regression models, and the Error metrics are calculated for each model. The Gradient Boosting Regressor showed the most promising results, achieving the lowest prediction errors (MAE of 76.50 kW and RMSE of 114.06 kW) and the highest R-squared value of 0.9918, indicating strong predictive performance and reliability. Future work could focus on integrating more complex machine learning and deep learning algorithms to further enhance prediction accuracy. The use of real-time data streaming and adaptive learning methods could also help in creating dynamic models that adjust to changing conditions, making wind power forecasting more efficient and responsive to operational needs.

REFERENCES

1. Y. Wang, Q. Hu, D. Srinivasan and Z. Wang, "Wind Power Curve Modeling and Wind Power Forecasting With Inconsistent Data," in IEEE Transactions on Sustainable Energy, vol. 10, no. 1, pp. 16–25, Jan. 2019, doi: 10.1109/TSTE.2018.2820198.
2. D. P. Cardoso Filho, D. N. da Silva Ramos, M. d. Carvalho Filho, R. C. Medrado, A. L. Cotta Weyll and T. R. Maritan," Wind Power Curve Modeling Through Data-driven Approaches: Evaluating Piecewise Linear Fitting and Machine Learning Applications in a Real-Unit Case," 2022 Workshop on Communication Networks and Power Systems (WCNPS), Fortaleza, Brazil, 2022, pp. 1–6, doi: 10.1109/WCNPS56355.2022.9969684.
3. C. Wang, L. Tengfe and F. Haiyan," Wind Power Prediction Based on Hybrid Neural Network and Wind Power Curve," 2023 China Automation Congress (CAC), Chongqing, China, 2023, pp. 4382–4386, doi: 10.1109/CAC59555.2023.10450176.
4. I. Mansoury, D. E. Bourakadi, A. Yahyaouy and J. Boumhidi," Wind Power Forecasting Model Based on Extreme Learning Machine and Time series," 2021 Fifth International Conference on Intelligent Computing in Data Sciences (ICDS), Fez, Morocco, 2021, pp. 1–6, doi:10.1109/ICDS53782.2021.9626737.
5. R. Veena, V. Femin, S. Mathew, I. Petra and J. Hazra, "Intelligent models for the power curves of small wind turbines," 2016 International Conference on Cogeneration, Small Power Plants and District Energy (ICUE), Bangkok, Thailand, 2016, pp. 1–5, doi: 10.1109/COGEN.2016.7728965.
6. M. Al-Gabalawy, H. S. Ramadan, M. A. Mostafa and S. A. Hussien, "Power Curve Estimation of a Wind Turbine Considering Different Weather Conditions using Machine Learning Algorithms," 2022 23rd International Middle East Power Systems Conference (MEPCON), Cairo, Egypt, 2022, pp. 1–8, doi: 10.1109/MEPCON55441.2022.10021759.

Note: All the figures and table in this chapter were made by the author.

Intelligent and Sustainable Power and Energy Systems – Dr. M. Premkumar et al. (eds)
© 2026 Taylor & Francis Group, London, ISBN 978-1-041-10314-1

15

Cost Optimization of Microgrid in EV Application using PSO Algorithm

**Manoj Nayak[1], Kiran R. Patil[2],
Satwik Mathad[3], Anoopkumar Patil[4]**
Department of Electrical and Electronics Engineering,
KLE Technological University,
Hubli, Karnataka

ABSTRACT: Fossil fuel prices rise, and families use more energy because of contemporary equipment. These factors underscore the widening imbalance between supply and demand for electricity. Microgrid operations optimisation becomes crucial as the transition to renewable energy continues, especially with the growing popularity of electric vehicles. To balance battery storage, grid supply, and renewable energy in microgrids with sizable EV loads, this study suggests a cost-optimization approach utilising the PSO algorithm. Two-wheeler and four-wheeler charging profiles at different times are considered, along with differences in home and EV demand. Considering limitations like battery capacity, state-of-charge limits, battery degradation costs, and penalties for unmet demand, the technique maximises energy output and storage while guaranteeing sustainable, affordable energy management.

KEYWORDS: Electric vehicle micro-grid, Battery storage system, Particle swarm optimisation (PSO) technique, Renewable source energy

1. Introduction

Microgrids (MGs) and electric vehicles (EVs) are key components of the shift to a clean and sustainable future because they lower carbon emissions and the use of fossil fuels. The disparity between supply and demand for electricity has worsened due to growing fossil fuel prices and increased household energy usage. Microgrids, self-sufficient energy systems that use solar panels, wind turbines, batteries, and generators, are essential to this change because they encourage clean

[1]manojnayak1662@gmail.com, [2]krpatil@kletech.ac.in, [3]satwik.mathad@kletech.ac.in, [4]anupkumar@kletech.ac.in

DOI: 10.1201/9781003654469-15

energy and lower greenhouse gas emissions [1]. Although they are cleaner than traditional cars, EVs have transformed transportation. However, they also present problems, such as grid instability when charging during peak hours. To overcome these obstacles, MGs must find economical energy management methods. Particle Swarm Optimization (PSO) is used in this study to reduce microgrid expenses while maintaining a steady power supply while considering renewable energy sources. Devices for Storing Batteries [2–4].

2. System Architecture

The system architecture is a microgrid model consisting of 400 houses, 180 two-wheelers, and 80 four-wheelers. This design comprises a solar panel, wind turbine, BSS, and grid, as illustrated in Fig. 15.1.

2.1 Proposed Methodology

A fundamental methodology is needed to carry out the microgrid cost optimisation, which comprises establishing the cost function and the constraints, creating a flow chart, and using optimisation techniques. When designing a microgrid that supports

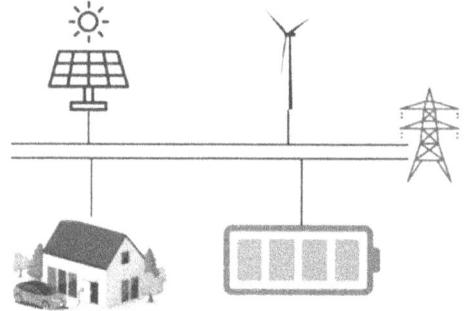

Fig. 15.1   Design of a microgrid system

EV charging, cost optimisation strategies must consider several variables, including the availability of renewable energy, load demand predictability, battery management, scalability, dependability, environmental effect, and operating expenses [5]. By optimising resource allocation, balancing grid and renewable power, adjusting to dynamic demands, and reducing battery degradation costs, PSO successfully addresses these issues.

2.2 Defining the Cost Function

To optimally minimise the total cost of the microgrid by integrating all the resources, the equation is given as [4].

$$Total\ Cost = \sum_{i=s,w,g} Pi \cdot (Ci + C_{OM,i}) + P_{battery} \left(\frac{c_{battery} + C_{OM.battery} + C_{degradation}}{efficiency} \right) \tag{1}$$

The total demand D must be satisfied by all sources of power [5]

$$P_{solar} + P_{wind} + P_{battery} + P_{grid} \geq D \tag{2}$$

Each energy source has a maximum capacity. Let C_{max} be the maximum capacity for each [6]

$$P_{solar} \leq C_{max,\,solar}; P_{wind} \leq C_{max,\,wind} \tag{3}$$

The state of charge (SOC) of a battery must be within the limits from maximum to minimum [7]

$$SOC_{min} \leq SOC \leq SOC_{max} \tag{4}$$

Throughout the day, solar and wind energy availability may fluctuate [8]

$$P_{solar} = 0\ (high\ duirng\ peak\ hour) \tag{5}$$

$$P_{wind} \leq A_{wind}\ (Avaliable\ wind) \tag{6}$$

$$P_{solar}, P_{wind}, P_{battery}, P_{grid} \geq 0 \tag{7}$$

2.3 Optimisation Technique

Particle Swarm Optimization (PSO) is a valuable method for reducing microgrid operating expenses, especially in light of the growing demand for EV charging. It effectively handles issues like dynamic EV charging schedules, battery energy storage optimisation, and the unpredictability of renewable energy outputs [8]. Because of its adaptability, PSO can incorporate particular restrictions, such as prioritising two-wheelers in the morning and four-wheelers in the evening, while maintaining cost-effectiveness and user needs. Through integrating load profiling, projections for renewable energy, and real-time energy prices, PSO successfully handles the nonlinear complexity of microgrids. Large-scale and complex microgrid systems benefit significantly from their capacity to integrate renewable energy sources and optimise battery consumption and dynamic loads [9].

2.4 PSO Algorithm and Processes

The process begins with establishing the PSO (Particle Swarm Optimization) parameters, including the number of particles and maximum iterations, and specifying each energy source's cost and operating parameters—wind, solar, and grid. The simulation is configured to run for 24 hours, with solar energy availability defined between 6 a.m. and 6 p.m. The initial positions and velocities of the particles are also set during this step. Once initialised, the main PSO loop is executed for each particle. At every iteration, the total energy from all sources is summed to determine if it meets the hourly demand. If the allocated energy exceeds the maximum limit, it is proportionally reduced, and if it falls short of the demand, penalties are applied. Next, the cost of each energy source—solar, wind, grid, and battery—is analysed based on predetermined parameters, including operational and maintenance (O&M) costs. The costs and particle positions are updated accordingly to reflect better solutions. The positions and velocities of the particles are then updated by adding the velocity to the current position while ensuring that boundary limits are enforced so that energy allocations remain within permissible limits. The iteration continues by tracking and displaying the progress of the best cost for each iteration. Finally, the optimised cost is presented after completing the iterations, representing the most cost-effective energy allocation strategy. Table 15.1 shows the parameters and their values for the PSO.

Table 15.1   Parameters and their values for the PSO

Parameter	Value/Description
Number of Particles	30
Maximum iteration	100
Inertia Weight (W)	0.7 (controls exploration vs. exploitation)
Cognitive Parameter (C1)	1.5 (personal learning factor)
Social Parameter (C2)	1.5 (global learning factor)

3. Results and Discussion

It is essential to understand the variance in load demand and to create a specialised model that calculates the overall average EV load demand of a particular area for the day, later on taking into account the 400 households with 180 two-wheelers 3 kW each (charging time 3 hrs) and 80 four-wheelers 13 KW each (charging time 6 hrs) as shown in Fig. 15.2 and Fig 15.3. Similarly, as illustrated below in the EV charging Schedule (Table 15.2), most four-wheelers are charged in the

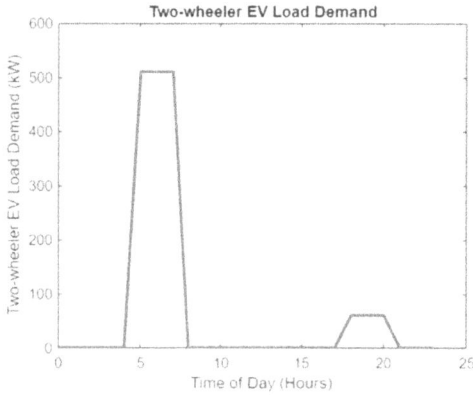

Fig. 15.2 Two-wheeler load demand per day

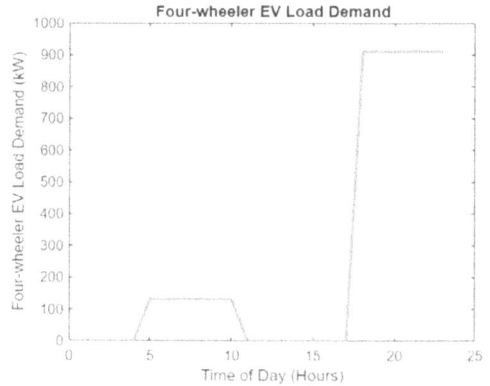

Fig. 15.3 Four-wheeler load demand per day

Table 15.2 EV charging schedule

Two-Wheeler charging	160 (charge between 6 am - 9 am) morning	20 (charge between 7 pm - 10 pm) Evening
Four-wheeler charging	10 (charge between 6 am - 12 pm) morning	70 (charge between 7 pm-midnight)

evening and rest in the morning. In contrast, most two-wheelers are charged in the morning and remain in the evening.

Figure 15.2 illustrates that most two-wheelers are charged in the morning, and the rest are charged in the evening. Figure 15.3 illustrates that most four-wheelers are charged in the evening. Consequently, as shown in Fig. 15.4, the average EV load demand for a particular area per day is 327.5 kW. by gathering each individual electric vehicle's charging data.

The average base load demand for a day, taking into account both residential and EV, is 1067 KW. Additionally, Figure 15.5 shows the locality's total load demand with and without EV. This graph demonstrates the overall pattern of local power usage over a 24-hour period. We can observe that greater electricity consumption occurs between 12 am and 11 pm. To capture a realistic environmental

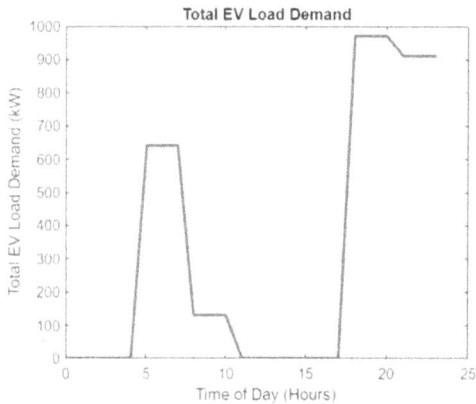

Fig. 15.4 Average EV load demand per day

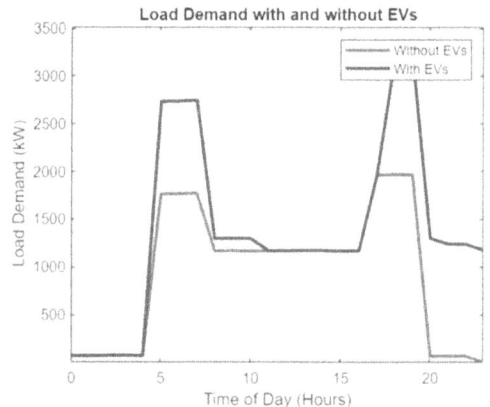

Fig. 15.5 Load demand of locality with and without EV

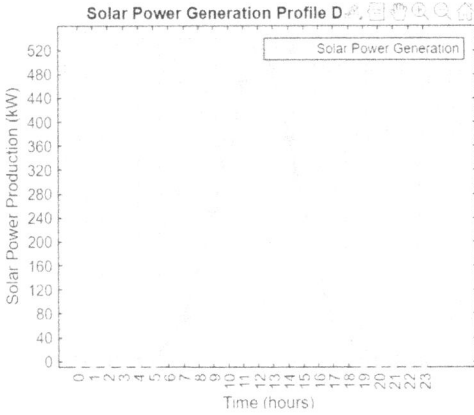

Fig. 15.6 Solar energy production profile

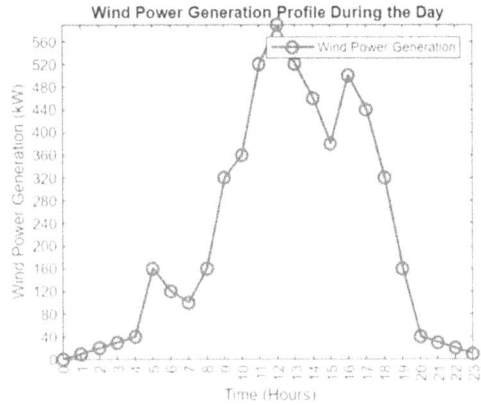

Fig. 15.7 Wind energy production profile

situation, load profiling determines the energy produced by solar, wind, and the grid. Figure 15.6 displays the amount of electricity generated by a grid using solar radiation, Figure 15.7 illustrates the energy generated by wind over time, and Fig. 15.8 displays the amount of power generated by a grid per day.

Taking the average of summer, winter, and rainy seasons, as shown in Fig. 15.7, we came up with the above graph since wind energy production is cyclical and may fluctuate. For example, in summer, energy will be moderate to low, while in winter, energy will be higher, and in the rainy season, energy will vary significantly. Figure 15.8 displays a grid's daily power generation from an existing power plant in a specific area. Operational and maintenance costs (O\&M) are separated into two categories: operational costs, the cost of energy per kilowatt-hour, and maintenance costs, which are fixed monthly expenses based on energy consumption. Based on the existing data, actual expenses are shown in Table 15.3 below.

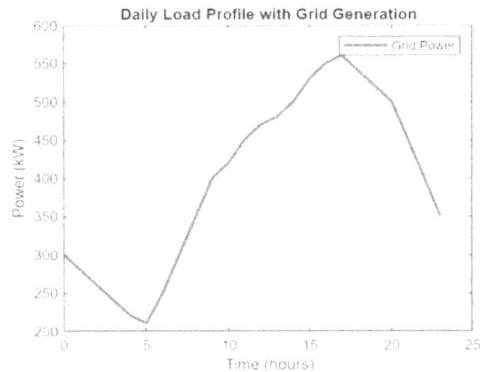

Fig. 15.8 Grid energy profile

Table 15.3 Summary of operational and maintenance costs

Energy source	Operational cost (per kWh)	Monthly maintenance cost
Solar	2.93	0.1
Wind	2.01	0.12
Battery	2.66	0.05
Grid	7	Included in OC

A penalty of 1000 INR will be imposed in every case where the load demand is not satisfied. Additionally, assume the battery degradation cycle is 1000. The monthly cost of battery degradation for a 90% % efficient and 0.5c rate battery would be 12 INR per kWh. Figure 15.9 depicts the battery

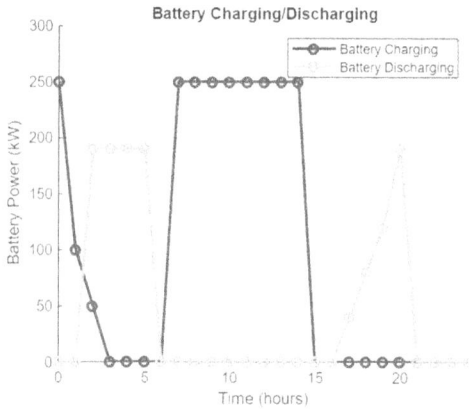

Fig. 15.9 Battery charging and discharging profile

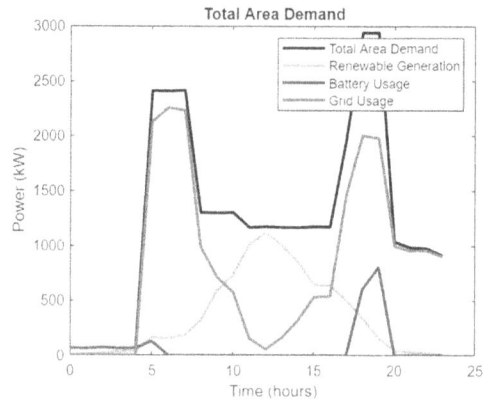

Fig. 15.10 Total area demand met per day

charging and discharging pattern. Limits on overcharging and over-discharging protect the battery from harm. The battery charges during off-peak hours when solar and wind power are adequate to meet load demands and discharges during peak hours.

The battery's power is 250 kWh, with a limit of 30 kWh. A battery's maximum and minimum SOC is 10% and 100%. Figure 15.10 above illustrates how the load demand is met from all the sources of micro-grid. the BSS (Battery storage system), along with the grid, is used to meet the demand between 4 and 6 am when the grid cannot meet the load demand of a locality. From 7 am to 5 pm, when the energy generated by renewable sources is high during peak time, gird usage is reduced to meet the load demand, and the battery charges until the evening when the renewable source fully meets the load. when the renewable source is low during off-peak time, the grid is used at that point, and the battery is discharged to meet the demand.

Before optimisation without EV, the microgrid's daily total cost was 164714 INR. After applying the PSO technique, the microgrid's overall cost was successfully lowered to 135935 INR, as shown in Fig. 15.11.

As shown in Fig. 15.12 the cost of a microgrid with EV before optimisation is 184994 INR, and after optimisation is seen to be 145948 INR.

```
Iteration 12, Best Cost: 142229.82 INR
Iteration 13, Best Cost: 142229.82 INR
Iteration 14, Best Cost: 142099.61 INR
Iteration 15, Best Cost: 142002.23 INR
Iteration 16, Best Cost: 141857.49 INR
Iteration 17, Best Cost: 140597.43 INR
Iteration 18, Best Cost: 140417.30 INR
Iteration 19, Best Cost: 139236.99 INR
Iteration 20, Best Cost: 138195.79 INR
Iteration 21, Best Cost: 136969.04 INR
Iteration 22, Best Cost: 136954.32 INR
Iteration 23, Best Cost: 136954.32 INR
Iteration 24, Best Cost: 136731.38 INR
Iteration 25, Best Cost: 136731.38 INR
Iteration 26, Best Cost: 136731.38 INR
Iteration 27, Best Cost: 136531.15 INR
Iteration 28, Best Cost: 136531.15 INR
```

Fig. 15.11 Optimization iteration without EV

```
Iteration 8, Best Cost: 151630.90 INR
Iteration 9, Best Cost: 151630.90 INR
Iteration 10, Best Cost: 148802.50 INR
Iteration 11, Best Cost: 148519.36 INR
Iteration 12, Best Cost: 146072.20 INR
Iteration 13, Best Cost: 146028.63 INR
Iteration 14, Best Cost: 146028.63 INR
Iteration 15, Best Cost: 145266.89 INR
Iteration 16, Best Cost: 145266.89 INR
Iteration 17, Best Cost: 145266.89 INR
Iteration 18, Best Cost: 145266.89 INR
Iteration 19, Best Cost: 144200.78 INR
Iteration 20, Best Cost: 143895.86 INR
Iteration 21, Best Cost: 143832.68 INR
Iteration 22, Best Cost: 143645.47 INR
Iteration 23, Best Cost: 143475.41 INR
Iteration 24, Best Cost: 143399.91 INR
Iteration 25. Best Cost: 143319.80 INR
Optimized Total Microgrid Cost: 141168.68 INR
```

Fig. 15.12 Optimisation iteration with EV

4. Conclusion and Future Work

This research aims to reduce and improve the cost-effectiveness of microgrids, particularly in EV applications, as part of load demand while also utilising renewable energy sources like solar and wind power. This will help maintain a locality's electrical supply consistency by meeting load demand with additional energy sources like the grid and BSS. Maximising the usage of renewable energy sources and lowering energy costs are the primary goals of this system. We adopt the particle swarm optimisation technique to accomplish this goal. Solar, wind, battery, grid, EV charging schedules, and the microgrid's overall functioning will all be considered in this method. This technique seeks to minimise carbon emissions to establish a cleaner environment and lower energy costs by maximising the use of renewable energy to preserve sustainability and to develop advancement in the field of microgrid EV application.

REFERENCES

1. Koli, S., Patil, S., Koti, A., M, A., & Patil, K. R. (2023). Cost optimization for integrating microgrid with electric vehicles using stochastic optimization technique. International Journal of Creative Research Thoughts (IJCRT), 11(12), b455–b462.
2. Jamehbozorg, A., Shahverdi, M., Serrato, C., & Flores, N. (2022). Optimal size of energy storage systems in microgrids under rapid growth of EV charging demands. In Proceedings of the 2022 IEEE Green Energy and Smart Systems Conference (IGESSC) (pp. 1–6). IEEE.
3. Wang, Y., Liu, Y., Zhao, K., Deng, H., Wang, F., & Zhuo, F. (2023). PEDF (Photovoltaics, Energy Storage, Direct Current, Flexibility) microgrid cost optimization based on improved whale optimization algorithm. In Proceedings of the 2023 IEEE 14th International Symposium on Power Electronics for Distributed Generation Systems (PEDG) (pp. 598–603). IEEE.
4. Xie, X., Xiong, H., Gao, C., & Hu, S. (2024). Microgrid economic and environmental cost optimization scheduling based on MO-PSO. In Proceedings of the 2024 6th International Conference on Energy Systems and Electrical Power (ICESEP) (pp. 1009–1014). IEEE.
5. Zhang, P., & Wu, X. (2024). Research on multi-objective microgrid operation optimization based on improved particle swarm optimization. In Proceedings of the 2024 4th International Conference on Electrical Engineering and Mechatronics Technology (ICEEMT)(pp.161–165).IEEE.
6. Zhang, P., & Wu, X. (2024). Research on multi-objective microgrid operation optimization based on improved particle swarm optimization. In Proceedings of the 2024 4th International Conference on Electrical Engineering and Mechatronics Technology (ICEEMT) (pp. 161–165). IEEE.
7. Sabzehgar, R., Kazemi, M. A., Rasouli, M., & Fajri, P. (2020). Cost optimization and reliability assessment of a microgrid with large-scale plug-in electric vehicles participating in demand response programs. International Journal of Green Energy, 17(2), 127–136.
8. Bijukchhe, P. M., & Wang, X. (2024). Optimization of economic power dispatch in microgrid featuring renewable resources. In Proceedings of the 2024 International Conference on Control, Automation and Diagnosis (ICCAD) (pp. 1–6). IEEE.
9. Phommixay, S., Doumbia, M. L., & Lupien St-Pierre, D. (2020). Review on the cost optimization of microgrids via particle swarm optimization. International Journal of Energy and Environmental Engineering, 11, 73–89.

Note: All the figures and tables in this chapter were made by the author.

Intelligent and Sustainable Power and Energy Systems – Dr. M. Premkumar et al. (eds)
© 2026 Taylor & Francis Group, London, ISBN 978-1-041-10314-1

16

Real-Time Cell Monitoring and Parameter Estimation for EV Applications

Manasa P. Gokarn[1],
Kiran R. Patil[2], Anoopkumar Patil[3],
Satwik Mathad[4], Sangmesh Melinmani[5]
Department of Electrical and Electronics Engineering,
KLE Technological University,
Hubballi

ABSTRACT: The automotive industry is shifting its focus from the gasoline to the electric and hybrid sectors. However, in an electric vehicle, the battery is the only source. Electric vehicles (EVs) employ Li-ion batteries because of their small size, feasibility, rechargeability, and low self-discharge rates. Despite all these benefits, there are drawbacks. Because of its high energy density, which is attained by employing more reactive chemicals, it tends to be unstable, demanding installing a battery management system (BMS) as an extra safety measure to prevent damage to the battery. To monitor the battery's overall state and minimise hazards, EVs have been installed with a BMS. The sensors offer the data needed to monitor the battery to the BMS, reducing the risks of an overcharge or over-discharge. In this paper, real-time monitoring of a Li-ion cell is carried out. As a result, the experimental setup for implementing this work integrates several sensors, such as a temperature sensor, voltage sensor, and current sensor, with an ESP32 controller. The data for the above-mentioned parameters is continuously monitored and logged through sensors. Data is synced to Google Sheets for analysis after these metrics are collected every second and sent to a cloud-based IoT platform in real time. These real-time sensor data are collected for varied battery loads, i.e. 25%, 50%, 75%, and 100%, to understand the cell's behaviour for different loads. As the load increases, the study shows that the temperature stays almost constant, but the voltage decreases from 3.06 V to 2.58 V, and the current ascends from 0.29 A to 1.02 A. Furthermore, by Li-ion charging characteristics, the State of Charge (SoC) shows a nonlinear relationship with voltage, increasing sharply above 3 V. When data exceeds a certain threshold, the user is alerted to inform about the potential danger.

KEYWORDS: Battery management system, Current sensor, Electric vehicles, Internet of things, State of charge, Temperature sensor, Voltage sensor

[1]gokarnmanasa@gmail.com, [2]krpatil@kletech.ac.in, [3]anupkumar@kletech.ac.in, [4]satwik.mathad@kletech.ac.in, [5]sangmesh.melinmani@kletech.ac.in

DOI: 10.1201/9781003654469-16

1. Introduction

Electric Vehicles (EVs) have heralded a new era of environmentally friendly transportation, lowering greenhouse gas emissions and dependency on fossil fuels [1]. Fully electric vehicles (EVs) have zero tailpipe emissions and are much better for the environment [2]. As the EVs are emission-free, the problem of air pollution is solved, and carbon footprints are reduced. So, EVs are known as green transportation. The battery is a critical component of EVs, providing the necessary power for the vehicle's motion. Most EVs use Li-ion batteries due to their higher energy density, ability to operate at higher voltages than other rechargeable batteries, low self-discharge rates, and long shelf life. Lithium is very reactive, and its batteries can hold high voltage and exceptional charge, making for an efficient, dense form of energy storage [3]. Additionally, Li-ion batteries perform effectively in the operating conditions required for EVs, making them competitive with diesel internal combustion (IC) engines for on-road use. However, managing the battery efficiently is significant for optimising its performance and increasing its shelf life in the long run. Li-ion batteries are unstable as they have high energy density, which is attained using more reactive chemicals. Li-ion batteries are susceptible to abuse conditions, such as overcharging, over-discharging, high temperatures, low temperatures, and overcurrent. Thus, battery monitoring is needed to manage these conditions and protect the battery from them. Battery monitoring systems monitor the battery's voltage, current, and temperature to maintain efficiency and safety. Much work has been laid out in the literature about battery monitoring. In this study [4], authors created an intelligent battery management system for electric vehicles (EVs) by integrating temperature and voltage sensors to identify anomalous situations and a fire sensor to identify potential threats such as battery overheating or fire. When an imminent risk is detected, a buzzer is alerted to set off alarms and show caution on the LCD. In the event of an emergency or danger, the system also has a relay attached to the DC motor of the car that disconnects it. Moreover, IoT can access battery status information from any location. In [5], the authors developed a system that integrates an RFID tag for the battery compartment and monitors battery voltage and charging status. A keypad allows users to select charging costs, activate charging, and deduct payments. After charging, battery voltage is measured by a sensor, and IoT notifies users. A PIC microcontroller monitors and controls the temperature with a cooling system (Peltier device and DC fan). Every detail is displayed on the LCD screen. The RF encoder-decoder system tells how far the nearest charging station is using push buttons. However, in [6], a system is developed to incorporate IoT and AI. IoT is for cloud connectivity and enabling real-time battery monitoring via mobile phones, as the battery is monitored through the temperature, voltage, and current sensors integrated with the NodeMCU microcontroller. AI estimates the SoC by training the model. All these data can be accessed by users using mobile apps. In [7], the authors proposed a system integrating voltage, current, and temperature sensors with Arduino UNO to monitor the Li-ion battery. If any parameters exceed the limit, actions are taken, i.e., the buzzer is triggered to alert, and all the data monitored is displayed on the LCD. If the temperature exceeds 50 degrees C, the system turns off the battery's power supply, and the fan is turned on. Data can be monitored through a mobile app through the ESP8266 WIFI module. In one of the works [8], authors developed an experimental setup to monitor the battery using a setup comprising a power circuit, i.e. a back-to-back converter to control both charging and discharging currents and a control circuit, i.c. achipKIT Max32 micro-controller to manage data acquisition via CAN to TTL and TTL to RS232 conversion circuits. Current, voltage, and temperature are displayed through PLC and computer. This paper explores the real-time monitoring of a Li-ion cell incorporating an ESP32 controller with temperature, voltage, and current sensors. To solve the difficulties in preserving the safety and health of EV batteries, the system enables continuous data collection and monitoring using Google Sheets and IoT platforms [9, 10]. With this configuration, users may use the TP4056 module to estimate battery percentages and examine how Li-ion cells react to different loads (25%, 50%, 75%, and 100%). In the case of a malfunction, the suggested system notifies users, disconnects the load using a relay, and sounds a buzzer to guarantee safety. By combining real-time analysis, failure avoidance, and user alerting systems, this effort attempts to close research gaps and offer a reliable, scalable battery monitoring solution. The II section details the experimental setup, data collection process, and safety mechanisms.

2. Proposed Methodology

The demonstrated real-time cell monitoring system uses an ESP32 controller, voltage sensor, temperature sensor (DHT11), and current sensor (ASC712) to obtain real-time battery data. First, all the sensors are calibrated, and then they are integrated with the controller. Then, the cell is continuously monitored to spot undesirable circumstances and implement safety precautions in response. Tracking the cell parameters is essential to extend its life, improve its performance, and shield it from potential hazards. The ASC712 current sensor measures the amount of current entering and leaving the cell using the Hall effect as the basis for measurement. It is employed to identify instances of excessive cell current. Upon detection, it generates a voltage commensurate to its magnetic field and has an output sensitivity of 66 to 185 mV/A. The voltage sensor monitors the battery's voltage using the voltage divider principle. It is also employed to identify over and under-voltage situations in the battery. The temperature sensor tracks the battery's temperature to identify instances of overuse or other dangerous circumstances that might cause the battery to overheat or explode.

According to the flowchart in Fig. 16.1, the following stages are outlined:

1. Initialize and integrate the current sensor, voltage sensor, temperature sensor, LCD, buzzer, and TP4056 charging module with the ESP32 controller.

2. Correspondingly, The sensor measures the temperature, voltage, and current.

3. The ESP32's integrated WiFi module facilitates IoT applications. The cell's real-time data is transferred to the cloud server. Real-time data, such as voltage, current, and temperature, is tracked and displayed via the IoT platform, i.e., in the Blynk app dashboard. The real-time voltage, temperature, and current figures are also shown on LCD.

4. Real-time data logging from the sensors synchronises on Google Sheets.

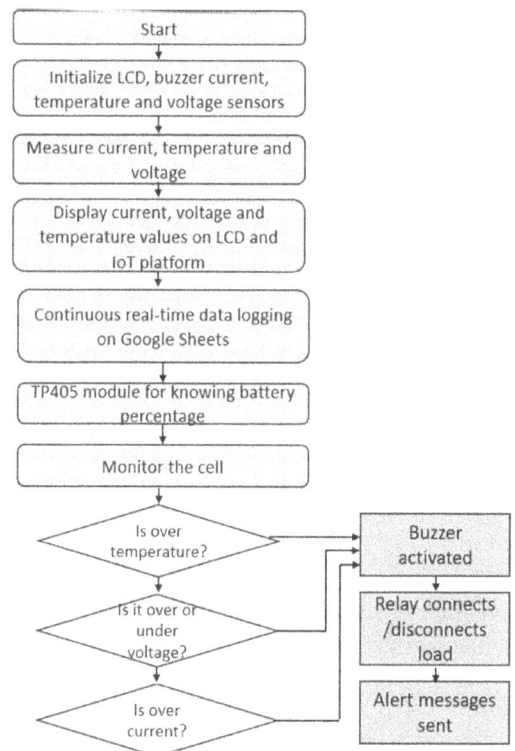

Fig. 16.1 Flow chart of the proposed methodology

5. The Li-ion cell is charged using the TP4056 charging module, providing the battery percentage in correspondence with the voltage.
6. The cell is monitored, and all safety conditions are examined to ensure it is within the permitted limits and that no imminent hazards, such as overvoltage, undervoltage, overtemperature, overcurrent, fire, or explosion, are present.
7. Three main safety measures are implemented if the battery is abused or comes across hazardous conditions:
   a. A buzzer is set off to notify the user of the abuse.
   b. The relay is connected to the DC motor as a safety feature that regulates turning the motor on and off. In the event of abuse conditions—undervoltage, overvoltage, overtemperature, and overcurrent—it is used to cut off the DC motor from the cell. The IoT platform also controls the relay's on and off functions.
   c. Users receive messages alerting them to a possible problem and urging them to take precautions.

Eventually, real-time data is logged to profile the battery and gain insight into its health and performance. These real-time data are entered into Google Sheets. One crucial component of the battery monitoring system is the State of charge (SoC). The SoC provides information on the amount of remaining charge in the battery, which aids in knowing when to charge and when to discharge the battery to avoid overcharging and undercharging, which can affect the battery's longevity, health, and effectiveness. The cell is charged using the TP4056 module, which also protects it from undercharging and overcharging. It is a constant current constant voltage (CC/CV) linear charger that indicates the remaining charge based on the voltage.

3. System Architecture

The circuit diagram for the real-time monitoring model is shown in Fig. 16.2, where all of the sensors, namely temperature, current, and voltage sensors, are connected to an ESP32 controller. The ESP32 controller powers the sensors, and the data output pins of the sensors are connected to the ESP32's analogue pins. The ESP32 has an integrated 12-bit ADC (Analog to Digital conversion) that can convert analogue signals into digital with a 12-bit resolution, meaning that it can represent analogue input values as digital values between 0 and 4095. The current sensor is connected in series with the battery, relay, and DC motor to measure the Li-ion battery's current. The voltage sensor is connected directly to the battery terminals to measure the voltage across the battery.

4. Result and Discussion

This section discusses the successful implementation of the proposed experimental setup and the findings derived from real-time monitoring of a Li-ion cell under varying load conditions. To ensure uniform testing under all circumstances, the experimental approach comprised discharging the Li-ion cell for 10 seconds for each applied load. The Blynk app continually monitors and logs real-time sensor data for voltage, current, and temperature on the IoT platform, as shown in Fig. 16.4. The software has a user-friendly interface for real-time information and a toggle button for controlling the relay. The system delivered alerts to users under dangerous situations, as seen in Fig. 16.3. This feature guarantees that the user will act quickly to stop more battery damage.

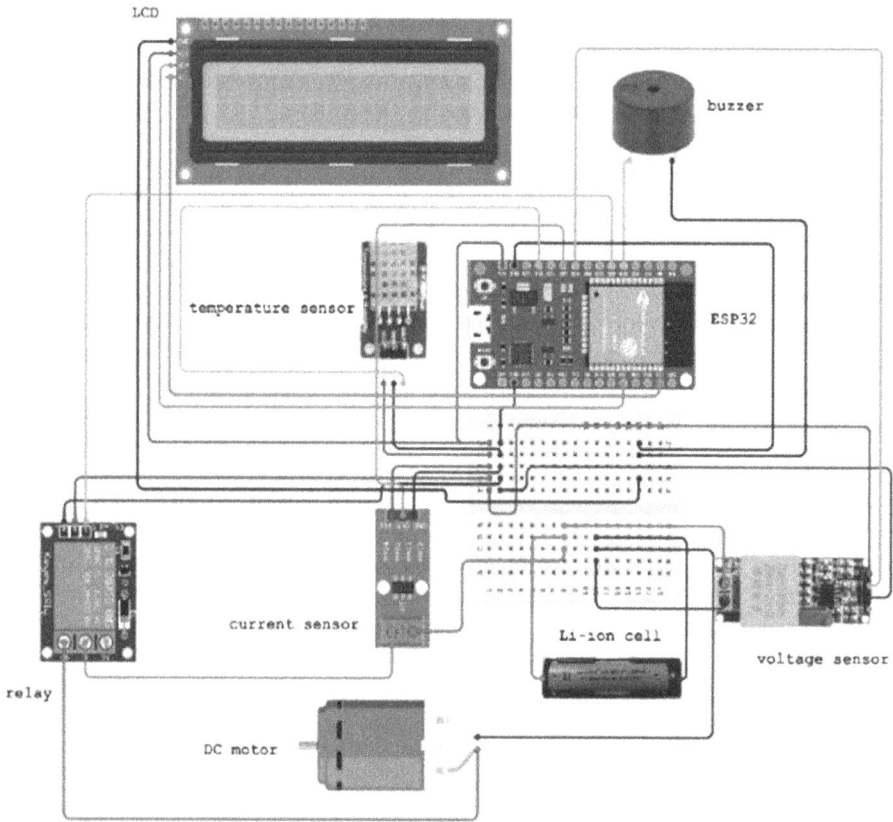

Fig. 16.2   Circuit diagram of proposed model

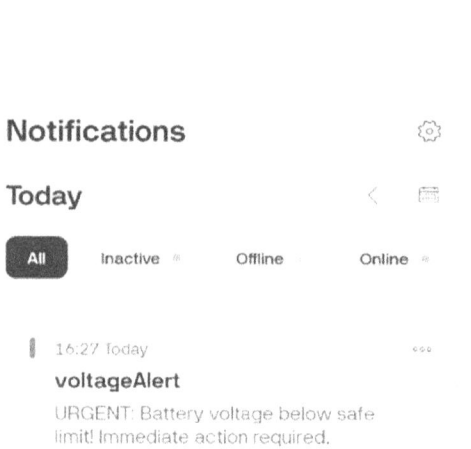

Fig. 16.3   Notifications received on the Blynk platform

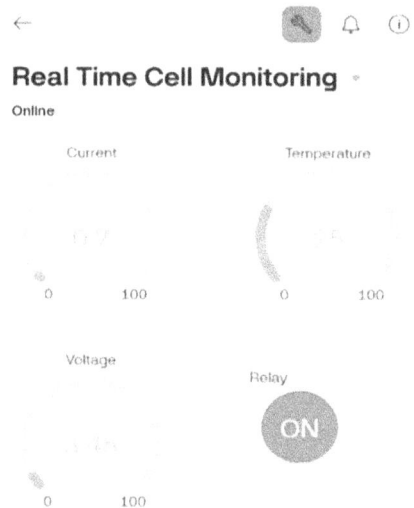

Fig. 16.4   Blynk app platform display

Tables 16.1, 16.2, and 16.3 display real-time sensor data for various load conditions: 25%, 50%, 75%, and 100%. The data is logged and presented using Google Sheets. The DC motor shaft was subjected to holding torque to replicate load circumstances. The combined representation of 50% and 75% loads in Table 16.2 was justified by the observation that they showed comparable patterns in current, temperature, and voltage values, with only slight deviations. Graphs demonstrate further how the battery behaves under these load circumstances. The relationship between voltage and time for various loads is shown in Fig. 16.5.

Table 16.1   Real-time data for 25% load

S. No.	Current (A)	Voltage (V)	Temperature (°C)
1	0.29	3.06	29.3
2	0.40	3.00	29.3
3	0.50	2.93	29.3
4	0.39	2.84	29.3
5	0.54	2.89	29.3
6	0.48	2.95	29.3
7	0.46	2.94	29.3
8	0.47	2.96	29.3
9	0.44	2.97	29.3
10	0.44	2.97	29.3

Table 16.2   Real-time data for 25% load

S. No.	Current (A)	Voltage (V)	Temperature (°C)
1	0.67	2.83	29.3
2	0.62	2.77	29.3
3	0.72	2.87	29.3
4	0.69	2.80	29.3
5	0.74	2.82	29.3
6	0.71	2.82	29.3
7	0.69	2.84	29.3
8	0.67	2.91	29.3
9	0.66	2.92	29.4
10	0.69	3.03	29.4

Table 16.3   Real-time data for 100% load

S. No.	Current (A)	Voltage (V)	Temperature (°C)
1	1.01	2.65	29.4
2	1.04	2.64	29.4
3	0.98	2.65	29.4
4	0.94	2.67	29.4
5	0.99	2.69	29.4
6	1.02	2.62	29.4
7	1.03	2.59	29.4
8	1.06	2.58	29.3
9	1.04	2.58	29.4
10	1.02	2.60	29.4

The voltage dropped dramatically from 3.06 V to 2.58 V as the applied torque increased from 25% to 100%. This voltage decrease aligns with internal battery resistance and Ohm's Law, which states that larger current draws result in more significant voltage loss. The current drawn over time under various loads is displayed in Fig. 16.6. The current increased from 0.29 A to 1.02 A as the torque increased from 25% to 100%, matching the power requirement. The motor shaft's torque increases the required electrical power, raising the battery's current draw.

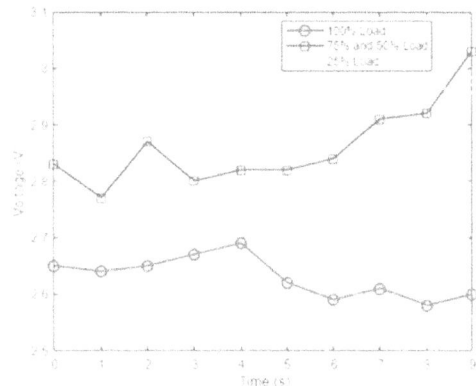

Fig. 16.5   Voltage vs time for different loads

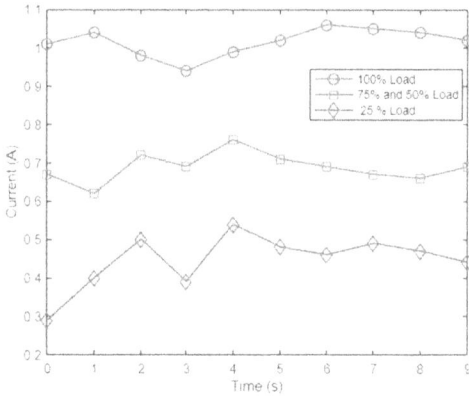

Fig. 16.6 Current vs time for different loads

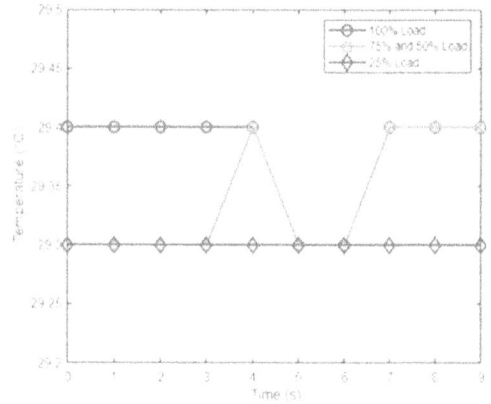

Fig. 16.7 Temperature vs time for different loads

Figure 16.7 shows the temperature changes for various loads displayed against time; with slight temperature variations between 29.3°C and 29.4°C, the battery operated within a safe thermal range. The controlled testing environment and effective heat dissipation during operation are responsible for this stability. However, extended high-load operations may result in more noticeable temperature fluctuations in real-world scenarios, underscoring the significance of thermal control in EV battery systems. The relationship between the battery's voltage and State of Charge (SoC) during a no-load charging procedure with the TP4056 module is shown in Fig. 16.8. The graph

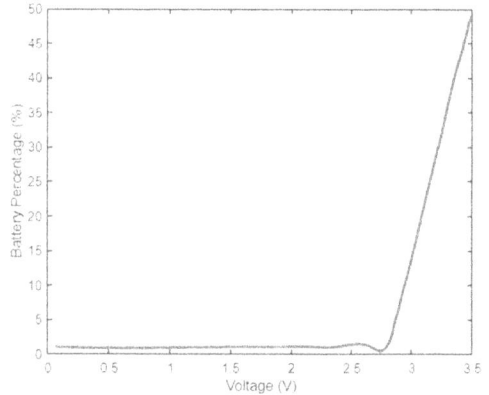

Fig. 16.8 Battery percentage vs voltage

confirms the nonlinear relationship between SoC and voltage, where the battery SoC is critically low for voltages below 3 V and grows dramatically when the voltage overcomes this threshold. This behaviour is consistent with Li-ion cells' charge curve, which shows a slower rise in SoC during the constant voltage (CV) phase of charging and a faster increase during the constant current (CC) phase.

5. Conclusion

This research emphasises the importance of the real-time cell monitoring system for Li-ion cells in EVs. This is accomplished by employing sensors to monitor the Li-ion cell in real-time, saving the data in the cloud and presenting it on the IoT platform, and synchronous data capture on Google Sheets. To safeguard the battery, prolong its life, and maintain its health to protect it from abusive conditions, every effort is made to alert the user through multiple methods in the case of any faults or safety concerns. Despite difficulties with sensor calibration and accuracy maintenance, the system effectively monitored vital battery characteristics across a range of load circumstances, including 25%, 50%, 75%, and 100% loads. As the load increased, the temperature stayed constant, but the voltage decreased from 3.06 V to 2.58 V, and the current ascended from 0.29 A to 1.02

A. Furthermore, by Li-ion charging characteristics, the SoC showed a nonlinear relationship with voltage, increasing sharply above 3 V. To further improve accuracy, future developments may incorporate sophisticated sensors with greater precision. Proactive battery management may also be possible by integrating machine learning algorithms for defect detection and prediction analysis. This would include anticipating possible problems like capacity decline or thermal runaway, which would improve remote monitoring and battery safety. In conclusion, the suggested real-time battery monitoring system meets present demands and upcoming developments in battery management technologies while offering a strong basis for guaranteeing the best possible performance and safety of Li-ion cells in EV applications.

References

1. K. I., V. V., A. B. K., and L. S. T. (2024). Smart battery management system for electric vehicles using IoT. 2024 5th International Conference on Intelligent Communication Technologies and Virtual Mobile Networks (ICICV), Tirunelveli, India, pp. 720–725.
2. Ramani, U., Thilagaraj, M., Vadivelan, T., Manikandan, P., and Ranjith, G. (2023). RFID-based real-time battery monitoring system for electric vehicle charging. 2023 5th International Conference on Inventive Research in Computing Applications (ICIRCA), Coimbatore, India, pp. 1574–1577.
3. Bharathi, M., Geetha, K., Mani, P. K., Vijayakumar, G. N. S., Srinivasan, K., and Kumar, K. R. (2022). AI and IoT-based electric vehicle monitoring system. 2022 Sixth International Conference on I-SMAC (IoT in Social, Mobile, Analytics, and Cloud) (I-SMAC), Dharan, Nepal, pp. 722–727.
4. Dwiyaniti, M., Kusuma, L. R. N., Silawardono, Kusumastuti, S. L., Wiguna, A. R., and Tohazen. (2022). A real-time performance monitoring of IoT-based lithium-ion battery pack. 2022 International Conference on Informatics Electrical and Electronics (ICIEE), Yogyakarta, Indonesia, pp. 1–4.
5. Hommalai, C., and Khomfoi, S. (2015). Battery monitoring system by detecting dead battery cells. 2015 12th International Conference on Electrical Engineering/Electronics, Computer, Telecommunications and Information Technology (ECTI-CON), Hua Hin, Thailand, pp. 1–5.
6. Adhikary, S., and Biswas, P. K. (2022). Bi-directional EV charging for grid to vehicle and vehicle to grid and battery monitoring system using IoT in Simulink. 2022 International Conference on Disruptive Technologies for Multi-Disciplinary Research and Applications (CENTCON), Bengaluru, India, pp. 173–178.
7. Supriya, M., P. K., P. S., N. S., and M. P. S. (2024). Design of EV control monitoring of multiple approach fault detection using multi-sensor IoT system. 2024 Ninth International Conference on Science Technology Engineering and Mathematics (ICONSTEM), Chennai, India, pp. 1–6.
8. Haq, I. N., et al. (2014). Development of battery management system for cell monitoring and protection. 2014 International Conference on Electrical Engineering and Computer Science (ICEECS), Kuta, Bali, Indonesia, pp. 203–208.
9. Calinao, H. A., Bandala, A., Gustilo, R., Dadios, E., and Rosales, M. (2020). Battery management system with temperature monitoring through fuzzy logic control. 2020 IEEE Region 10 Conference (TENCON), Osaka, Japan, pp. 852–857.
10. Ehteshaam, M., Amir, M., and Haque, A. (2023). Investigation of battery health monitoring system for electric vehicles using IoT-cloud based Arduino controller. 2023 Second International Conference on Smart Technologies for Smart Nation (SmartTechCon), Singapore, Singapore, pp. 1249–1254.

Note: All the figures and tables in this chapter were made by the author.

Intelligent and Sustainable Power and Energy Systems – Dr. M. Premkumar et al. (eds)
© 2026 Taylor & Francis Group, London, ISBN 978-1-041-10314-1

17

Sound-Sensor-Based Emergency Vehicle Priority Detection at Urban Traffic Intersections

Kv. Karthik Reddy*, Aashutosh Swami
Dept of ECE, MS Ramaiah Institute of Technology,
Bangalore, India

Kevin Andrew Francis
Dept of ISE, MS Ramaiah Institute of Technology,
Bangalore, India

Prateek B. N.
Dept of EEE, California State University Fullerton,
California, USA

ABSTRACT: Obstructed junctions are a major cause of ambulance response time delays in urban areas with high traffic congestion. Existing emergency traffic control technologies like RFID, GPS, and image recognition work well, but they are expensive and difficult to implement in settings with limited resources. This study offers a scalable and reasonably priced method for automating traffic signal regulation in response to ambulance sirens using a sound sensor-based system. Our technology uses acoustic sensors that are tuned to detect siren frequencies. When an emergency vehicle is identified, traffic signals are switched to green, allowing ambulances to have real-time precedence. The prototype is appropriate for high-density traffic suburbs and big cities across the nation since it exhibits faster reaction times and fewer false alarms when evaluated in traffic simulation scenarios. The design, execution, and practicality of the system are covered in this study, which offers insights into a cost-effective strategy for improving emergency response in urban environments.

KEYWORDS: Emergency vehicle detection, Automated traffic signal control, Ambulance priority, Sound sensor, Urban traffic management, Real-time traffic response, Cost-effective traffic solution

*Corresponding author: 1ms19ec058@gmail.com

DOI: 10.1201/9781003654469-17

1. Introduction

Traffic congestion is a growing challenge in urban areas, and with the rising number of vehicles on the roads, cities are struggling to cope with the increased volume of traffic. This issue significantly impacts emergency services, particularly ambulances, which need to navigate congested streets to deliver patients to hospitals quickly. Studies indicate that delays in ambulance response times, especially in busy urban areas, can have a serious impact on patient outcomes, particularly in critical situations where immediate medical attention is required. Efficient traffic management systems that prioritize emergency vehicles at intersections are crucial for ensuring their timely passage during peak traffic hours when traditional traffic control systems may not adequately address the urgent requirements of emergency services.

Currently, several methods are employed to give priority to emergency vehicles at intersections, including RFID (Radio Frequency Identification) systems, GPS tracking, andimage recognition technology, in addition to manual intervention. In RFID-based systems, ambulances are equipped with RFID tags that communicate with RFID readers installed at intersections. GPS-based systems track the real-time location of ambulances and adjust the traffic signals accordingly to prioritize their routes. While these technologies have proven effective, they come with several limitations, especially for cities facing financial constraints. The high costs of installation, maintenance, and the infrastructure required for these systems make them impractical for many urban areas, particularly those that lack the financial resources to deploy such advanced technologies on a wide scale. Another alternative is image recognition systems, which use cameras to detect the presence of emergency vehicles and adjust the traffic signals accordingly. However, these systems are highly susceptible to environmental factors such as poor weather conditions, low lighting, and physical obstructions, which can impair their ability to detect vehicles accurately. These limitations, combined with the significant costs associated with installation and maintenance, make image recognition systems less effective in certain conditions and less accessible for cities with budgetary constraints.

Given these challenges, there is a clear need for an alternative solution that isaffordable, easy to implement, and scalable. The solution proposed in this research is a sound sensor-based systemfor traffic signal management, specifically designed to prioritize ambulances at intersections. This system utilizes sound sensors(KY-038 & MAX9814), that are capable of detecting the distinct frequencies of ambulance sirens. By focusing on the unique characteristics of the siren sound, the system reduces the possibility of interference from other background noises, ensuring that emergency vehicles are accurately prioritized. Unlike other technologies that require costly infrastructure, such as GPS units or cameras, a sound sensor-based system is a low-cost solution that only requires the installation of a few sensors at key intersections. This makes it particularly suitable for urban cities where the demand for better emergency traffic management is high, but financial constraints limit the ability to implement more expensive alternatives. In this approach, the sound sensorsdetect the siren of an approaching ambulance and sends the signal to an Arduino microcontroller, which processes the sound and activates the traffic signal to give priority to the ambulance by turning the light green in the direction of the approaching vehicle. The simplicity of the hardware setup, which includes the sound sensors (KY-038 & MAX9814), microcontroller, traffic signal controller, LEDs, breadboard, jumper wires, and power supply, makes the system easy to deploy and maintain. The Arduino microcontroller is programmed using the Arduino IDE, and libraries such as the Fast Fourier Transform (FFT)are used to analyse the sound frequencies and filter out background noise. Testing during the research phase ensures that the system reliably detects ambulance sirens, even in noisy environments. The system has demonstrated its potential to improve emergency vehicle response times in urban intersections, while being cost-effective and easy to implement.

This research aims to demonstrate the feasibility and reliability of a sound sensor-based system for emergency vehicle prioritization. By focusing on scalability, cost, andease of installation, the proposed system offers an effective solution for enhancing traffic management in urban areas. The system's simple hardware setup and reliable performance make it an ideal choice for cities looking to improve emergency vehicle response times without incurring significant costs. In the following sections, the system architecture implementation, procedure,and the results of real-world testing will be presented to showcase how this technology can be integrated into urban traffic control systems to improve emergency response outcomes.

2. Literature Survey

Techniques based on the Global Positioning System (GPS) and Radio Frequency Identification (RFID) are frequently employed for emergency vehicle detection. RFID systems entail attaching RFID tags to ambulances, which are then picked up by intersection readers. When an emergency vehicle is spotted, these systems activate automatic green light signals [2]. However, RFID technology necessitates the installation of RFID devices on both ambulances and intersections, resulting in significant setup costs that are sometimes unaffordable in metropolitan areas with tight budgets [1]. Using GPS-based techniques, traffic lights are adjusted based on the real-time location of ambulances. Despite their excellent accuracy, they also necessitate a large infrastructure investment and intricate integration with real-time traffic control systems [4]. This intricacy and expense underscore the constraints of RFID and GPS methodologies for extensive urban use. Image-based systems recognize emergency vehicles using visual cues like flashing lights or vehicle forms by using cameras and image processing algorithms. To increase detection accuracy, particularly in crowded metropolitan environments, these systems frequently use machine learning techniques [6]. But environmental factors can affect image identification, and problems like fog, rain, and low light levels can make detection less accurate [7]. Furthermore, it is challenging to deploy such systems widely, especially in cities with tight budgets, because of the high cost of cameras and the computational resources required for real-time video processing [8]. In order to change traffic signals, sound sensor-based systems concentrate on identifying the distinctive siren sounds of emergency vehicles. Sound sensors are less expensive and require less infrastructure than RFID, GPS, and image-based technologies. According to a study, sound-based systems can successfully identify sirens even in loud settings, proving that sound recognition algorithms are capable of efficiently removing irrelevant noise [3]. Because of its low cost and simplicity of use, this technology is perfect for towns with tight budgets. But issues like false positives from other loud road noises still exist, underscoring the need for more improvement [17]. Although image-based, RFID, and GPS systems are accurate under certain circumstances, their use is restricted in cost-sensitive, congested urban settings like Metropolitan cities because of their high implementation and maintenance costs. However, because sound sensor systems are easy to install at junctions and reasonably priced, they offer a viable, scalable substitute. Research shows that sound-based systems, especially those with algorithms that eliminate non-siren noises, can offer a dependable way to detect emergency vehicles [5]. Despite its effectiveness, RFID, GPS, and image recognition have limited real-world applications in urban environments with little funding. Systems based on sound sensors present a viable substitute since they are inexpensive and simple to implement.

3. Block Diagram and Implementation

In Fig. 17.1 above, the sound sensor-based traffic signal management system is designed to prioritize emergency vehicles, such as ambulances, at intersections using a straightforward configuration. The primary components of this system are the power supply, sound sensor, signal processing unit (microcontroller), traffic signal controller, and traffic lights. The power supply (either 5V or 12V) provides the necessary power to all components in the system, including the Arduino uno (used as the microcontroller), the sound sensors (KY-038 & MAX9814),, and the traffic lights. The sound sensor is responsible for detecting the siren sound from the approaching ambulance. When the sound sensor detects the siren, it converts the sound waves into an electrical signal and sends this signal to the signal processing unit (microcontroller). The microcontroller processes the signal and identifies whether the incoming sound corresponds to an ambulance siren. If it does, the microcontroller sends a command to the traffic signal controller unit, which, in turn, changes the traffic light to green in the direction of the ambulance. This gives the ambulance priority and allows it to pass through the intersection quickly, reducing delays during critical situations.

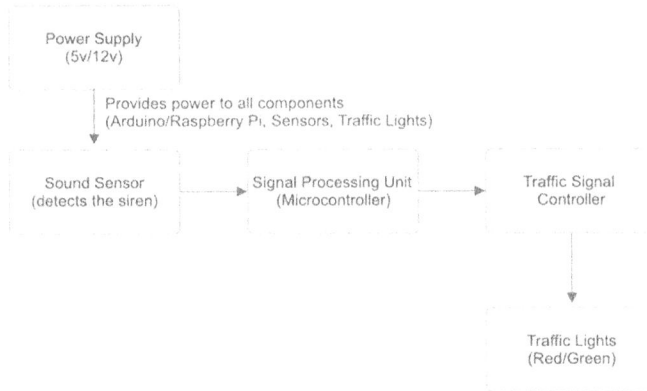

Fig. 17.1 Block diagram of the proposed model

Finally, the trafficlights, which are controlled by the traffic signal controller, switch between red and green depending on the signal received from the microcontroller. The power supply ensures that all components operate continuously without interruption. This system design illustrates how automated traffic signal control can be used to improve emergency response times by giving priority to emergency vehicles at intersections, all while minimizing infrastructure changes and cost.

3.1 Flowchart

The soundsensor-based traffic control system, as shown in Fig. 17.2, is designed to dynamically prioritize emergency vehicles, such as ambulances, at intersections. The system starts with the initialization phase, where all components, including the sound sensors (KY-038 & MAX9814), microcontroller (Arduino UNO), LED's are powered on and set for operation. Once initialized, the system enters a continuous listening state, where it detects sirens from approaching emergency vehicles. If no siren is detected, it continues monitoring; however, when a siren is detected, the system validates the frequency to ensure it matches the specific profile of an ambulance siren. If the sound does not match the expected frequency, the system returns to the listening state, but if it matches, the system processes the signal and proceeds to adjust the traffic signal based on its current state.

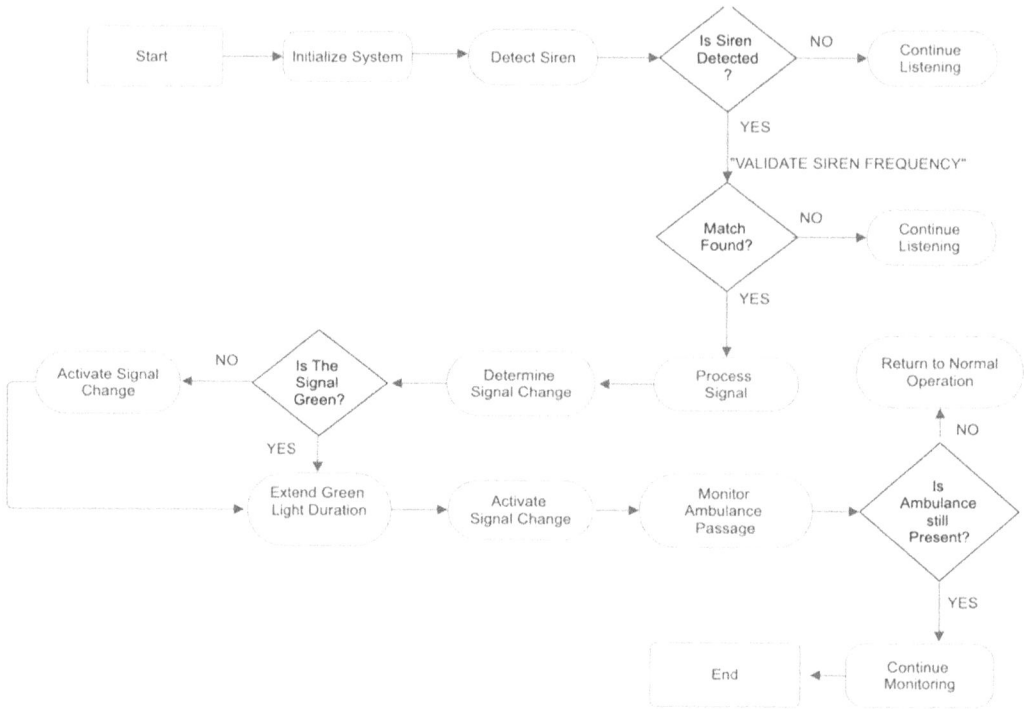

Fig. 17.2 Flowchart and implementation

In the case of a single ambulance, if the traffic signal is already green, the system will extend the green light duration to allow the ambulance to pass through the intersection. During this time, all other traffic signals are turned red to prevent conflicting traffic. If the signal is red, the system immediately changes it to green for the ambulance's direction. Once the ambulance passes, the system returns to normal operation and continues monitoring for further sirens. In scenarios where two ambulances arrive from the same direction, the system prioritizes both vehicles within the same green light cycle. Upon detecting the first ambulance, the system extends the green light for both vehicles to pass. Similarly, the system ensures that all other traffic signals are turned red to avoid conflicting vehicles. After both ambulances have cleared the intersection, the system resumes normal traffic control. In the case of two ambulances arriving from different directions, the system will prioritize the first arriving ambulance. The system turns the signal green for the direction of the first ambulance and extends the green light duration to ensure it passes without delay. After the first ambulance clears the intersection, the system waits for a 20-second window to confirm the vehicle's safe exit. If a second ambulance arrives during this window from a different direction, the system turns the signal green for the second ambulance's direction, turning all other signals red. Once the second ambulance clears the intersection, the system returns to normal operation.

The system's flexibility in handling multiple ambulance scenarios, whether from the same or different directions, is key to its efficiency. It prioritizes emergency vehicles by extending green light durations and managing the safe passage of multiple ambulances while maintaining normal traffic flow. The integration of the Arduinomicrocontroller, FFT (Fast Fourier Transform) algorithm, and modular hardware components like breadboards, jumper wires, and LEDs makes the system adaptable, reliable, and easy to test and deploy across various urban intersections. The Arduino IDE is used to program the microcontroller, where the core logic for sound detection, signal validation, and traffic light control is implemented. The LEDs simulate the traffic light system, representing green and red signals that change based on the sound sensor's detection of ambulance sirens.

3.2 Schema Diagram

In Fig. 17.3. The schema diagram outlines the flow of signals and power between the components of the sound sensor-based traffic control system, ensuring that the traffic signals are dynamically adjusted based on real-time ambulance siren detection.

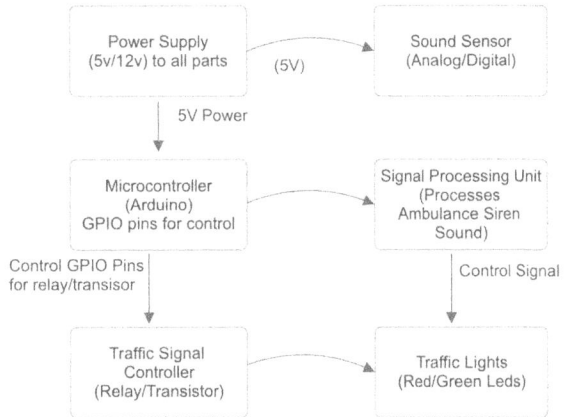

Fig. 17.3   Schema diagram and implementation

- Power Supply: The 5V or12V power supply is responsible for powering the entire system. It provides 5V power to components like the Arduino microcontroller, the sound sensor, and the traffic signal controller. The 12V supply is used when other components like relays or motors require higher voltage. The consistent power ensures that the system operates smoothly and continuously.

- Sound Sensor: The sound sensor is responsible for detecting the ambulance sirens. It can be either Analog or Digital, depending on the sensor type. The sensor receives sound input and, upon detecting the specific frequency of an ambulance siren, sends this data to the Arduino for further processing. The sensor is connected to the Arduino and continuously monitors the environment for relevant sound frequencies, like sirens. The sensors used are 2'(KY-038) & 2'(MAX9814)

- Microcontroller (Arduino UNO): The Arduino serves as the central processing unit. Once the sound sensor detects a siren, the Arduino processes the signal and validates whether the detected sound corresponds to an ambulance siren. The Arduino handles the GPIO (General Purpose Input/Output) pins used to control the traffic signal controller. Based on the processed input, the Arduino triggers the appropriate response, such as switching the traffic signal to green for the ambulance's direction.

- Signal Processing Unit: The signal processing unit inside the Arduino processes the raw data from the sound sensor. The FFT (Fast Fourier Transform) algorithm helps to analyse the sound and filter out any irrelevant noise. If a valid ambulance siren frequency is detected, the signal is passed to the traffic signal controller to initiate a change in the traffic signal.

- Traffic Signal Controller: The traffic signal controller is responsible for physically changing the state of the traffic lights. The Arduino sends a signal to the traffic signal controller (using relays or transistors), which controls the LED traffic lights (represented by red/green LEDs) in each direction. If an ambulance is detected, the controller changes the corresponding traffic signal to green and ensures all other signals turn red to prevent traffic interference.

- Traffic Lights (LEDs): The LEDs represent the traffic signals—green for the ambulance's direction and red for other directions. The Arduino controls these LEDs based on the signals received from the traffic signal controller. The green light allows the ambulance to pass safely, while the red light ensures other vehicles are stopped during this emergency.

In this schema diagram, the system effectively shows how the components work together: the sound sensor detects the ambulance, the Arduino processes the signal, and the traffic light controller changes the signal to green for the ambulance, ensuring that the intersection is cleared for emergency vehicles

3.3 Use Case Diagram

In Fig. 17.4 The system's interactions with the main players are depicted in the use case diagram for a sound sensor-based traffic signal control system that gives priority to emergency vehicles. The main actor at the beginning is the Emergency Vehicle, which sounds a siren that the System picks up via a sound sensor. To give the emergency vehicle a clear way, the system changes the traffic signal as soon as it detects the siren. The system's function in enhancing traffic flow for emergency responders is highlighted in this interaction, which illustrates the main use case. Interactions between the Traffic Management System and the main system allow for more extensive synchronization of traffic signals across intersections, guaranteeing the emergency vehicle's safe and effective passage. Furthermore, Maintenance Personnel are involved in keeping an eye on the system's condition, makingsure all of its parts are operating as intended, and repairing or adjusting it as necessary. This exchange highlights how crucial system maintenance is to preserving dependable operation.

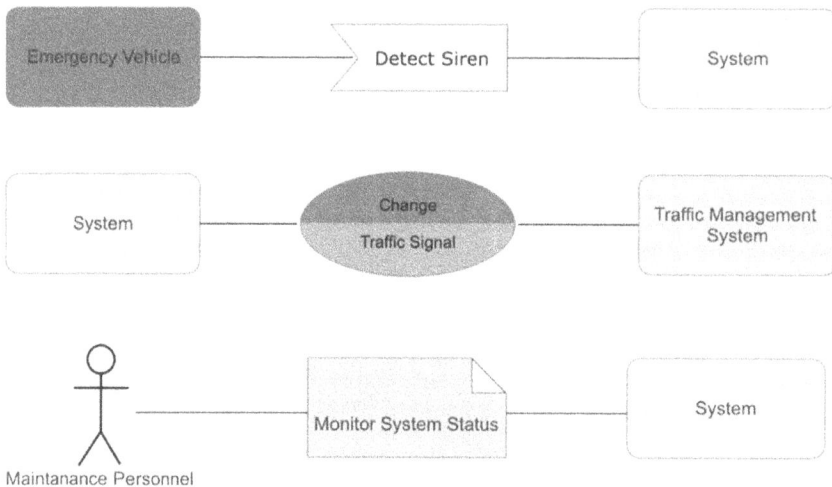

Fig. 17.4   Use case (UMl) diagram

Overall, the figure shows a simple yet effective sequence of contacts, emphasizing how the system enhances emergency response times at traffic crossings by integrating seamlessly with emergency vehicles, traffic control systems, and human operators.

3.4 System Architecture and Connections

As Shown in Fig. 17.5, The system architecture of the sound sensor-based ambulance detection system combines various components to detect ambulance sirens and trigger traffic control. The central unit of the system is the Arduino Uno, which processes inputs from two sound sensors: the KY-038 sound sensors and two MAX9814 microphone amplifiers. The KY-038 provides a quick digital signal for detecting loud sounds, while the MAX9814 delivers a cleaner analog signal. This analog signal is processed using FFT (Fast Fourier Transform) to analyze frequency components, helping distinguish the ambulance siren's frequency (typically between 500 Hz and 2 kHz) from background noise. Once the siren is detected, the Arduino controls the traffic lights by activating the green LED in the corresponding direction, allowing the ambulance to pass. The green light stays on for 20 seconds, after which the signal switches to red. The red LEDs are used to stop the traffic in the direction of the incoming ambulance, ensuring the vehicle's safe passage. Resistors connected to the LEDs help protect the components by limiting current. The Arduino reads the digital output (DO) pin from the KY-038 for fast detection and the analog output from the MAX9814

COMPONENT	PIN ON ARDUINO	CONNECTION DETAILS
Sound Sensor 1 (Direction 1)	VCC	5V rail on breadboard
	GND	GND rail on breadboard
	DO (Digital Output)	Pin 2 on Arduino
Sound Sensor 2 (Direction 2)		5V rail on breadboard
		GND rail on breadboard
		Pin 3 on Arduino
2 MAX9814 Microphones	VDD	5V rails on both breadboards
	GND	GND rails on both breadboards
	OUT (ANALOG OUTPUT)	A0 (Sensor 1) & A1 (Sensor 2) on Arduino
	GAIN	5V rails on both breadboards
	AR	GND rails on both breadboards
Green LED 1 (Direction 1)	Pin 13	Anode (long leg) → Pin 13, Cathode (short leg) → Negative rail via 220Ω resistor
Red LED 1 (Direction 1)	Pin 12	Anode (long leg) → Pin 12, Cathode (short leg) → Negative rail via 220Ω resistor
Green LED 2 (Direction 2)	Pin 11	Anode (long leg) → Pin 11, Cathode (short leg) → Negative rail via 220Ω resistor
Red LED 2 (Direction 2)	Pin 10	Anode (long leg) → Pin 10, Cathode (short leg) → Negative rail via 220Ω resistor
Arduino Power	5V pin	Positive rail on breadboard
	GND pin	Negative rail on breadboard

Fig. 17.5 System architecture and connections table

for frequency analysis, adjusting the traffic lights based on the detected sound. The system is set up on two breadboards with jumper wires connecting the components, allowing for flexible testing and adjustments. The simplicity of this architecture, with its modular design, ensures easy installation, both fast response and precise sound analysis for real-world traffic control applications.

4. Comparative Study

4.1 Cost-Effectiveness

In practical urban scenarios, cities often face budget constraints when it comes to upgrading or maintaining traffic management systems. Traditional systems like RFID and GPS require significant upfront costs for infrastructure, including RFID readers, GPS trackers, and data processing systems, which can be prohibitively expensive, especially for developing cities. Image recognition systems also require high-cost cameras and data processing equipment. On the other hand, the sound sensor system is significantly more affordable to install and maintain. Its cost-effectiveness is especially apparent in low-budget urban environments, where municipalities are often looking for low-cost solutions that can scale without draining resources.

4.2 Simplicity of Implementation

In practical deployment scenarios, GPS and image recognition systems often require substantial integration with existing infrastructure and traffic management systems. The need for sophisticated hardware setups, such as GPS tracking systems on vehicles or high-resolution cameras at every intersection, makes these technologies complex to implement and more time-consuming to set up. Sound sensor systems, in contrast, can be installed quickly and with minimal disruption to existing traffic setups. This makes them highly practical for cities looking for quick fixes to improve emergency vehicle response times. They can be installed at key intersections without requiring major infrastructure changes, making them ideal for rapid deployment.

4.3 Robustness in a Range of Weather Conditions

In real-world scenarios, traffic management systems are often put to the test under adverse weather conditions like rain, fog, or snow, which can severely affect camera-based systems. Cameras rely on clear visibility to detect objects, and poor weather can reduce their effectiveness, leading to detection failures. Sound sensors, however, are much less impacted by weather conditions because they depend on auditory signals rather than visual cues. In cities with unpredictable or extreme weather, such as foggy coastal cities or snow-prone areas, sound sensor systems provide a more reliable solution compared to image-based systems, which struggle in low-visibility conditions.

4.4 Less Dependency on Infrastructure

In practical scenarios, especially in remote or underdeveloped areas, establishing a strong infrastructure base for systems like RFID or GPS can be a major challenge. RFID systems, for instance, require readers to be installed at every intersection and RFID tags on emergency vehicles, which increases both the cost and complexity of deployment. Sound sensor systems, on the other hand, function independently and only require the installation of sound sensors at key locations, such as intersections. This minimal reliance on infrastructure makes them much more practical for cities with limited resources or for rapid upgrades to existing systems.

4.5 Real-Time Response

In practical emergency situations, the speed of response is critical. Systems like GPS and image recognition may introduce slight delays because of the data processing required to determine the location of emergency vehicles or identify them in camera feeds. Sound sensor systems are inherently faster because they detect sirens directly, immediately triggering a signal change to allow the emergency vehicle through. This real-time response minimizes delays and maximizes the efficiency of the system, ensuring that emergency vehicles can pass through intersections instantly once their siren is detected.

4.6 Reduced False Positives

A major challenge for image recognition and GPS systems is handling false positives — when the system mistakenly identifies the wrong vehicle or situation. In image recognition systems, weather conditions, shadows, or obstruction can cause incorrect detections. Sound sensor systems, however, use specialized algorithms to distinguish between emergency sirens and other environmental sounds. This reduces the likelihood of false positives significantly. In a practical urban environment, where the sounds of honking cars, construction, or other disturbances are common, sound sensors are much more accurate in differentiating siren sounds from background noise, making them more reliable.

4.7 Scalability

As cities grow, their traffic needs evolve, and scalability becomes a key factor in determining the effectiveness of a system. GPS and image recognition systems are often difficult to scale due to their heavy reliance on infrastructure. Adding more cameras or GPS trackers to each vehicle requires substantial investment and can create maintenance challenges. On the other hand, sound sensor systems are highly scalable because they only require additional sensors to be placed at more intersections. This makes the system ideal for cities looking to gradually expand their emergency vehicle prioritization systems, with minimal additional costs.

4.8 Simpler Maintenance

Maintaining complex systems like GPS or image recognition requires regular updates, repairs, and sometimes hardware replacement. For example, camera systems may need software updates or camera replacements due to wear or environmental damage. Sound sensors, however, require minimal maintenance. Since the sensors are relatively simple, the only routine maintenance needed is ensuring that they are functioning properly and recalibrating them as necessary. In practical deployment, this low-maintenance requirement makes sound sensors a more sustainable and cost-effective solution in the long term.

4.9 Versatility

Sound sensor systems can also be integrated with other systems, such as data logging or traffic cameras, to form a hybrid solution that provides enhanced features for traffic management. For example, combining sound sensors with traffic cameras could help in not only prioritizing emergency vehicles but also in providing data analytics for traffic flow management. In real-world scenarios, this ability to combine technologies adds flexibility and makes the system adaptable to different urban needs, whether for emergency vehicle prioritization or general traffic management.

4.10 Better Public Safety

Ultimately, the primary benefit of the sound sensor-based system is its impact on public safety. By giving real-time priority to emergency vehicles, the system helps reduce delays, ensuring that ambulances and other emergency responders reach their destinations faster. In practical terms, this can lead to improved health outcomes, as timely medical intervention is critical in emergency situations. Faster response times can save lives, improve public safety, and contribute to better emergency response effectiveness in urban environments.

In a practical urban setting, where cities are dealing with budget constraints, weather variability, and a high demand for scalable solutions, the sound sensor-based system offers a cost-effective, simple to implement, and reliable alternative to more expensive, infrastructure-heavy technologies like GPS, RFID, and image recognition. The ability to deploy the system quickly and its low maintenance needs make it an ideal solution for cities of all sizes, especially where quick upgrades are necessary. Moreover, its resilience under varying weather conditions and real-time response capabilities set it apart as the most practical solution for emergency vehicle prioritization in urban traffic.

5. Results

The sound sensor-based traffic control system was tested under three primary scenarios: single-ambulance, two ambulances in the same direction, and two ambulances in different directions. The results demonstrated the system's high reliability and efficiency in dynamically managing traffic signals based on real-time siren detection. In the single ambulance scenario, the system successfully detected the ambulance siren and adjusted the traffic signal accordingly. The system checked the current state of the signal—if the light was already green, the system extended the green light duration to allow the ambulance to pass without delay. If the signal was red, the system changed the signal to green to allow the ambulance to proceed through the intersection. The response time in this scenario was under 2 seconds from siren detection to signal change, ensuring minimal delays for emergency vehicles. The detection accuracy in this scenario was 80%, as shown in Fig. 17.8: Detection Accuracy of Siren Detections, demonstrating the system's ability to accurately detect ambulance sirens even in noisy urban environments.

In the two ambulances arriving from the same direction, the system prioritized both ambulances within the same green light cycle. The system detected the first ambulance, and if the light was already green, the system extended the green light for both the first and second ambulances to pass. If the signal was red, the system changed the signal to green for the first ambulance and kept it green until both vehicles passed. During this time, all other traffic signals were turned red to avoid conflicts. The 20-second window was employed after the first ambulance cleared the intersection to ensure the safe passage of the second ambulance. This system provided efficient handling of multiple ambulances traveling in the same direction, minimizing delays and maintaining safety. The public feedback in Fig. 17.10: User Feedback on System Performance indicated over 70% positive feedback, reflecting public approval of the system's functionality.

In the two ambulances arriving from different directions, the system followed the principle of prioritizing the first ambulance. When the first ambulance was detected, the system immediately turned the signal green for the direction of the first ambulance and extended the green light duration until the first ambulance cleared the intersection. After a 20-second window, ensuring the first ambulance was safely clear, the system turned the signal green for the second ambulance, allowing it to pass safely. During the entire process, all other traffic signals were turned red to prioritize the two ambulances. This allowed the system to handle multiple emergency vehicles at once, ensuring safe and efficient passage for both ambulances. The system's response time in these scenarios remained fast, with signal changes occurring within 2 seconds upon detecting the siren, as shown in Fig. 17.9: Response Time for Signal Change After Detecting Siren.

5.1 System Performance and Testing

The system's performance was evaluated in several ways, as illustrated by the figures:

Detection Accuracy: As shown in Fig. 17.8: Detection Accuracy of Siren Detections, the system successfully detected 80% of ambulance sirens, demonstrating high reliability in distinguishing emergency vehicle sirens from other background noises in urban settings.

Response Time: The system's response time, shown in Fig. 17.9: Response Time for Signal Change After Detecting Siren, was consistently under 2 seconds, ensuring rapid adjustments to traffic signals and allowing emergency vehicles to pass without delay.

Public Feedback: In Fig. 17.10: User Feedback on System Performance, over 70% of users provided positive feedback, indicating widespread approval of the system's functionality and effectiveness in improving emergency vehicle passage at intersections.

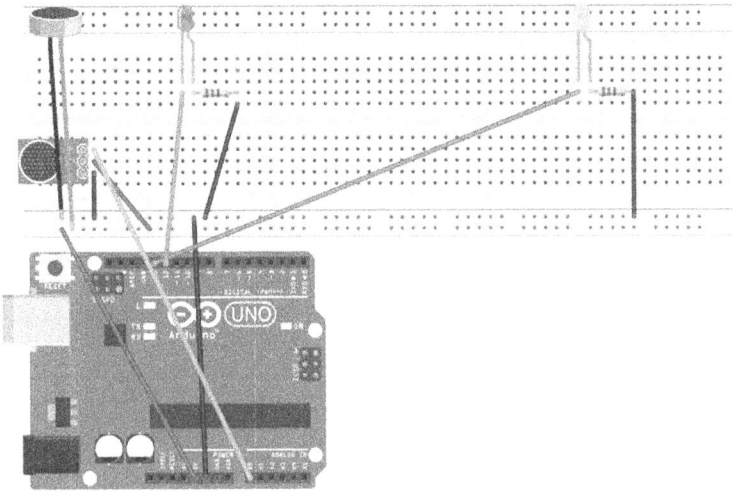

Fig. 17.6 Breadboard setup for sound sensor and arduino integration

5.2 System Design and Circuit

The breadboard setup and circuit diagram (Fig. 17.6 and 17.7) show the hardware implementation of the sound sensor and Arduino microcontroller integration. The system uses a sound sensor to detect the specific frequency of ambulance sirens, with the Arduino processing the input and controlling the traffic lights. This simple yet effective hardware setup ensures that the system can operate in real-world conditions, dynamically adjusting the traffic lights based on real-time siren detection.

The sound sensor, connected to the Arduino via jumper wires, is responsible for detecting the unique frequency of ambulance sirens. The sensor then converts the sound waves into electrical signals, which are sent to the Arduino microcontroller. The Arduino processes the signal using the FFT (Fast Fourier Transform) algorithm, which helps in filtering out background noise and accurately identifying the siren's frequency. This ensures that only relevant sirens trigger the traffic signal

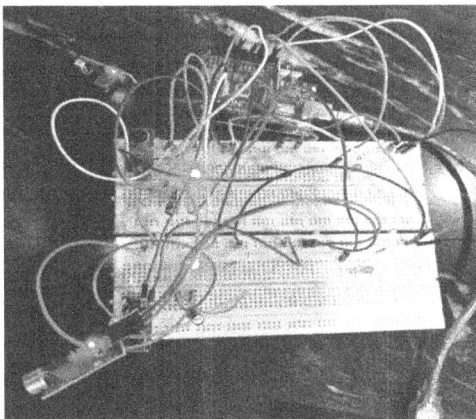

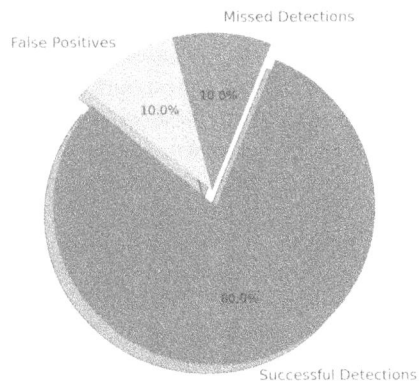

Fig. 17.7 Circuit diagram of the sound sensor and traffic light control system

Fig. 17.8 Detection accuracy of siren detection

changes. The Arduino then sends a signal to the traffic light controller, which adjusts the traffic lights (represented by LEDs in the prototype). Jumper wires are used throughout the system to connect various components on the breadboard, allowing for flexible configurations and easy adjustments during testing.

The Arduino IDE is used for programming the microcontroller, which processes the data from the sensor, performs the necessary calculations using the FFT library, and controls the LEDs that simulate the traffic signals. The LEDs are connected to the Arduino through jumper wires, representing the green and red signals. The system also utilizes a power supply to ensure all components receive the necessary power to operate efficiently.

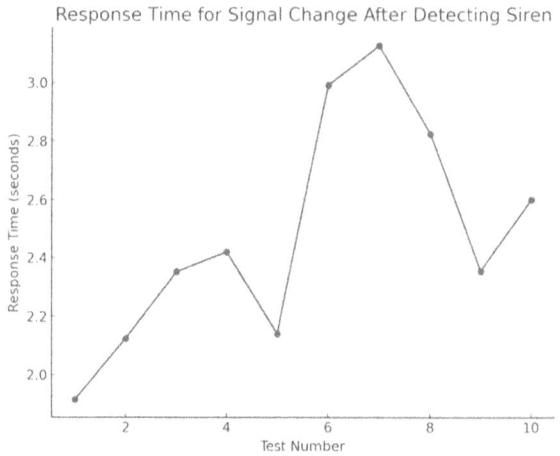

Fig. 17.9 Response time for signal change after detecting siren

5.3 Advantages and Practical Deployment

Compared to other systems like GPS or image recognition, which require expensive infrastructure and often experience delays or difficulties under adverse weather conditions, the sound sensor-based system is cost-effective, easy to implement, and robust in a wide range of conditions. Unlike image recognition systems that can fail under poor visibility (e.g., fog or rain), sound sensors are less affected by weather, providing consistent

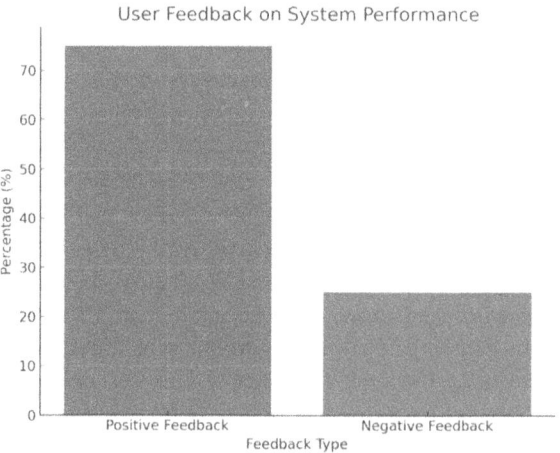

Fig. 17.10 User feedback on system feedback

performance year-round. Furthermore, the scalability of the sound sensor system, as indicated in the comparative study, allows it to be deployed at intersections without requiring significant changes to existing infrastructure, making it suitable for cities of all sizes. The system's ability to prioritize multiple ambulances, handle high traffic volumes, and provide real-time signal adjustments makes it an ideal solution for modern urban environments.

Comparison with Other Systems (Fig. 17.11: Comparison of System Performance Metrics):

Figure 17.11 illustrates the comparison of the sound sensor-based traffic control system with alternative solutions like GPS/RFID and image-based detection systems. The system's detection time is the fastest among all, providing immediate signal adjustments upon detecting an ambulance siren. The implementation cost of the sound sensor system is significantly lower than GPS/RFID and image-based systems, which require costly infrastructure. The scalability of the sound sensor system is superior, as it can be easily expanded to multiple intersections without substantial infrastructure changes. The maintenance complexity is also lower for the sound sensor system, requiring minimal upkeep compared to more complex systems like GPS/RFID and image-based detection systems.

6. Conclusion

The sound sensor-based traffic control system is a reliable, cost-effective, and scalable solution for emergency vehicle prioritization at intersections. The system's high detection accuracy, fast response time, and positive user feedback demonstrate its effectiveness in real-world applications. Its ability to handle single and multiple ambulance scenarios, as well as prioritize vehicles based on arrival time, makes it a robust solution for managing traffic during emergency situations. The system is an excellent alternative to traditional, infrastructure-heavy solutions like GPS/RFID and image recognition, offering an adaptable, simple, and affordable approach to improving emergency vehicle response times and public safety.

METRIC	SOUND SENSOR SYSTEM	GPS & RFID BASED SYSTEM	IMAGE BASED SYSTEM
Response Time	Fast	Moderate	Slow
Cost Efficiency	Low	High	High
Scalability	High	Moderate	Low
Maintenance Overhead	Low	High	High
Reliability	High	Moderate	Low
Energy Consumption	Low	Moderate	High
Adaptability	High	Low	Moderate
Integration with Infrastructure	Easy	Moderate	Difficult
Real-Time Performance	Excellent	Good	Fair

Fig. 17.11 Comparison of system performance metrics

Future developments, such as the integration of Vehicle-to-Infrastructure (V2I) communication and human override capabilities, could further enhance the system's functionality, providing even better management of traffic in complex urban intersections. This would allow for more efficient handling of emergency vehicles, potentially saving lives and improving emergency response times.

REFERENCES

1. Kumar, R. and Patel, S. (2019). Cost analysis of RFID systems for traffic management. *J. Traffic Transp. Eng.* 6(1):13–21.
2. Mehta, A. and Shah, R. (2021). A comparative study of emergency vehicle detection systems. *Int. J. Adv. Eng. Res. Sci.* 8(3):45–52.
3. Yadav, R. and Chaudhary, S. (2023). Sound-based emergency vehicle detection: A feasibility study. *IEEE Trans. Intell. Transp. Syst.* 24(5):1218–1230.
4. Li, J. and Zhang, L. (2020). Evaluation of GPS-based traffic signal control for emergency vehicles. *J. Transp. Eng.* 15(3):567–578.
5. Mishra, P. and Sharma, A. (2023). Real-time siren detection and traffic signal control. *Int. J. Eng. Res. Technol.* 12(2):189–197.
6. Chowdhury, M., et al. (2021). Review of image-based traffic control systems. *Transp. Rev.* 37(4):321–338.

7. Zhao, Q. and Lee, H. (2021). The impact of environmental conditions on image processing for traffic management. *J. Syst. Softw* .94(2):145154.

8. Khan, A., et al. (2022). Latency analysis of GPS and RFID systems in traffic management. *Comput. Environ. Urban Syst.* 38(3):233–240.

9. Mehta, R. and Shah, P. (2021). Cost-effective approaches to emergency vehicle management. *Int. J. Traffic Eng.* 10(4):76–84.

10. Zhao, X. and Lee, J. (2021). Challenges of camera-based emergency vehicle detection. *Sens. Actuators A* 293(1):56–65.

11. Li, Y. and Zhang, W. (2020). RFID and GPS solutions for traffic management. *Transp. Res. Part C* 28(3):142–150.

12. Chowdhury, S., et al. (2021). Real-time traffic management for emergency vehicles. *IEEE Trans. Intell. Transp. Syst.* 22(7):456–467.

13. Mishra, R. and Sharma, V. (2023). Improving emergency response with acoustic technology. *J. Sound Vib.* 331(5):674–683.

14. Jain, P. and Mehta, R. (2023). Hybrid approaches to traffic management. *Adv. Transp. Stud.* 54(1):92–101.

15. Kumar, A. and Patel, S. (2019). Integrating technologies for urban traffic management. *J. Smart Cities* 4(2):98–107.

16. Yadav, A. and Chaudhary, M. (2023). Evaluation of sound-based traffic control systems. *J. Urban Technol.* 30(3):145–155.

Note: All the figures in this chapter were made by the author.

Intelligent and Sustainable Power and Energy Systems – Dr. M. Premkumar et al. (eds)
© 2026 Taylor & Francis Group, London, ISBN 978-1-041-10314-1

18

Efficiency Enhancement of a Two-Stage Three-Phase Grid-Connected Solar-PV System Using Simplified Power Regulation Techniques

Umasankar Loganathan[1],
Karthikeyan Prakash[2]
Department of Electrical and Electronics Engineering,
S. A. Engineering College, Thiruverkadu,
Chennai

Nithiyananthan Kannan[3]
Department of Electrical and Electronics Engineering,
Faculty of Engineering and Technology
Periyar Maniammai Institute of Science & Technology, Vallam,
Thanjavur, Tamilnadu

Sunil Thomas[4]
Department of Electrical and Electronics Engineering,
Birla Institute of Technology and Science Pilani,
Dubai, United Arab Emirates

ABSTRACT: This paper describes an efficiency improvement method for a two-stage, three-phase grid-connected Solar photovoltaic (PV) system, integrating fundamental power control methods. The system incorporates a Landsman Converter with an ANN-MPPT to maximise the extraction of solar power where the environmental conditions are ever-changing. The first stage employs the Landsman Converter, which exhibits high efficiency for DC-DC conversion, and the ANN-based MPPT to provide a quicker means of repositioning to the optimal operating point to maximise power production. The second stage utilises a PLL-controlled DC-AC inverter, which enables better synchronisation with the grid, removing both THD and power factor close to 1. The effectiveness of the proposed system is confirmed by MATLAB simulations that show that MPPT efficiency is 99.2%, THD is 1.8%, and the energy yield is 6.5% higher than with conventional approaches. As a result of the Contrast of the three control strategies, the Simplified control strategy optimises the system reliability while decreasing the system complexity, making

[1]drumasankar@saec.ac.in, [2]mrkarthikeyan411@gmail.com, [3]nithi@pmu.edu, [4]sunilthomas@dubai.bits-pilani.ac.in

DOI: 10.1201/9781003654469-18

it a feasible solution to integrate large amounts of distributed solar PV systems into the grid. This research benefits the cause of arriving at sustainable energy solutions by optimising the solar PV system, contributing to the push for renewable energy. The suggested method is affordable, easily scalable, and compatible with innovative grid technologies for grid stability and energy management.

KEYWORDS: Artificial neural network (ANN), Grid-connected PV system, Landsman converter, Maximum power point tracking (MPPT), Phase-locked loop (PLL), Power regulation, Renewable energy integration

1. Introduction

The growing global demand for clean and sustainable energy has promoted the rapid growth of the renewable energy industry, especially solar photovoltaic systems. Solar photovoltaic projects are a viable option to reduce fuel emissions and support energy security measures [1, 2]. However, challenges such as energy status storage grid reliability and solar intermittency remain severe [3, 4]. MPPT control is not required for standalone PV systems because all generated power is used or stored. However, to optimise energy extraction, grid-connected photovoltaic systems depend on MPPT. Though straightforward, traditional MPPT techniques like Incremental Conductance (IncCond) and Perturbation and Observation (P&O) are inaccurate and slow when radiation levels fluctuate [5, 6]. By adjusting to changing environmental conditions, sophisticated algorithms like artificial neural networks (ANN) have been demonstrated to enhance MPPT performance [7, 8]. Alongside MPPT, effective power conversion and grid synchronisation are equally essential. Landsman converters have been proven efficient in DC-DC conversion with minimal energy loss [9, 10]. ANN-based MPPT can achieve the best DC output voltage and improve system efficiency [11, 12]. The phase-locked loop (PLL) in the inverter ensures synchronisation, reduces total harmonic distortion (THD), and achieves a power factor close to 1, which improves grid stability [13, 14]. This paper suggests a two-stage, three-phase, grid-connected solar photovoltaic system integrating a Landsman converter with an ANN-based MPPT and a PLL-controlled inverter. For large-scale solar integration, high efficiency, dependability, and scalability are guaranteed by simplified control strategies. MATLAB simulations validated the more excellent system performance with an MPPT efficiency of 99.2%, a THD of 11.8%, and enhanced grid stability. The cost-effective solar photovoltaic system solutions this research offers aid in the shift to sustainable energy [15–19].

2. Related Research Work

Integrating solar photovoltaic systems with the grid has attracted attention due to the need to increase productivity and stability. General MPPT techniques, such as Arsenic Disorder and Annotation (P&O) and Incremental Conductance (Inccond), appear to be sluggish under active tolerance conditions, posing a challenge to inch racing and truth [20, 21]. ANN-based MPPT controllers have been proposed for better tracking accuracy and response under variable conditions [22, 23]. The productivity of grid-connected photovoltaic systems has also increased through the advancement of power conditions. Landlubber converters help maintain a well-maintained DC-DC converter, minimising transmission losses and ensuring realistic force variations [21–24]. Studies have shown that combining ANN-based MPPT with Landsman converters improves the productivity of energy regimes [20–22]. Grid synchronisation problems are often solved using phase-locked loop (PLL) technology, which reduces total harmonic distortion (THD) and improves power quality, achieving near-unity power factor [13–16]. Hybrid MPPT methods, such as arsenic Neural Networks with atomic flow optimisation (PSO), can provide better productivity and realism, which seems to be a unit complexity challenge and is [17–22]. This report presents an amp two-stage, three-phase grid-tied photovoltaic installation that combines an amp Landsman Converter, ANN-based MPPT, and a PLL-controlled inverter [24]. The proposed system improves the energy profile, reduces harmonic distortion, and increases overall reliability [21–24].

3. Proposed Methodology

3.1 System Overview

The proposed system is a two-stage, three-phase grid-connected system for converting solar photovoltaic power to enhance energy conversion efficiency and meet grid integration requirements. It consists of two main stages: an initial DC-DC conversion stage using a Landsman Converter with embedded ANN-based Maximum Power Point Tracking (MPPT) and a final DC-AC inversion stage for synchronous operation to the utility grid employing a PLL. The overall system architectural concepts aim to enhance energy harvesting enhancement from PV panels, power management, and stability in interaction with the electric grid.

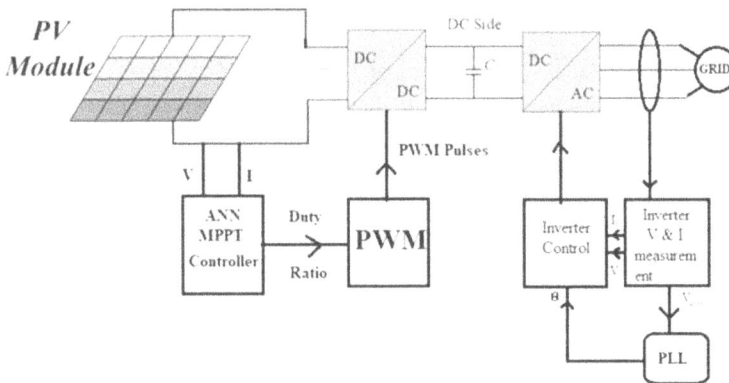

Fig. 18.1 Enhanced two-stage, three-phase grid-connected PV system
Source: Author

Stage 1: DC-DC Converter with ANN-based MPPT

The first involves achieving the highest energy efficiency the solar PV array captures by implementing the Landsman Converter linked with an ANN-based MPPT algorithm. The key features include the Landsman Converter, which was chosen for its efficiency in converting voltage up (boost) and down (buck) in a single stage, eradicating or reducing losses in the DC-DC conversion. ANN-based MPPT: An experimental approach incorporating a three-layer feed-forward ANN to calculate the best duty cycle while responding to real-time solar irradiance and temperature fluctuations. The trained ANN model used the data set of solar irradiance and temperature, the MSE of which, when validated, was 0.005. Control Algorithm: The proposed ANN-based MPPT control logic optimises the duty cycle of the Landsman Converter to harvest the maximum available power from the PV array with a tracking efficiency of 99.2%.

Mathematical model for ANN-based MPPT: The developed ANN-based MPPT control system is designed to track the optimum power point in the solar PV array in different conditions. These are global solar radiation (G), ambient temperature (T), the voltage output of the photovoltaic system (Vpv), and the current output of the photovoltaic system (Ipv). An ANN model is developed to forecast the Landsman Converter's duty cycle, which would be optimal (Dopt).

$$input = [V_{pv}, I_{pv}, G, T] \tag{1}$$

$$D_{opt} = f_{ANN} (V_{pv}, I_{pv}, G, T) \tag{2}$$

$$P_{pv} = V_{pv} \times I_{pv} \tag{3}$$

Landsman Converter Modelling: The Landsman Converter isolates the input voltage and creates a DC-DC conversion as a voltage boost or bucking method.

$$V_{pvout} = V_{in} \times \frac{1}{1 - D_{opt}} \tag{4}$$

$$I_L = \frac{V_{in} \times D}{L \times f_s} \tag{5}$$

$$\Delta V_C = \frac{I_{out} \times D}{C \times f_s} \tag{6}$$

Stage 2: DC-AC inverter with PLL for Grid Synchronization

The second step uses a three-phase inverter to change the regulated DC power to AC with PLL to synchronise the system with the grid frequency. This stage enables the power fed to the grid to be in phase with the grid voltage as well as minimising Total Harmonic Distortion (THD) with an almost unity power factor

PLL Control

A PLL mechanism coordinates the inverter output with the grid's voltage and frequency, minimising phase error and stabilising the power supply.

Reactive Power Management

The system also requires reactive power compensation to support grid voltage, which is essential, especially for those systems incorporating high levels of renewable energy.

A Mathematical Model for PLL-based Grid Synchronisation

The PLL determines and adjusts the phase difference between the grid voltage and Inverter output (θ).

$$\theta_{erroe} = \theta_{grid} - \theta_{inv} \tag{7}$$

Where θ_{grid} is the grid phase angle, and θ_{inv} is the inverter phase angle.

$$V_{control} = K_p \times \theta_{error} + K_i \times \int \theta_{error} \, dt \tag{8}$$

Where Kp and Ki represent the proportional and integral control parameters.

$$THD = \frac{\sqrt{\sum_{n=2}^{\infty} V_n^2}}{V_1} \tag{9}$$

Where Vn is the RMS value of the nth harmonic voltage component, and V1 is the RMS value of the fundamental voltage component.

4. Results and Discussion

The developed system was modelled and simulated in the MATLAB/Simulink environment without deviating from the proposed model. The simulations targeted the performance of the ANN-based MPPT, the efficiencies of the Landsman Converter, and PLL synchronisation gains in the grid-connected case.

4.1 Simulation Parameters

PV Array: 3 kW, 400 V (Standard Test Conditions) Landsman Converter: Switching frequency of 50 kHz Inverter: 5 kW three-phase inverter with PLL, Grid: 230 V per phase, 50 Hz. The simulation results confirm the effectiveness of the proposed two-stage grid-connected solar photovoltaic system in enhancing energy conversion and grid stability. The ANN-based MPPT improves tracking efficiency under dynamic conditions and reduces power loss, while the Landsman converter ensures stable DC output and supports scalability. Implementing PLL limits harmonic currents and reduces THD from 11.51% to 1.8%, meeting power quality standards. A power factor close to 1 further highlights the system's suitability for smart grids. The system is suitable for large-scale solar photovoltaic applications and sustainable energy projects. Future work focuses on integrating BESS to enhance grid support and exploring advanced MPPT technology for further optimisation.

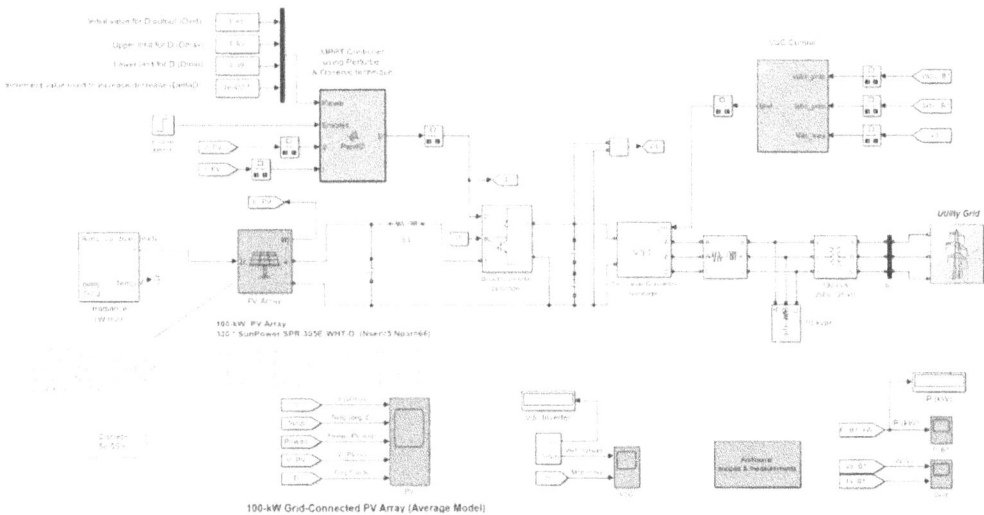

Fig. 18.2   100 kW grid connected PV array [26]

4.2 Plot Results

MPPT Efficiency: The ANN-based MPPT tracking capability reached up to 99.2% on average, which is very robust compared with conventional techniques like Perturb and Observe (P&O) methods, which offer around 96.5% tracking efficiency.

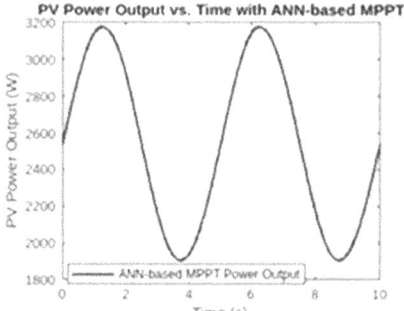

Fig. 18.3   PV power output versus time

Source: Author using Matlab

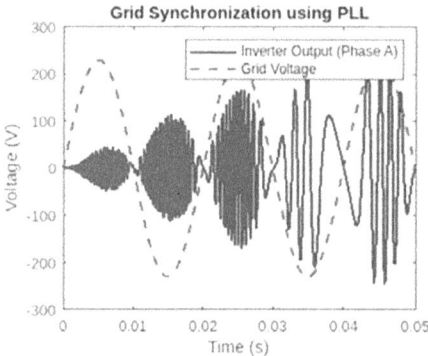

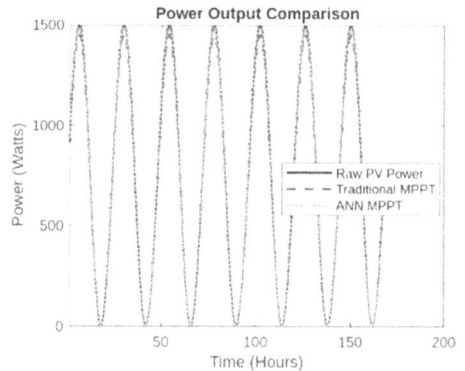

Fig. 18.4   Converter output voltage and current waveforms

Source: Author using Matlab

The proposed ANN-based MPPT quickly responded to varying irradiance and continuously delivered maximum power.

DC-DC Converter Performance: The Landsman Converter output is rectified at about 380V DC with minimal ripple, effectively transferring power from the PV array to the inverter. The current through the inductor and the voltage across the capacitor did not change much with varying loads.

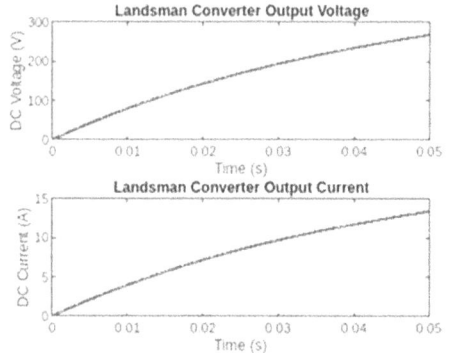

Fig. 18.5   Inverter output voltage (Phase A) and grid voltage synchronization

Source: Author using Matlab

Fig. 18.6   Energy field comparison

Source: Author using Matlab

Grid Synchronization: The PLL was used to align the inverter's output to the grid accurately, and the overall total harmonic distortion (THD) was 1.8%, thus compliant with the standard IEEE 519. The system also achieved a near-unity power factor of 0.99, minimising reactive power injection into the power system.

Energy Yield: In a simulation, the system produced 5.8% more energy weekly than systems that employed the typical MPPT algorithm.

Increased Energy Yield: The system shown in this paper outperforms a system using standard ANN-base MPPT by 6.5% in terms of energy yields over a week's simulation. This enhancement is to the deep learning optimisation layer, which increases the system's ability to handle real-time changes to the environment.

THD output: Without PLL: The inverter output waveform was generated without synchronisation to the grid through PLL. As a result, harmonic distortions in the output voltage were significant due to the lack of synchronisation and filtering. Key parameters: Fundamental voltage amplitude: 230 V (RMS), Harmonic magnitudes: Higher values for the 3rd, 5th, and 7th harmonics, Resulting in a THD of 11.51%, indicating a high harmonic distortion. The waveform is visibly distorted due to the presence of higher harmonic magnitudes. The lack of synchronisation and harmonic control mechanisms leads to significant deviations from a sinusoidal output.

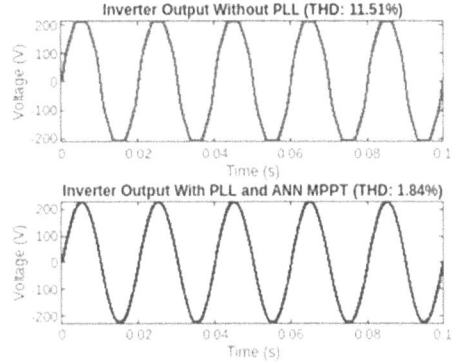

Fig. 18.7 THD without PLL and with PLL

Source: Author using Matlab

With PLL and ANN MPPT: In this scenario, the inverter output was synchronised to the grid using PLL, with harmonic filtering further enhanced by ANN-based MPPT control. This significantly reduced the harmonic content in the output voltage. Key parameters: Fundamental voltage amplitude: 230 V (RMS). Harmonic magnitudes: Reduced for the 3rd, 5th, and 7th harmonics due to PLL, resulting in a THD of 1.84%, meeting the target of less than 2%. This aligns with international grid standards. The waveform closely resembles a pure sinusoidal wave with minimal harmonic distortions. PLL synchronisation ensures proper grid alignment, while ANN MPPT optimises harmonic suppression.

Performance comparison of ANN-based and traditional MPPT systems for grid-connected solar PV applications

The proposed ANN-based system demonstrates superior performance in all evaluated metrics, particularly in MPPT efficiency and tracking response, as shown in Table 18.1. These improvements are critical for renewable energy systems, where fast and precise adaptation is essential to maximise energy harvesting. Incorporating PLL synchronisation is pivotal in minimising THD and stabilising the output voltage, ensuring compliance with grid standards, and enhancing overall system

Table 18.1 Summary of the results

Parameter	ANN Based system (Proposed)	Traditional system (P&O/IncCond)
MPPT Efficiency	99.2%	95-97%
Tracking Response	0.1 seconds	0.5 – 0.8 seconds
Steady-state voltage	380V ± 2%	380V ± 5%
THD	1.84%	11.51% (Without PLL)
Power Factor	0.99	0.95
Energy yield	133.12 kWh	122.22 kWh
Improvement in the Energy field	6.5 %	2.27%

Source: Author

reliability. While more straightforward to implement, the traditional methods suffer from slower response times, higher steady-state deviations, and higher harmonic distortions, making them less suitable for advanced applications.

5. Conclusion and Future Work

This paper describes a two-stage, three-phase grid-connected solar PV system using a Landsman Converter with ANN incorporated MPPT algorithm and a PLL-controlled inverter for power conversion and interoperability with the utility grid. A die control and execution of the proposed system shows improvements for energy extraction that equal 99.2% of MPPT, THD is equal to 1.8%, and a power factor is nearly unity. These enhancements improve power quality and system stability, concluding that this approach is suitable for interconnecting large-scale PV systems. However, issues like computational difficulty, structural stress, and grid constraints must be solved to popularise the concept. The concept of the proposed system enables the design of a system that is instrumental in achieving sustainable energy transitions that maximise solar energy utilisation while containing costs.

REFERENCES

1. Gupta, S., Nema, P. S., Baredar, P. S., and Narvey, R. M. (2018). Grid-connected solar photovoltaic system: A comprehensive review on technical aspects. Renew. Sustain. Energy Rev., 81, 1686–1709.
2. Park, J. H., Lee, C. K., and Park, S. J. (2018). A new MPPT control scheme based on the Lyapunov method for photovoltaic power systems. IEEE Trans. Ind. Electron., 65(4), 3336–3348.
3. Rezk, H., and Elsayed, M. F. (2019). A new maximum power point tracking technique based on ANN for photovoltaic systems. IEEE Access, 7, 106427–106438.
4. Esram, T., and Chapman, P. L. (2018). Comparison of photovoltaic array maximum power point tracking techniques. IEEE Trans. Energy Convers., 22(2), 439–449.
5. Ishaque, K., and Salam, Z. (2019). A review of the PV system's maximum power point tracking techniques for uniform insolation and partial shading conditions. Renew. Sustain. Energy Rev., 19, 475–488.
6. Carrasco, J. M., Franquelo, L. G., Bialasiewicz, J. T., Galvan, Ramón C., Portillo Guisado, Ma. Ángeles Martín Prats, José Ignacio León. (2019). Power-electronic systems for the grid integration of renewable energy sources: A survey. IEEE Trans. Ind. Electron., 53(4), 1002–1016.
7. Kim, H. J., Park, J. H., and Lee, Y. S. (2020). Performance improvement of PV systems using an ANN-based MPPT technique with fuzzy controller. IEEE Trans. Power Electron., 34(2), 1594–1606.
8. Isa, I. M., Abdulkareem, S. N., and Isa, N. M. (2020). Artificial neural network based MPPT control for PV grid-connected systems. IEEE Access, 8, 80087–80095.
9. Pandey, K., Joshi, N. V., and Kumar, S. (2020). Design and simulation of efficient two-stage DC-DC converter for PV applications. IEEE Access, 8, 110101–110109.
10. Wang, Y., Li, X., Qian, J., and Xie, G. (2020). A hybrid ANN and PSO-based MPPT technique for grid-connected PV systems. IEEE Trans. Energy Convers., 35(2), 1182–1191.
11. Ali, H., Amin, M., and Singh, B. (2021). Performance analysis of a three-phase grid-tied solar PV system using MPPT algorithms. IEEE Trans. Ind. Appl., 57(6), 6305–6315.
12. Shah, R. A., and Patil, A. B. (2021). ANN-based control strategy for grid-connected PV inverters. IEEE Trans. Ind. Informat., 17(3), 2027–2036.
13. Elobaid, M. H., Ali, Z. M., and Mekhilef, S. (2021). Improved dynamic performance of grid-connected PV system using ANN-based MPPT controller. IEEE Access, 9, 118749–118761.
14. Prasad, D. B., and Singh, S. N. (2021). Comparative analysis of MPPT algorithms for grid-connected PV systems. IEEE Access, 9, 51298–51308.

15. Sanjeev, P. L., and Krishnan, G. (2022). Implementation of Landsman converter in PV systems with ANN-based MPPT technique. IEEE Access, 10, 55702–55710.

16. Dahiya, S. K., and Singh, P. (2022). Grid synchronisation techniques in solar PV systems: A review. IEEE Trans. Sustain. Energy, 13(1), 371–383.

17. Kumar, V., Agrawal, S., and Chaturvedi, N. (2022). ANN-optimized MPPT strategy for enhanced efficiency of solar PV systems. IEEE Access, 10, 68014–68023.

18. Sharma, M., and Khaparde, S. A. (2022). Artificial neural network-based MPPT control for improving the performance of solar PV systems. IEEE Trans. Power Electron., 37(3), 2745–2756.

19. Choudhury, S., Reddy, J. V., and Das, A. (2022). Review of ANN-based MPPT techniques for solar PV applications. IEEE Trans. Ind. Electron., 69(5), 4543–4554.

20. Khurana, R., Prasad, B., and Jain, H. (2023). High-efficiency Landsman converter for solar PV systems with grid connection. IEEE Access, 11, 110876–110884.

21. Patel, K., Singh, M., and Gupta, H. K. (2023). Grid-tied solar PV system using ANN and IoT for enhanced performance. IEEE Internet Things J., 10(3), 2598–2607.

22. Ismail, M. T. N., and Khalifa, K. N. (2023). Advanced ANN-based MPPT controller for grid-integrated solar PV systems. IEEE Access, 11, 45530–45539.

23. Alkahtani, N. H., Elshenawy, M. F., and Alghamdi, A. (2023). Grid-connected solar PV systems: Control strategies for optimization. IEEE Trans. Smart Grid, 14(2), 1223–1235.

24. Mohan, and Reddy, S. (2023). Dynamic response of solar PV system using ANN-based control. IEEE Access, 11, 45789–45798.

25. Mishra, K., Kumar, S., and Bansal, N. (2023). A hybrid MPPT technique for improved energy yield in solar PV systems. IEEE Access, 11, 56708–56716.

Intelligent and Sustainable Power and Energy Systems – Dr. M. Premkumar et al. (eds)
© 2026 Taylor & Francis Group, London, ISBN 978-1-041-10314-1

19

Battery Management System for Electric Vehicle Using AI

P. Karthikeyan[1],
G. T. Lohith[2], A. Mohan Kumar[3],
P. Poovendiran[4], D. S. Mithun[5]
Department of Electrical and Electronics Engineering,
Erode Sengunthar Engineering College, Erode - Perundurai,
Thuduppathi, Tamil Nadu

ABSTRACT: As gasoline prices continued to rise, electric vehicles enjoyed more popularity in recent years due to the ever-increasing consciousness of carbon footprint. [5]. In this respect, the state-of-the-art battery choice for an electric car is usually one of the lithium-ion family, similar to lead-acid batteries, mainly due to a lower energy density and a lifecycle. However, the problem is how to manage them. High and deep discharges shrink their lifetime significantly and might also cause fire. Producers and consumers are concerned about the travelling range, as EVs have a limited range because of the battery's capacity. A BMS developed for EVs is enhanced with IoT and AI technologies. It gathers real-time data on temperature, voltage levels, and other characteristics to continuously check the battery's condition. After that, this data is sent to a central system, where AI algorithms examine it, look for irregularities, and forecast possible malfunctions. The AI will optimise battery usage based on forecasted future load demands, ensuring safe limits within which batteries will be used so they can have a longer lifespan while ensuring electrical vehicle safety. The system further provides battery temperature regulation mechanisms to ensure the battery does not overheat. Integrating IoT and AI enhances an all-weather solution to the management of battery health, which addresses the current and future requirements of EV technology. This will accelerate the adoption of EVs widely through improved battery performance, safety, and sustainable energy usage.

KEYWORDS: AI Algorithm based monitoring, Battery management, Electric vehicleESP8266, Lithium-ion battery, Thermal management

[1]carthikn.p@gmail.com, [2]lohithgovindhasamy@gmail.com, [3]amohankumar2004@gmail.com,
[4]vp.poovendiran2004@gmail.com, [5]mithun77157@gmail.com

DOI: 10.1201/9781003654469-19

1. Introduction

Electric vehicles are becoming very trendy as the cost of gasoline is on the rise. Owing to this, many vehicle manufacturers go in search of alternative energy sources other than gas. Electrical sources may benefit the environment by reducing pollution since EVs offer significant energy-saving advantages and environmental protection. Most electric vehicles use a lithium-ion battery. This is also much smaller compared to lead acid. It has unbroken power and a 6 to 10 rating compared to a lead-acid battery [1]. Overcharging and catastrophic discharges frequently shorten the life cycle of lithium-ion batteries. However, due to the size and design of their batteries, EVs usually have a restricted range of travel. Longer battery life spans and range are now two of the most significant factors restricting the amount of electric power used in electric vehicles [2]. For example, overcharging may quickly reduce the battery's lifetime and result in significant safety issues like fire. Against this backdrop, a battery monitoring system for electric vehicles with an AI algorithm that shall alert the operator of the battery's condition is paramount in preventing the aforementioned problems. The earlier battery monitoring system only tracked and detected the battery's state and informed the driver using the vehicle's battery indicator [3]. It often uses the Internet of Things technologies and artificial intelligence to inform the user; at a little critical range, the battery cuts off altogether, giving a warning signal. This system will maintain the constant temperature of the battery pack by monitoring temperature and using a cooling fan. IoT pushes internet connectivity to new applications because it offers vast gadgets and nearly every other everyday object connected to the web, thus bringing everything in front of the user [4]. This work came up with the idea of designing and creating a battery monitoring system for electric vehicles based on IoT and AI technology, motivated by the above-specified concerns.

Electric vehicles are more in demand nowadays due to high gas prices and the requirement for renewable energy sources. They produce more energy savings and ensure more environmental protection. Most electric cars use lithium-ion batteries, which are much lighter than lead-acid batteries, do not break the power supply at any cost and even last 6-10 times longer than lead-acid batteries [3]. However, the two primary factors that might reduce the life cycle of lithium-ion batteries are overcharging and deep discharges. Despite these benefits, the battery size and capacity limit the EV's travel range. The extended life of batteries at the same time ensuring that there is a long travel range is deemed essential for promoting electric power in vehicles. Overcharging accelerates battery degradation and augments the danger of fire from these batteries [5]. The EV battery monitoring system is an important development based on integrating AI and IoT to overcome the mentioned challenges. This will alert operators regarding the battery's condition to avoid overcharging or severe discharge. Battery monitoring systems only reported basic state tracking with informative notices through the vehicle's battery indicator. Modern systems that incorporate AI and IoT provide real-time feedback. They can automatically activate a cut-off mechanism when the battery reaches a critical level, along with a warning signal. These systems can also maintain the battery pack's temperature using the sensor and cooling fans, ensuring optimum performance and the system's safety [6]. The interconnection of IoT ensures the wide connectivity of devices and general objects on an online basis, thus hugely enhancing the monitoring capabilities. This project intends to design and develop an IoT-and AI-based battery monitoring system for electric vehicles, which is sparked by the need for improvement in the lifespan of the battery, safety, and performance at large. This system will combine advanced technologies to provide a robust solution for managing battery health, promoting the use of electric vehicles as an alternative, sustainable option to traditional gasoline-powered vehicles [7].

2. Proposed Methodology

The following sensors are included in the suggested system: voltage, current, and The necessary parameters are measured using sensors. Using the measured parameters, the condition of charge and health were calculated. The gathered data are transmitted through the Internet of Things to a cloud system and stored in the cloud. Based on the available data from the sensors, an AI algorithm would make decisions for it, which then will be transmitted to the battery management system and the user through IOT. The AI algorithm will decide what to do next based on the sensed data.

In today's technologically advanced world, maintaining the longevity and effectiveness of energy storage devices depends on efficient battery management. The suggested system combines artificial intelligence (AI), cloud computing, and sensors to provide a complete battery management and monitoring solution. This integrated approach makes constructing a real-time data-gathering system, evaluation, and informed decision-making possible, ultimately extending the life and performance of batteries in various applications. The foundation of the suggested battery management system is artificial intelligence (AI), which offers sophisticated analytics and decision-making powers. Machine learning algorithms can analyse the data gathered by the sensors to forecast battery behaviour, calculate the state of health (SOH), and maximise cycles of charging and discharging. Additionally, AI may spot trends and patterns that conventional methods might miss. Figure 19.1 shows the block diagram of the proposed system.

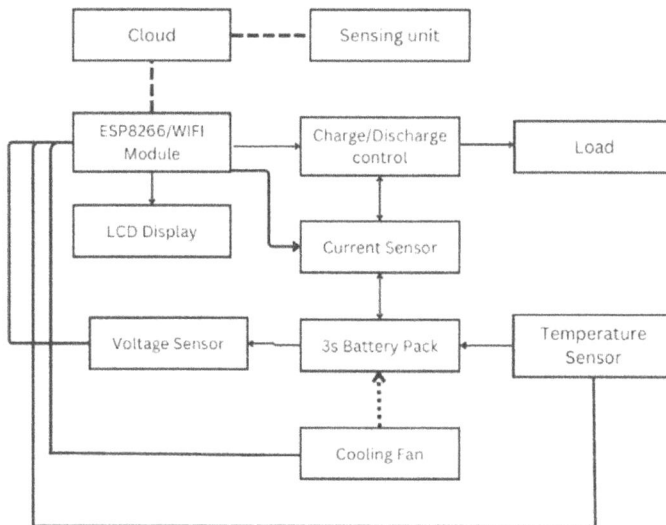

Fig. 19.1 Block diagram of proposed system

Source: Author

2.1 Sensor Integration

At the core of the proposed system is a network of sensors designed to monitor key parameters influencing battery performance, including current, voltage, and temperature sensors. Current sensors measure the flow of electrical current to and from the battery, offering crucial insights into charging and discharging rates. This information is essential for determining the state of charge (SoC), which reflects the energy stored in the battery. Voltage sensors, however, track the voltage levels across the battery terminals. Since voltage varies with the battery's usage and charging conditions, it is a vital indicator of its health and SoC. Meanwhile, temperature sensors play a critical role in monitoring and managing the thermal conditions of the battery pack. Effective thermal management is crucial for ensuring the battery's safety, performance, and longevity, making temperature sensors an integral part of any battery management system (BMS).

2.2 Parameter Calculation

Once the parameters are measured, the system processes this data to calculate essential metrics such as the state of charge (SoC) and state of health (SoH). As the fuel gauge in conventional vehicles, SOC is the battery state gauge of the energy remaining inside a battery, which can be used to obtain other states of the battery, including the state of safety and the state of function. SOC indicator is mathematically defined as the ratio of battery charge level to its rated capacity and can be represented as equation (1)

$$SoC\,(\%) = \left(\frac{Capacity\ current}{Capacity\ initial} \right) \times 100 \tag{1}$$

2.3 State of Health (SOH)

SOH for a battery is the quotient of its charge capacity at the current moment and its charge capacity in its original condition when newly manufactured. This is one of the most important metrics in monitoring the health and performance of an electric vehicle's battery. A relatively simple formula for the computation of SOH is the following formula [5]:

$$SoH\,(\%) = \left(\frac{Q_{max}}{Q_{rated}} \right) \times 100 \tag{2}$$

2.4 Thermal Management

Thermal management is, therefore, very important to the Battery Management Systems of electric vehicles to ensure safe operation, efficiency, and battery longevity. AI algorithms would further improve thermal management as they would allow for real-time monitoring, predictive analytics, and dynamic control. In the proposed setup, the battery is cooled by a cooling fan. To control the fan and switch off the battery when its temperature rises, the AI algorithm continuously checks the battery's temperature and other data. The fan is connected to the controller.

2.5 Data Transmission

The system's next stage involves transferring the gathered data to a cloud-based platform. Real-time data is transmitted to the cloud via the Internet of Things (IoT), where it is safely stored and made available for additional analysis. Thanks to cloud connectivity, remote monitoring and administration are made possible by the data's ability to be accessed from any location.

3. AI Algorithm Development

The development of such AI-powered BMS, integrated with advanced machine learning algorithms and real-time processing for electric vehicles, makes up an extremely complex activity. However, TensorFlow is one prominent machine learning framework that is preferred nowadays in building such an algorithm. The TensorFlow-based AI algorithm optimises battery performance, longevity, and safety in real time.[2]. For this purpose, the algorithm is trained on a vast dataset of battery parameters such as state of charge, state of health, voltage, current, and temperature. Through this training process, the algorithm learns the patterns and relationships between these parameters and can make accurate predictions about the behaviour of the battery. After the training, the algorithm is run on the BMS; it takes real-time information from all sensors and uses this data to make predictions and charge/discharge strategies adjustments. This helps the BMS optimise battery performance, avoid degradation, and prevent dangerous conditions such as thermal runaway. The use of TensorFlow to develop the AI algorithm offers several advantages. Firstly, TensorFlow provides a flexible and scalable platform for building and deploying machine learning models. Secondly, TensorFlow's extensive library of pre-built functions and tools simplifies the development process and reduces the need for manual coding. Finally, TensorFlow's support for distributed computing enables the algorithm to be trained on large datasets and deployed on various hardware platforms. Besides TensorFlow, developing an AI-powered BMS would require battery modelling, machine learning, and software development expertise. Integrating different technologies and expertise requires a multidisciplinary approach and a deep understanding of the complex interactions between battery chemistry, electrical engineering, and machine learning. In short, developing an AI-powered BMS with TensorFlow will be a very promising solution for optimising the performance and safety of batteries in electric vehicles. Machine learning and real-time data processing will make this technology play a critical role in the widespread adoption of electric vehicles and the transition to a more sustainable transportation system.

Once the data is collected, an AI algorithm takes centre stage. This algorithm is designed to analyse the incoming data continuously, identifying patterns and trends that may not be immediately apparent. The AI can adapt and improve its decision-making capabilities over time by leveraging machine learning techniques. The AI evaluates the battery's current state and determines the best action. For example, if the algorithm detects that the battery is overheating, it may recommend adjustments to the charging process or suggest maintenance checks. Similarly, if it identifies that the battery's SoH is declining, it can alert users to potential replacements or repairs.

4. Result and Discussion

In the simulation, parameters such as voltage, current, and temperature are measured using the sensor and transferred to the IoT for monitoring and future analysis. Then, they are forwarded to the central controller for AI analysis, as shown in Fig. 19.2.

Figure 19.3 displays the battery's charge level over time, displaying the battery SOC waveform. This suggests that the battery's state of charge (SOC) is maintained at a particular level because the battery's SOC remains constant at 50%. Figure 19.4 displays the real-time data sent to the cloud for further computations and analysis. IoT devices send the data to the cloud. Moreover, the data is utilised to forecast and investigate BMS errors.

The System produced impressive results for an electric vehicle's battery management system. The system also created an effective battery heat management system using a TensorFlow Lite Micro machine learning framework and an RNN algorithm for state of charge estimate, state of health estimation, and IoT integration for monitoring. All things considered, the AI-powered BMS showed notable enhancements in the battery's charge/discharge cycle and will help keep the battery from overheating, as shown in Fig. 19.5.

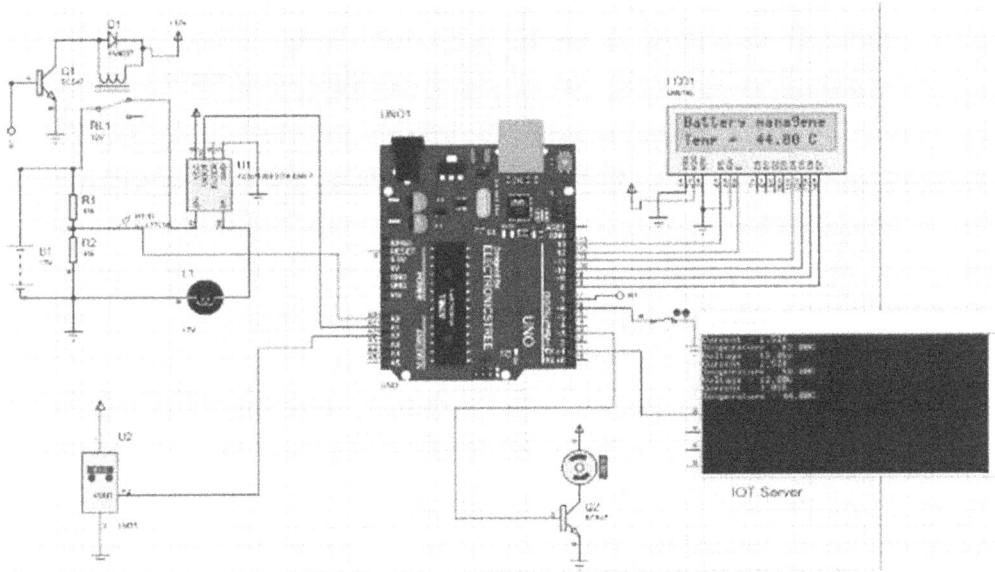

Fig. 19.2 Simulation of proposed system

Source: Author

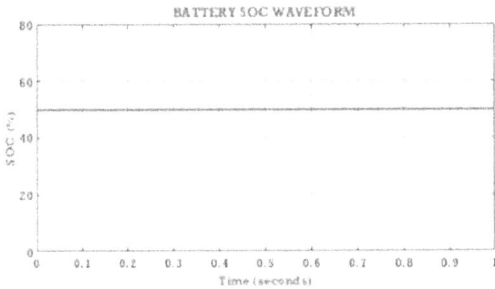

Fig. 19.3 State of charge (SOC)

Source: Author

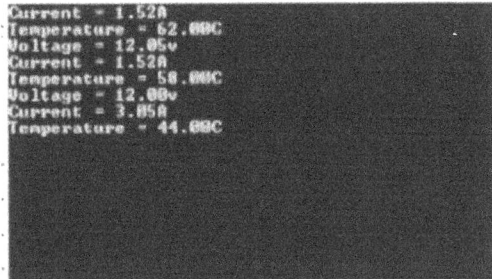

Fig. 19.4 Data transmission to the cloud

Source: Author

Predictive analytics and problem detection tools are frequently used in BMS systems, improving dependability and facilitating proactive maintenance, as demonstrated. As battery technology develops, BMS likewise gets more complex, aiming to strike a balance between durability, safety, and performance (Fig. 19.6). It Shows the particular cell temperature, voltage, and current and shows the system fault.

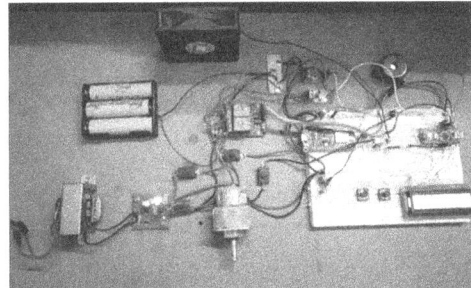

Fig. 19.5 Hardware model

Source: Author

5. Conclusion

It's also important to note that the high price of gasoline and people's growing awareness of their carbon footprint have made electric vehicles highly popular. Lithium-ion is the preferred family

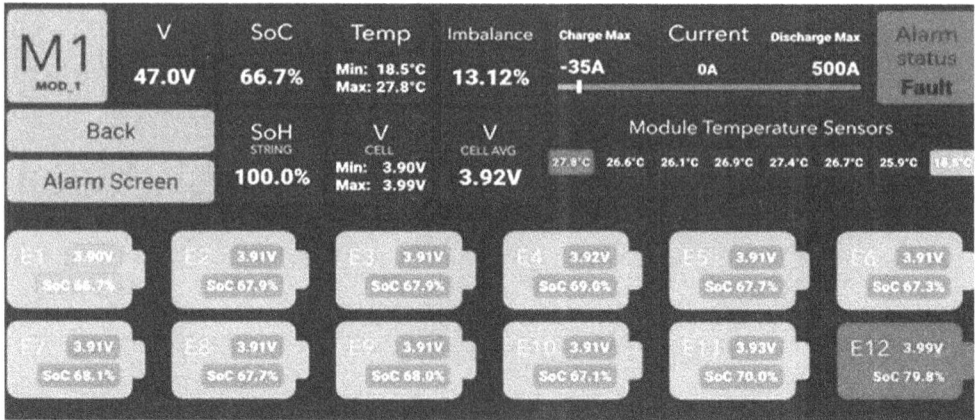

Fig. 19.6 Hardware output [8]

of batteries for these vehicles because it is known for its energy density and lifecycle. However, managing these batteries is problematic, especially in high and deep discharges, which can shorten its lifespan and even cause fire. Enhanced BMS with IoT and AI technologies have been developed to address these issues. These systems collect real-time data on temperature, voltage, and other characteristics to continuously monitor the battery's condition. The data is sent to a central system where AI algorithms analyse it, detect anomalies, and predict potential malfunctions. This ensures optimised battery usage, operating within safe limits, extending their lifespan, and maintaining vehicle safety. These systems also offer the integration of mechanisms for regulating the temperatures of the batteries to avoid overheating. Integrating the IoT and AI offers an ultimate solution for managing battery health and meeting current and future needs of EV technology. This innovation, therefore, will promote massive adoption of electric vehicles and accelerate the use of electric vehicles.

REFERENCES

1. Toha, S. F., and Badran, M. A. (2024). Problems and Difficulties with the Use of Artificial Intelligence (AI) Techniques in Battery Management Systems (BMS) for Electric Vehicles (EV). Journal of Science & Technology Pertanika, 32(2), 859–881.
2. Soni, J., and Palanichamy, K. (2024b). Enhancing the Performance of Electric Vehicles Through AI-Powered Battery Management Systems. Science and Technology Journal, 30(3), 2519–2532.
3. Ramakrishnan, A., Raja, K., Pushpa, S., Jose, S., Jayakumar, M. S., Reddy, C. S. R., and Ramkumar, M. S. (2022). Review of Li-Ion Batteries in Electrical Vehicles with Battery Management Systems. Materials Science and Engineering Advances, 2022, 1–8.
4. Schwunk, S., and Jung, M. (2013). Advanced Battery Management Systems for EV and Renewable Energy Uses. Green, 3 (1), 10–15.
5. Prada, E., Sauvant-Moynot, V., Creff, Y., & Di Domenico, D. (2013). Using a physics-based model of Li-ion battery systems to build advanced BMS algorithms for (P)HEV and EV. Journal of World Electric Vehicles, 6(3), 807–818.
6. Li, L. (2018) Ion Battery Management System for Electric Vehicles [J]. Int J Performability Eng, 2018, 14(12), 3184–3194.
7. Kumar, T. P., Rajendra, U., Varaprasad, G., Suryaprakash, G., Sadanand, T., Awaar, V. K., & Gairola, S. P. (2023). EV BMS With Temperature and Fire Protection. E3S Web of Conferences, 430, 112–125.

Intelligent and Sustainable Power and Energy Systems – Dr. M. Premkumar et al. (eds)
© 2026 Taylor & Francis Group, London, ISBN 978-1-041-10314-1

20

Real-Time Monitoring and Alert System for Earthing Integrity in Electrical Installations

Johny Renoald A.[1],
Thirulakshmipathi T.[2], Tamilmani S.[3],
Muthu Esakki M.[4], Dhinesh A.[5]
Department of Electrical and Electronics Engineering,
Erode Sengunthar Engineering College (Autonomous),
Erode, Tamilnadu, India

ABSTRACT: This paper presents a novel IoT-enabled real-time monitoring and alert system explicitly designed to maintain the earthing integrity of streetlight installations. This system significantly enhances safety measures within municipal street lighting networks by proactively addressing potential electrical hazards. The proposed system employs an ACS712 current sensor to monitor current flow along the earthing line continuously. Upon detecting an anomaly, such as an increase in leakage current indicative of an earth fault, the system's ESP32 microcontroller initiates a response by actuating a relay. This relay isolates the faulty streetlight, redirecting its connection to an alternative earthing line, effectively protecting the network from further risk and isolating the compromised unit. To achieve precise fault identification, the system incorporates an 8-bit ADC to convert the analog output from the ACS712 sensor into digital data. Each streetlight is assigned a unique identifier, enabling the ESP32 microcontroller to accurately locate and address specific faults in real time. Additionally, integrating a GPS module enhances the system's fault localisation capabilities by providing precise geographic coordinates of the affected streetlight. This geolocation data, combined with the unique streetlight identifier, is transmitted to a cloud platform via the ESP32's Wi-Fi module, allowing for remote monitoring and data logging. Once a fault is detected and isolated, the cloud platform automatically sends an SMS alert to maintenance personnel containing details about the specific fault location and streetlight identifier. This immediate notification system ensures rapid response to potential electrical hazards, minimising downtime and risk to public safety. The remote monitoring capabilities allow for efficient data analysis and preventative maintenance scheduling, contributing to long-term infrastructure reliability.

KEYWORDS: ACS712 current sensor, ESP32 microcontroller, GPS tracking for streetlights, Relay-based fault isolation

[1]jorenoeee@esec.ac.in, [2]thirulakshmipathi03@gmail.com, [3]thirulakshmipathi03@gmail.com, [4]muthuudr95@gmail.com, [5]dhineshalan086@gmail.com

DOI: 10.1201/9781003654469-20

1. Introduction

Street lighting is a critical component of urban infrastructure, enhancing visibility, deterring criminal activities, and ensuring the safety of pedestrians and drivers [1–4]. However, traditional street lighting systems often operate continuously, regardless of traffic conditions or environmental factors, leading to significant energy wastage and inefficiencies [5]. Maintaining and detecting faults in such systems, particularly in large urban areas, also presents a considerable challenge. Faults in streetlight earthing systems, often caused by wear, tear, or improper installation, can pose severe electrical hazards to humans and animals [6–10].

To address these challenges, this project proposes a real-time monitoring and alert system for maintaining earthing integrity in streetlight installations. By leveraging IoT technology, the system aims to mitigate electrical hazard risks, reduce energy consumption, and minimise maintenance costs through continuous monitoring and prompt fault detection [11–13]. The system architecture integrates several components to achieve its objectives. At its core, an ACS712 current sensor continuously monitors the earthing line current, detecting potential leakages that could indicate electrical issues [14–16]. The ESP32 microcontroller processes the sensor data, evaluates current levels, and triggers appropriate anomaly responses. When a fault is detected, the ESP32 activates a relay to isolate the malfunctioning streetlight by disconnecting it and redirecting it to an alternate earthing line. The system offers additional functionality to enhance its efficiency. Faulty streetlights are precisely identified, allowing maintenance personnel to address issues promptly. An 8-bit ADC digitises the sensor data for accurate monitoring, and the processed data is transmitted to a cloud-based platform via Wi-Fi, enabling real-time remote monitoring [17–19]. Fault detection automatically generates SMS alerts with specific streetlight IDs, significantly reducing response times. Furthermore, an integrated GPS module provides precise location data, streamlining fault localisation and ensuring swift repairs. This system improves public safety and optimises energy use by targeting and isolating only the faulty streetlights. Such an approach eliminates the need for large-scale repairs while ensuring efficient use of existing infrastructure [2, 3, 12, 14, 15].

2. Related Work

The increasing demand for sustainable and energy-efficient urban infrastructure has led to a significant focus on developing innovative street lighting systems in recent years. Numerous systems incorporate Internet of Things (IoT) technologies to enhance streetlight monitoring and control, lowering maintenance costs and energy consumption. [14–16] offer a noteworthy method for bright street lighting. An Internet of Things-based system monitors and regulates lamps according to environmental variables and traffic situations. Similar to the goals of the current effort, optimising lighting settings based on real-time requirements through wireless sensors is a big step toward attaining energy efficiency and strongly emphasises combining wireless connectivity and environmental sensors to control streetlight brightness dynamically. Building on this idea, our work focuses on detecting earthing faults and uses an ACS712 current sensor to track leakage currents in real-time. It is a more complete approach to urban safety since this extra layer of monitoring guarantees that issues are found before they become dangerous. [9] discussed an automated system with defect detection and automated system warnings that uses IoT and GSM for smart street lighting [10–13]. In their analysis of different innovative street lighting options, [2, 3, 12] emphasise the incorporation of IoT, energy-saving algorithms, and remote monitoring. The current effort considers safety by identifying earth faults in streetlight installations, even if their primary focus is operational optimisation and energy savings. Much study has been done on fault detection in power systems, especially about earthing and leakage. For instance, [14] emphasises the significance of fault identification for system dependability by proposing a novel protection mechanism for single-phase earth faults in distribution systems. Similar to our system's focus on locating and isolating faults at the streetlight level, [20] investigate fault feeder identification in networks with secondary earth faults. According to [12], our system's real-time earth fault detection capability significantly improves over conventional fault detection systems.

3. Methodology

The Real-Time Monitoring and Alert System for Earthing Integrity in Electrical Installations aim is to control and monitor the earthing line in streetlight circuits to provide safety while isolating a fault condition. It constantly observes, isolates, and alerts the maintenance staff as required. It has a current sensor, ACS712, to monitor the current flowing through the earthing line. The current sensor output has to be monitored by the ESP32 microcontroller, which is programmed for a continuous operation to measure the current. Suppose it is beyond a set limit, considering it is the earth's fault. In that case, it has to isolate the fault streetlight by opening a relay and isolating the earthing line for secured operation. ESP32 procures the name of the Streetlight from the ADC once the fault is identified. The latitude and longitude coordinates of the fault location are recorded by the GPS module even before the fault. Both these details are sent to a cloud platform for remote monitoring with an SMS alert to the charge facilitating the exact fault position, and the details of the streetlight failed to enable real-time data. The street light operation can be restored manually within seconds once the fault is found at the site, as the relay contact should be closed after fault clearance. Figure 20.1 shows the block diagram of the streetlight earthing integrity monitoring system, comprising the important components of electrical safety, whereby fault current is routed directly to the earth. The ACS712 current sensor would watch the current continuously flowing in the earthing line. Upon detecting any abnormality in the current (earth leakage), the relay will be activated to switch off and isolate the faulty streetlight, thus preventing further hazards.

The activation of the relay is detected by the ADC, converting this analog signal into high-resolution digital data. Each streetlight has a unique binary pattern, aiding in accurate identification in the event of a fault. Data from the ADC is routed to the ESP32 microcontroller, which further communicates with the Cloud Platform to allow remote monitoring by maintenance personnel, as depicted in Fig. 20.1. Local real-time alerts, made possible through guidance from the display unit, facilitate prompt identification of faulty streetlights. Upon detecting abnormal currents, the ACS712 directs the ESP32 to retrieve latitude and longitude data from the GPS module and relates this information to the streetlight under fault. The maintenance team sends this fault and location data to the cloud for remote access. An SMS alert, relaying the streetlight's ID and precise GPS location, is sent to those responsible, ensuring immediate action is taken, as depicted in Fig. 20.5, even if active monitoring of the Cloud Platform does not occur. The overall system, shown in Fig. 20.1, adds safety and reliability improvements for streetlight installations through rapid fault detection and precise isolation, combined with effective remote monitoring.

4. ESP32 Master Control Workflow

As can be seen from Fig. 20.2, the ESP32 microcontroller is a pivotal block for the maintenance of earthing integrity within street lights and fault isolation. The ESP32 runs a continuous loop,

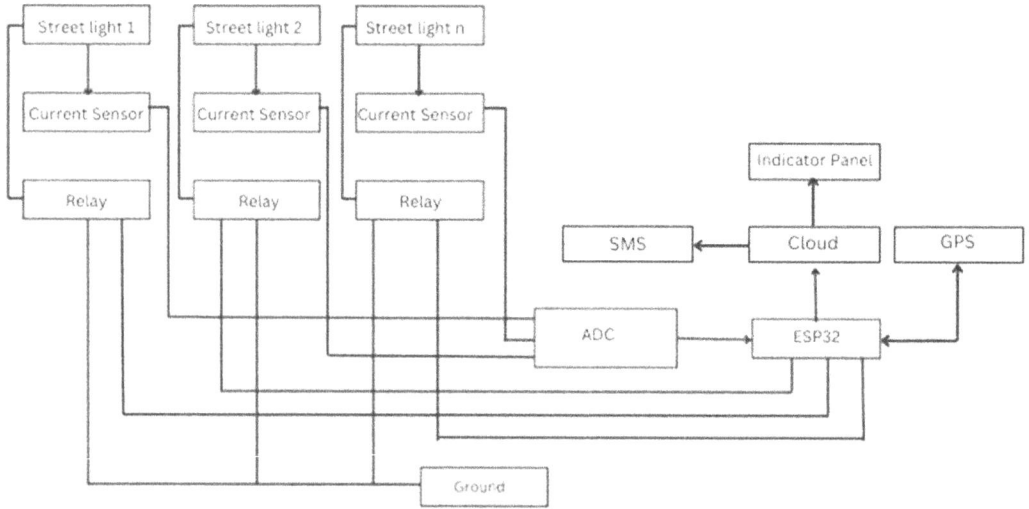

Fig. 20.1 Block diagram of proposed system

where the output of the ACS712 is monitored continuously for current. The ESP32 will continue monitoring until it detects a potential earth fault condition, at which point a reading greater than zero is determined. If the current is greater than 0, the ESP32 breaks the loop. Afterwards, it checks the ADC channel through which the current sensor feeds signals. This channel corresponds to a specific Streetlight Identification Number informing the ESP32 of that particular streetlight with the fault. This is indicated in Fig. 20.5 of the system overview. In this manner, the ESP32 is made to trigger a relay that is associated with the identified streetlight. The relay then switches the earthing line of the defective streetlight to a diversion line, thereby successfully isolating the faulty streetlight into circuit separation to prevent further hazards, as shown in Fig. 20.2. Its possible causes can be isolated by locating the fault, and then the GPS module is accessed by the ESP32 to acquire the latitude and longitude of the defective streetlight. With this GPS data and the identification number, the precise location details of the affected unit are known. Cloud Transmission and Alert: The ESP32 sends the streetlight identification number and GPS coordinates to the cloud platform via Wi-Fi. From the cloud, an SMS alert containing these details is dispatched to the maintenance personnel, ensuring immediate action, as demonstrated in Fig. 20.2.

This mode of operation integrates the processes into an automated, real-time system that isolates streetlight faults and alerts operators with accurate location-specific information. Underscored in this workflow, as represented in Figs. 20.2 and 20.4, is a demonstration of the system's effectiveness in enabling the remote management of streetlight safety and reliability.

5. Hardware Model

Testing concluded successfully with these key findings, and a section of an artificial earthing fault circuit was incorporated into the hardware prototype, as the use of 220V bulbs prevented direct integration of an actual earth line. This separate circuit created an artificial earth fault condition. As shown in Fig. 20.3, the fault current was detected by the ACS712 current sensor through this simulated earthing line. Upon detecting the fault, the ESP32 microcontroller activated a relay,

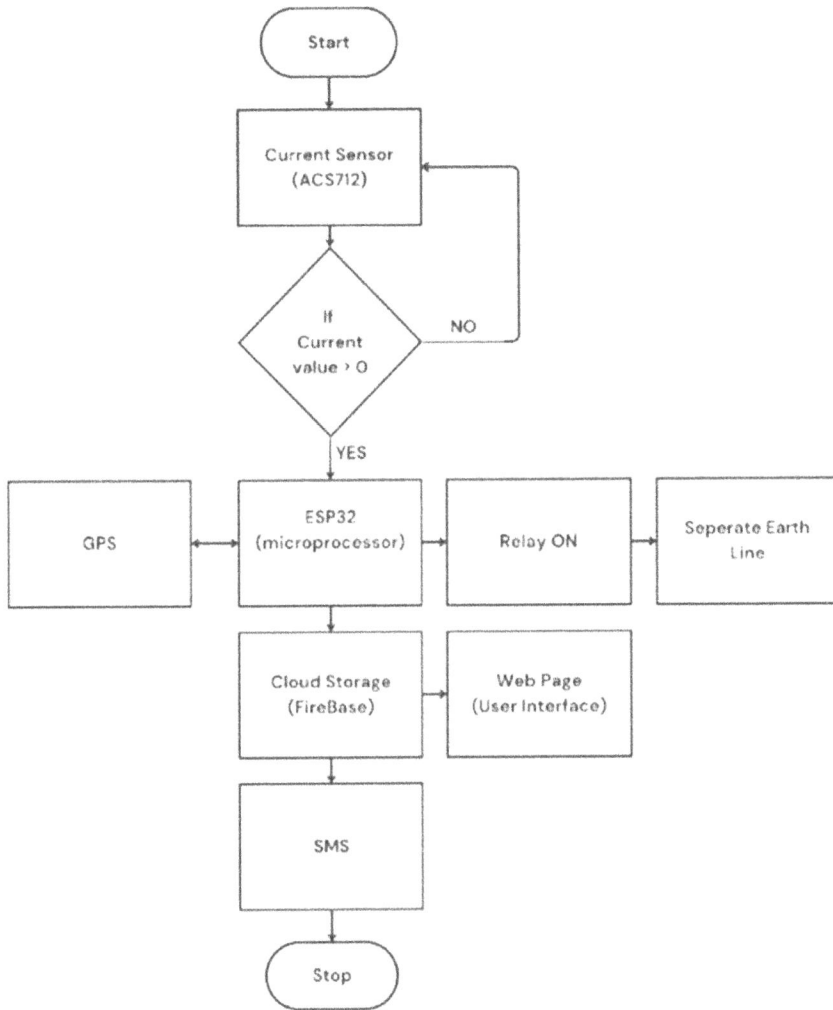

Fig. 20.2 Operations at the controller device

isolating the faulty streetlight's earthing line from the primary circuit, preventing further hazards. This prototype did not include an ADC, as only four bulbs were connected to the system. The ACS712 sensor identified fault occurrence based on changes in current. Upon fault detection, the ESP32 transmitted the error to a cloud platform for remote monitoring. An SMS alert, illustrated in Fig. 20.4, was sent to notify relevant personnel of the faulty streetlight (bulb) for immediate action, like "Lamp 2 Under Earth Fault."

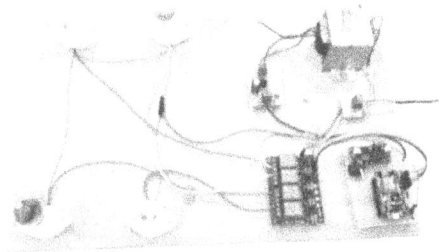

Fig. 20.3 Hardware model

6. Result Analysis

In this simulation, the lights act as typical streetlights, pointing to an actual scenario where the earth-leakage detection and isolation system would function. The limited-scale nature of this setup leads to the omission of many components pertinent to the realisation of a fully real-time system. Usually, the real-time system would use an ADC to monitor several streetlights. At the same time, in the simulation, only three LEDs act as mock representations of streetlights, making the testing more manageable and computational performance quicker. Each LED is a proxy for a streetlight, and this setup is indeed a launch pad for theoretical validation of the fault detection process, as illustrated in Fig. 20.4. A battery and switch configuration, shown in Fig. 20.4, provides an alternative current path to create an earth leakage condition. When the switch acts to bring about closure, an artificial earth fault condition is simulated, thus permitting the current sensor, ACS712, to detect the increase in current that is created, mimicking the actual earth leakage behaviour. The rise in current detected in this setting is shown in Fig. 20.4.

Fig. 20.4   Hardware result

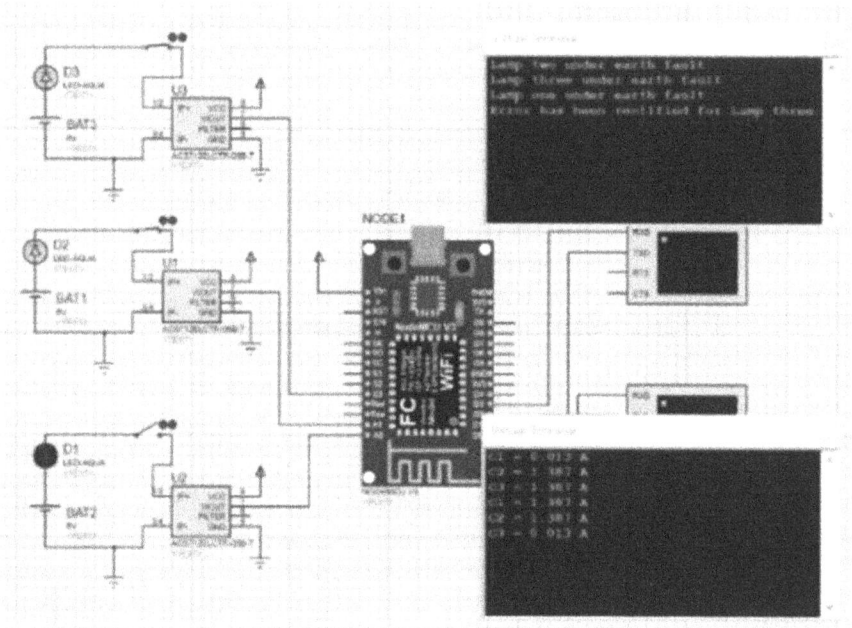

Fig. 20.5   Simulation of proposed work

The ACS712 current sensor monitors the current flowing to the LEDs under regular operation (switch off) and during faults (switch on). For instance, the usual stable current is approximately

0.013A during steady-state applications. Once the switch introduces the earth fault, the current peaks at approximately 1.387A, so the ESP32 can sense and react to the fault condition. Figure 20.3 demonstrates this setup and current sensing arrangement. However, difficulties arise in the development stage, such as managing several high-power street lamps. Real-time implementations of the system require an ADC extension to determine the accurate streetlight failure at detection time and establish each streetlight individual for the ADC as output. The output produced by the ADC in Fig. 20.4 is connected to each streetlight circuit, allowing the ESP32 to locate and act on the source of the fault accurately. The ACS712 commands the ESP32 to activate a relay to switch the faulty streetlight to another earth line for isolation. In practice, the power consumption was way too low due to LEDs. Therefore, the thresholds were adjusted according to the more realistic current levels found at the time of actual implementation, as per Fig. 20.3, where simulated current thresholds provide a workable approximation for testing purposes.

7. Conclusion

The Real-Time Monitoring and Alert System for Earthing Integrity in Electrical Installations demonstrates the importance of safety and proactive maintenance in streetlight installations. By integrating the ACS712 current sensor, ESP32 microcontroller, GPS module, and cloud-based monitoring, this system detects earth leakage, isolates faulty streetlights, and provides real-time SMS alerts to maintenance personnel, enabling immediate response to potential hazards. The system identifies faults through current variations detected by the ACS712 sensor, triggering the ESP32 to actuate a relay and disconnecting the faulty streetlight from the primary circuit. The SMS alerts also incorporate GPS location data, enhancing the system's utility by precisely informing maintenance staff of the fault location for efficient repair coordination (see Fig. 20.4 for an SMS alert sample).

REFERENCES

1. Krishnamoorthy, M. and Albert, J.R. (2024). Electricity theft detection in IoT-based smart grids using a parameter-tuned bidirectional LSTM with pre-trained feature learning mechanism. Springer publication. 106:5987–6001.
2. Albert Johny Renoald (2022). Design and Investigation of Solar PV Fed Single-Source Voltage-Lift Multilevel Inverter Using Intelligent Controllers. Journal of Control, Automation and Electrical Systems. Springer publication. 33(1).
3. Albert, J.R. and Stonier, A.A. (2020). Design and development of symmetrical super-lift DC–AC converter using firefly algorithm for solar-photovoltaic applications. IET Circuits Devices Syst., IET Publication. 14: 261–269.
4. Johny Renoald Albert, Aditi Sharma, B. Rajani, Ashish Mishra and Ankur Saxena, C. Nandagopal, and Shivlal Mewada. (2022). Investigation on load harmonic reduction through solar-power utilization in intermittent SSFI using particle swarm, genetic, and modified firefly optimization algorithms. Journal of Intelligent and Fuzzy Systems, IOS Press publication. 42(4):4117–4133.
5. Vanchinathan, K., Valluvan, K.R., Gnanavel, C., Gokul, C. and Albert, J.R. (2021). An improved incipient whale optimization algorithm based robust fault detection and diagnosis for sensorless brushless DC motor drive under external disturbances. Int. Trans. Electr. Energ. Syst., Wiley publication. 31(12):e13251.
6. Ramaraju, S.K., Kaliannan, T., Androse Joseph, Sheela, Kumaravel, Umadevi D., and Albert, J.R. (2022). Design and Experimental Investigation on VL-MLI Intended for Half Height (H-H) method to improve power quality using Modified PSO algorithm. Journal of Intelligent and Fuzzy Systems.42(6):5939–5956.
7. G. C., M.P., V.K., and J.R.A. (2021). Experimental Validation and Integration of Solar PV Fed Modular Multilevel Inverter (MMI) and Flywheel Storage System. IEEE Mysore Sub Section International Conference (MysuruCon). pp. 147–153.

8. Thangamuthu, L., Albert, J.R., Chinnanan, K., and Banu Gnanavel. (2021). Design and Development of Extract Maximum Power from Single-Double Diode PV Model for Different Environmental Conditions Using BAT Optimization Algorithm. Journal of Intelligent and Fuzzy Systems. 1–12.

9. Albert, J.R., (2022). Design and Investigation of Solar PV Fed Single-Source Voltage-Lift Multilevel Inverter Using Intelligent Controllers. Journal of Control, Automation and Electrical Systems. Springer publication. 33(1).

10. Palanisamy, R., Govindaraj, V., Siddhan, S., Saravanan, A. And Albert, J.R.(2022). Experimental Investigation and Comparative Harmonic Optimization of AMLI Incorporate Modified Genetic Algorithm Using for Power Quality Improvement. Journal of Intelligent and Fuzzy Systems.1–14.

11. Johny Renoald, Stonier Albert Alexander and Vanchinathan Kumarasamy (2022). Testing and Performance Evaluation of Water Pump Irrigation System using Voltage-Lift Multilevel Inverter. International Journal of Ambient Energy. 1–14.

12. Albert, J.R., Selvan, P., Sivakumar, P. and Rajalakshmi, R. (2022). An Advanced Electrical Vehicle Charging Station Using Adaptive Hybrid Particle Swarm Optimization Intended for Renewable Energy System for Simultaneous Distributions. Journal of Intelligent and Fuzzy Systems. IOS Press. 4395–4407.

13. Johny Renoald A., Kannan R., Karthick S., Selvan P., Sivakumar A. and Gnanavel C. (2022). An Experimental and Investigation on Asymmetric Modular Multilevel Inverter an Approach with Reduced Number of Semiconductor Devices. J. Electrical Systems. 18(3):318–330.

14. Babypriya, B. and Johny Renoald A. (2022). An Experimental Simulation Testing of Single-Diode PV Integrated MPPT Grid-Tied Optimized Control Using Grey Wolf Algorithm. Journal of Intelligent and Fuzzy Systems. IOS Press. 43(5):5877–5896.

15. Albert, J.R., Ramasamy, K., Joseph Michael Jerard, V., Rajani Boddepalli, G. S. and Anbarasu Loganathan. (2023). A Symmetric Solar Photovoltaic Inverter to Improve Power Quality Using Digital Pulse Width Modulation Approach. Wireless Pers. Commun. pp.1–14.

16. Gnanavel C., Muruganatham P., Vanchinathan K., and J.R. Albert.(2021). Experimental Validation and Integration of Solar PV Fed Modular Multilevel Inverter (MMI) and Flywheel Storage System. IEEE Mysore Sub Section International Conference. 147–153.

17. C. Gnanavel, A. Johny Renoald, S. Saravanan, K. and Vanchinathan. (2022). An Experimental Investigation of Fuzzy-Based Voltage-Lift Multilevel Inverter Using Solar Photovoltaic Application. Smart Grids and Green Energy Systems, Wiley publication. 59–74.

18. Albert, J.R., Premkumar, K., Vanchinathan, K., Nazar Ali, A., Sagayaraj, R. and Saravanan, T.S. (2022). Investigation of Super-Lift Multilevel Inverter Using Water Pump Irrigation System. Smart Grids and Green Energy Systems, Wiley publication, pp. 247–262.

19. Karthikeyan S. and Albert J.R. (2023). Optimizing Energy Utilization in the Weaving Industry: Advanced Electro-Kinetic Solutions with Modified Piezo Matrix and Super Lift Luo Converter. Electric Power Components and Systems. Taylor & Francis. pp. 1–14.

20. Dineshkumar, R., Anna Alphy, Kalaivanan, C., Bashkaran, K., Pattanaik, Balachandra, Logeswaran, T., Saranya, K., Deivasikamani, Ganeshkumar, and Johny Renoald, A. (2023). A Novel Hyperparameter Tuned Deep Learning Model for Power Quality Disturbance Prediction in Microgrids with Attention-Based Feature Learning Mechanism. Journal of Intelligent and Fuzzy Systems. 46(1):1–17.

Note: All the figures in this chapter were made by the author.

Intelligent and Sustainable Power and Energy Systems – Dr. M. Premkumar et al. (eds)
© 2026 Taylor & Francis Group, London, ISBN 978-1-041-10314-1

21

Photovoltaic Based Charging of 5V Smart Gadgets in Indian Household

Suraj S.[1]
Research Scholar,
Dept of EEE, BITS Pilani, K.K Birla Goa Campus

Narayan S. Manjarekar[2],
Sudarshan Swain[3]
Assistant Professor,
Dept of EEE, BITS Pilani, K.K Birla Goa Campus

ABSTRACT: The penetration of smart gadgets is increased in every one's day today life along with the advancement of technology and reduced cost. The commonly used smart gadgets in all households are smart phone, smart watch, wireless earphones, and portable speakers. These gadgets are recharged daily and contributes to the electricity bill paid to the grid operator. Moreover, a portable charging solution of these gadgets is also necessary. The energy consumption of these smart gadgets in Indian Household per day is analysed and the specification solar panel is derived. This paper proposes a solar based charging of 5V Smart gadgets with portable power bank as the battery storage providing portability feature too. The energy consumed by these gadgets is found out to be 100 Wh and solar panel of 30W and power bank with capacity of 20,000mAh is utilized. The LM2576 based simple buck converter is used to interface the solar panel and power bank and the smart gadgets. The proposed methodology is validated with the hardware results.

KEYWORDS: Smart gadget, Power bank, Solar photovoltaic, Buck converter

1. Introduction

The term Smart Gadget is used for a small hand-held or wearable battery powered portable electronic device which can be used for communication, entertainment, medical care, fitness and professional

[1]p20190057@goa.bits-pilani.ac.in, [2]narayan@goa.bits-pilani.ac.in, [3]sudarshans@goa.bits-pilani.ac.in

DOI: 10.1201/9781003654469-21

work in day-to-day life [1],[2],[3]. Smart gadgets use IoT, Bluetooth and wireless connectivity along with various sensors in order to improve the user's need of the hour. With technology evolving day by day smart gadgets will be part of smart homes, cities and lifestyle as well along with security and sustainability [4-5]. With the aid of battery, VLSI and embedded technologies these smart gadgets are available for affordable prices and has penetrated into Indian population at ease.

The rechargeable smart gadgets come with various battery charge capacities in the range of 100mAh to 10,000mAh. The charging of these smart gadgets is often done with a 5V charger with varying power output of 5W to 50W. The power of the charger indicates how fast the gadget can be charged [6-7]. The advancements in the battery technology of lithium ion and solod state batteries have helped the fast charging. These smart gadgets are often charged once per day. The smart gadgets even come with wireless charging option along with conventional USB based wired charging [8-9]. The size of the chargers has also reduced due to the advancement in power electronic converters.

With the push towards sustainability some gadgets these days does not come up with chargers. The power for charging is also shifting towards the green energy especially the solar [10-11]. The portable chargers, power banks powered by Solar Photovoltaic(SPV) are gaining a lot of interest by the researchers as well as the consumer. These portable chargers are handful during travelling and rural areas where it is difficult to find grid electricity and costlier.

2. Analysis of Power Consumption of 5V Smart Gadgets per Indian Household

The penetration of 5V Smart Gadgets of Table 21.1 in Indian Households is analysed and its power consumption is studied for implementation of Photovoltaic based Charging.

Table 21.1 Specification of smart gadgets, daily energy required per gadget and for Indian population

5V Smart Gadget	Battery Capacity (mAh)	Penetration Level (%)	E_d(Wh/day)	E_{dp}(MWh/day)
Smart Phone	4000	70%	20	21,777.77
Portable Bluetooth Speaker	1200	40%	6	3,733.33
Wireless Bluetooth Earphone	500	20%	2.5	777.77
Smart Watch	100	35%	0.5	450

Assuming every gadget is charged once per day, daily energy required per gadget, E_d is given as

$$E_d = V \times Q \tag{1}$$

Equation 1. Daily energy required per gadget formula

where V and Q are the gadget's battery voltage and capacity respectively.

Considering the smart gadgets penetration for 1.4 billion Indian Population P and charger efficiency, η_c of 90%, the daily consumption for total population per gadget, E_{dp} is given as

$$E_{dp} = 1.4 \ billiion \times P \times \frac{E_d}{\eta_c} \tag{2}$$

Equation 2. Daily Consumption for Total Indian Population per Gadget formula.

Thus E_d and E_{dp} for various gadgets considered in this framework is calculated and is depicted in Table 21.1. The combined daily energy consumption of all Smart Gadgets per household, CE_{dp} is calculated as

$$CE_{dp} = (21,777.77 + 3,733.77 + 733.77 + 450) = 26,739.31 \text{ MWh per day}$$

Considering four people per household H_P, the total number of households, N_H is calculated as

$$N_H = 1.4 \, billiion \times \frac{1}{H_P} = 350 \, Million \tag{3}$$

Equation 3. Number of Household formula

The Average Daily Power Consumption per Household by Smart Gadgets, AE_{dH} is given by

$$AE_{dH} = \frac{CE_{dp}}{N_H} = 76.4 \text{ Wh/day/Household} \tag{4}$$

Equation 4. Average Daily Power Consumption per household by Smart Gadgets

3. Literature Review

The various related work regarding the smart gadgets and their charging using solar photovoltaic are summarized in Table 21.2. The contribution and gaps are also identified and depicted in Table 21.2. charging of smart gadgets is also mentioned. The objective of this literature review is to understand the enhancements in the smart gadgets and its charging technology along with utilization of solar energy for charging for promoting the renewable energy-based charging.

4. Proposed Work

Figure 21.1 shows the proposed work. The block diagram of the proposed work, shown in Fig. 21.1(a) illustrates the utilization of solar photovoltaic power by charging the power bank and the smart gadget connected to the power bank. The rechargeable smart gadget can be a device with anything of a 5V rating and maximum charging current rating of 3A.

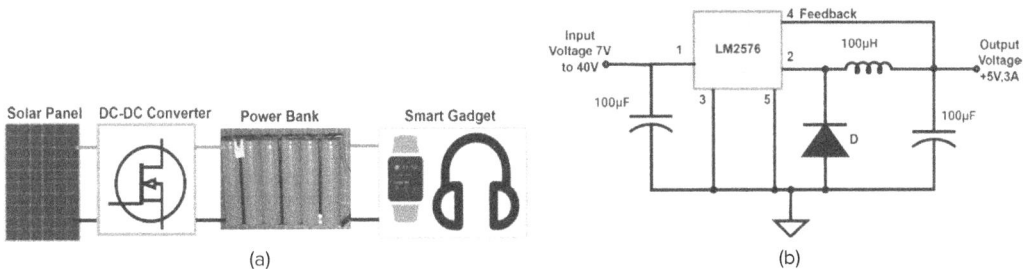

Fig. 21.1 Proposed work (a) Block diagram (b) Circuit diagram of LM2576 based buck converter

Circuit Description and Operation of LM2576 based 5V, 3A Buck Converter

The circuit diagram of LM2576[12] based 5V, 3A output buck converter for a varying input voltage of 7V to 40V is shown in the Fig. 21.1(b). In this proposed work the input to the buck converter is from the solar photovoltaic panel. LM 2576 is a 5 pin IC with DIP hole through package.

Table 21.2 Related work with contribution and gaps identified

Ref. No	Work	Contribution	Gaps Identified
[13]	Arduino Based Platform for Managing a PV Battery Charge	Hardware experiment of BMS for battery storage station.	The design and application were not discussed
[14]	Mini Solar Home System for Electricity Supply in Coastal Rural Area, Case Study: Yogyakarta, Indonesia	A case study on solar power system for Rural homes was discussed	No experimental validation provided
[15]	Design Low-Cost Battery Management System for Low Power Applications of Photovoltaic Systems	Design of BMS for low power and low-cost PV system was discussed	The PV system was not designed for experimental validation
[16]	Solar photovoltaic charging of lithium-ion batteries	Li-Ion batteries charging with solar power was introduced	Only Simulation based results were discussed
[17]	WPT: A Smart Magnetic Resonance Technology based Wireless Power Transfer System Design for Charging Mobile Phones	Wireless charging for Mobile Phones with magnetic resonance technology was demonstrated	The design of WPT was not discussed
[18]	Automatic Gadget Charger using Coin Detection	A public coin payment based Solar mobile charging is discussed	Costly and complex
[19]	Simulation of Solar Bag Using DC-DC Boost Converter	solar photovoltaic placed backpack to charge individuals' electronic gadgets was introduced.	Only MATLAB Simulink based results were discussed
[20]	Electrouter-An Automated Wireless Charging Gadget Zone	Wireless charging for the gadgets with IoT Interface was introduced.	No provision to charge Gadgets without wireless charging option
[21]	The Design of a Control System for an Automated photovoltaic Power Pack	Photovoltaic powered Power Pack was designed with sun tracking	Costlier and the photovoltaic charging circuit for power bank is not explained.

Design consideration of LM2576 based 5V, 3A Buck Converter

The buck converter operated in continuous conduction mode is designed for the following specifications shown in Table 21.3.

Table 21.3 Specification of LM2576 based buck converter

Parameter	Specification
Minimum Input voltage(V_{in} (min))	7V
Maximum Input voltage (V_{in} (max))	40V
Output voltage (V_o)	5V
Output current (I_0)	3A
Switching Frequency (F_{SW})	52KHz

The minimum duty cycle needed to get V_o is for $V_{in(max)}$ and is given as,

$$D_{min} = \frac{V_0}{V_{in(max)}} = \frac{1}{8}$$

(5)

Equation 5. Minimum Duty Ratio of LM2576 Buck Converter

where D is switching duty cycle and thus for varying input.

The inductor current is given by, $I_L = I_0 = 3\ A$

Considering 40% peak to peak ripple, $\Delta I_L = 1.2\ A$

The maximum inductor value L_{max} is obtained for $V_{in(max)}$ and D_{min} is given by

$$L_{max} = \left(\frac{(V_{in(max)} - V_0)D_{min}}{\Delta I_L * F_{SW}} \right) = 70..1\ \mu H$$

(6)

Equation 6. Inductance of LM2576 Buck converter

For 2% ripple voltage, the capacitance C value is given by,

$$C = \left(\frac{\Delta I_L}{8 * F_{SW} * \Delta V_0} \right) = 100\ \mu F$$

(7)

Equation 7. Capacitance of LM 2576 Buck converter

$1.25 \times L_{max}$ & C is taken for ensuring that the converter is working in Continuous Mode

Specification of Solar Panel Needed

Considering loses and 80% efficiency of solar photovoltaic system, the energy required to be produced by solar, E_S

$$E_S = \left(\frac{AE_{dH}}{0.8} \right) = 96\ Wh$$

(8)

Equation 8. Energy to be produced from Solar

With an average solar hours of 4 hours per day in India, the solar panel power needed, P_S

$$P_S = \left(\frac{E_S}{4} \right) = 24\ W$$

(9)

Equation 9. Solar Power Required

Considering cloudy days and efficiency losses a 30W solar panel is chosen to meet 96 Wh energy.

Specification of Power Bank Needed

The capacity of 5V power bank required for the daily load of 96 Wh, Q_P

$$Q_P = \left(\frac{E_S 1000}{5V} \right) = 19,200\ mAh$$

(10)

Equation 10. Capacity of Power Bank

Thus, a 5V,20,000mAh power bank is chosen in this proposed work

5. Hardware Prototype and Results

The complete hardware set, along with the solar output voltage and current readings for implementing the proposed work, is shown in Fig. 21.2. The hardware implementation is shown in Fig. 21.2(a). The measured open circuit voltage is 21.4V, while the short circuit current was found to be 1.16A for the same irradiance, as shown in Fig. 21.2(b) and Fig. 21.2(c), respectivelyhas a maximum power of 30W with an open circuit voltage of 21.78V and a short circuit current of 1.82A.

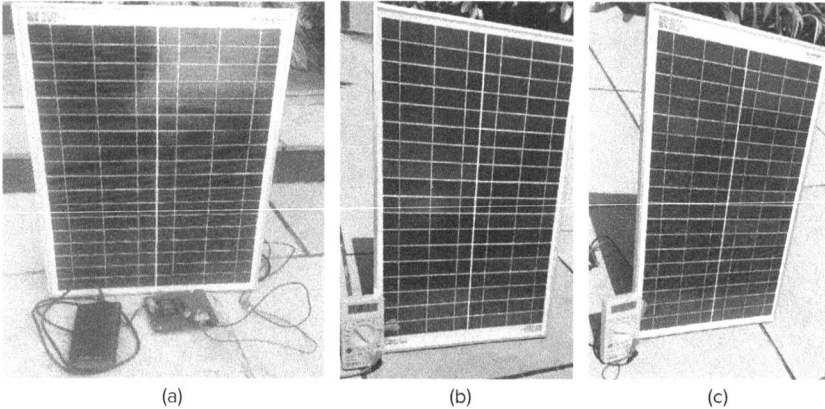

| (a) | (b) | (c) |

Fig. 21.2   Hardware results (a) Completed hardware setup (b) SPV open circuit voltage (c) SPV short circuit current

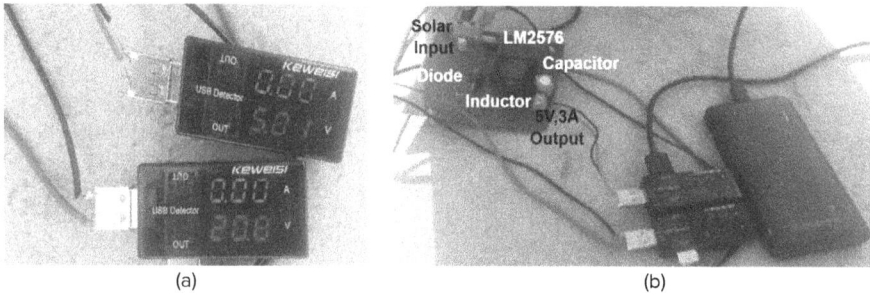

| (a) | (b) |

Fig. 21.3   Input output voltage and current (a) Without load (b) With load

The hardware setup of LM2576 of Fig. 21.1(b) is depicted in Fig. 21.3. The input-output parameters measured of LM2576 is depicted in Fig. 21.3 (a). The input voltage from the solar is measured to be 20.8Vand the corresponding output voltage from LM 2576 obtained is 5V. The output current obtained is 0A as the load power bank for the buck converter is not connected and thereby forcing the input current extracted from the solar panel to be 0A as well. The input-output voltage and current measurement from the USB multimeter when the Power bank is connected for charging is shown in Fig. 21.3(b). The input and output voltages have dropped down to 4.92V and 19.8V respectively. The output current absorbed by the load, i.e., the charging current of the power bank, is 0.54A, and as the converter is of buck topology, the input current is half of the output for a 50% duty cycle. The input current extracted from solar is thus 0.29A.

6. Conclusion

Smart Gadgets plays important part in the day-to-day life of all the people in the world for communication, entertainment and health monitoring. These rechargeable smart gadgets come with batteries of limited capacity and thereby there is need for charging these gadgets time to time. Though, the energy consumed by these smart gadgets per day is small compared to other household equipment's there is need for a portable charging as well as reducing the stress on the nonrenewable fuel-based energy production.

In this paper, the consumption of energy by smart gadgets is analyzed and a solar photovoltaic based charging system is designed. The existing work was studied to identify the gaps. The buck converter for interfacing the solar photovoltaic with the power bank and the smart gadget was designed and the effectiveness of the proposed work was validated with hardware results. The hardware prototype of 30W solar panel and 5V-20,000mAh power bank was shown.

REFERENCES

1. Fullerton, B. and Schmidt, A. (2005), Gadget Review: A Look at Devices for Work, Home and Play. *Library Hi Tech News*, Vol. 22 No. 4, pp. 13–17. https://doi.org/10.1108/07419050510604639

2. Singu, S. (2020). Socially distant smart gadget for COVID-19. *Transactions on Latest Trends in Health Sector, 12*(12). Retrieved from https://www.ijsdcs.com/index.php/TLHS/article/view/194

3. Gljušćić, P., Zelenika, S., Blažević, D., &Kamenar, E. (2019). Kinetic Energy Harvesting for Wearable Medical Sensors. *Sensors, 19*(22), 4922. https://doi.org/10.3390/s19224922

4. Xu, C., Song, Y., Han, M., & Zhang, H. (2021). Portable and wearable self-powered systems based on emerging energy harvesting technology. Microsystems & Nanoengineering, 7(1), 25. https://doi.org/10.1038/s41378-021-00248-z

5. Rong, G., Zheng, Y., &Sawan, M. (2021). Energy Solutions for Wearable Sensors: A Review. *Sensors, 21*(11), 3806. https://doi.org/10.3390/s21113806

6. Manoharan, S., Mahalakshmi, B., Ananthi, K., &Agalya, A. (2024, January 18). A comprehensive study on fast charging in smart phones. 2024 5th International Conference on Mobile Computing and Sustainable Informatics (ICMCSI), 530–535. Presented at the 2024 5th International Conference on Mobile Computing and Sustainable Informatics (ICMCSI), Lalitpur, Nepal. https://doi.org/10.1109/icmcsi61536.2024.00083

7. Liu, X., Ansari, N., Sha, Q., & Jia, Y. (2022). Efficient green energy far-field wireless charging for internet of things. IEEE Internet of Things Journal, 9(22), 23047–23057.https://doi.org/ 10.1109/jiot.2022.3185127

8. Kok, C. L., Fu, X., Koh, Y. Y., & Teo, T. H. (2024). A Novel Portable Solar Powered Wireless Charging Device. *Electronics, 13*(2), 403. https://doi.org/10.3390/electronics13020403

9. Akbari, H., Browne, M. C., Ortega, A., Huang, M. J., Hewitt, N. J., Norton, B., & McCormack, S. J. (2019). Efficient energy storage technologies for photovoltaic systems. Solar Energy (Phoenix, Ariz.), 192, 144–168. https://doi.org/10.1016/j.solener.2018.03.052

10. Lee, H.-S., & Yun, J.-J. (2019). High-efficiency bidirectional buck–boost converter for photovoltaic and energy storage systems in a smart grid. IEEE Transactions on Power Electronics, 34(5), 4316–4328. https://doi.org/10.1109/tpel.2018.2860059

11. Yu, X., Fan, J., Wu, Z., Hong, H., Xie, H., Dong, L., & Li, Y. (2024). Simulation and Optimization of a Hybrid Photovoltaic/Li-Ion Battery System. *Batteries, 10*(11), 393. https://doi.org/10.3390/batteries10110393

12. LM2576xx Series SIMPLE SWITCHER® Power Converter 3-A Step-Down Voltage Regulator, Ti, available: online from https://www.ti.com/lit/ds/symlink/lm2576.pdf

13. Hammoumi, K. E., Bachtiri, R. E., Boussetta, M., &Khanfara, M. (2019, November). Arduino based platform for managing a PV battery charge. 2019 7th International Renewable and Sustainable Energy Conference (IRSEC). Presented at the 2019 7th International Renewable and Sustainable Energy Conference (IRSEC), Agadir, Morocco. https://doi.org/10.1109/irsec48032.2019.9078303

14. Megantoro, P., Ma'arif, A., Priambodo, D. F., Iswanto, Kusuma, H. F. A., Perkasa, S. D., &Suhono. (2022, November 8). Mini solar home system for electricity supply in coastal rural area, case study: Yogyakarta, Indonesia. 2022 International Conference on Information Technology Systems and Innovation (ICITSI).. https://doi.org/10.1109/icitsi56531.2022.9970836

15. Ethman, Y., Elbastawesy, M., Emad, M., Younes, I., Al-Hosseny, B. I., Matar, O., … Badr, B. M. (2023, February 19). Design low-cost battery management system for low power applications of photovoltaic systems. 2023 IEEE Conference on Power Electronics and Renewable Energy (CPERE), 1–6. https://10.1109/CPERE56564.2023.10119604

16. Gibson, T. L., & Kelly, N. A. (2010). Solar photovoltaic charging of lithium-ion batteries. Journal of Power Sources, 195(12), 3928–3932.https://doi.org/10.1016/j.jpowsour.2009.12.082

17. Tamilselvi, M., Ramesh Babu Durai, C., Senthilkumar, B., Latha, B., SamuthiraPandi, V., & Lakshmi Priya, J. (2024, January 24). WPT: A smart magnetic resonance technology based wireless power transfer system design for charging mobile phones. 2024 International Conference on Intelligent and Innovative Technologies in Computing, Electrical and Electronics (IITCEE), 1–6. https://doi.org/10.1109/iitcee59897.2024.10467828

18. Chhabra, G., Kumar, S., & Badoni, P. (2015, September). Automatic gadget charger using coin detection. 2015 1st International Conference on Next Generation Computing Technologies (NGCT).https://doi.org/10.1109/ngct.2015.7375261

19. Deepikavalli, K., &Puviarasi, R. (2018, February). Simulation of solar bag using DC-DC boost converter. 2018 Fourth International Conference on Advances in Electrical, Electronics, Information, Communication and Bio-Informatics (AEEICB). https://doi.org/:10.1109/aeeicb.2018.8480943

20. Sonawane, A., Vinerkar, S., Thote, U., Suryavanshi, A., &Waykar, S. (2019, March). Electrouter-an automated wireless charging gadget zone. 2019 3rd International Conference on Computing Methodologies and Communication (ICCMC). https://doi.org/10.1109/iccmc.2019.8819704

21. Mayingi, A. B., Malatji, J. M., & Chowdhury, S. P. D. (2019, September). The design of a control system for an automated photovoltaic power pack. 2019 IEEE AFRICON. Presented at the 2019 IEEE AFRICON, Accra, Ghana. https://doi.org/10.1109/africon46755.2019.9133933

Note: All the figures and tables in this chapter were made by the author.

Intelligent and Sustainable Power and Energy Systems – Dr. M. Premkumar et al. (eds)
© 2026 Taylor & Francis Group, London, ISBN 978-1-041-10314-1

22

AI-Based Control System for Wireless EV Charging with Real-Time Battery Monitoring and Management

**P. Selvan[1], K. Anbarasan[2],
K. Kumaravel[3], C. Prakash[4], G. Vignesh[5]**
Department of Electrical and Electronics Engineering,
Sengunthar Engineering College, Erode

ABSTRACT: Wireless EV charging stations use electromagnetic fields to transfer energy between a charging pad and the vehicle without physical connectors, enhancing convenience and safety. However, existing systems face drawbacks such as lower energy transfer efficiency, slower charging speeds compared to wired solutions, and misalignment issues that reduce performance. Therefore, this paper proposes an advanced high-frequency inverter with an isolation transformer-based EV battery system. The high-frequency inverter configuration is designed for efficient energy conversion and storage. Starting with a DC supply, the inverter utilises a Pulse Width Modulation (PWM) generator to produce high-frequency AC signals. This output is then directed into an LC tank circuit, optimising resonance for improved performance. An isolation transformer ensures electrical safety and stabilises the system by decoupling the load. The parallel synchronous rectifier facilitates efficient power conversion, minimising losses during rectification. PWM pulses are generated to regulate the output, ensuring the energy delivered to the battery is stable and efficient. The Artificial Neural Network (ANN) controller monitors and adjusts the system's performance, maintaining the desired voltage levels. This system is particularly beneficial for renewable energy applications, enhancing the system's reliability and efficiency. This paper emphasises advanced control techniques and component integration for optimal energy management. The system's versatility makes it suitable for various applications, promoting sustainable energy practices. Finally, this paper is implemented using MATLAB 2021a simulink software.

KEYWORDS: Battery charging system, High-frequency inverter, LC tank network, Parallel synchronous rectifier, Wireless EV charging

[1]selvan_14@rediffmail.com, [2]anbarasananbarasankeee@gmail.com, [3]kk8307319@gmail.com, [4]prakash183131@gmail.com, [5]gopalvignesh03@gmail.com

DOI: 10.1201/9781003654469-22

1. Introduction

Electric Vehicles (EVs) have gained significant traction as automakers have driven advancements in EV technology, with growing acceptance due to their environmental benefits [1]. Shifting to EVs is crucial in reducing fuel dependency, enhancing transportation sustainability, and protecting the environment [2]. This led to substantial investments by the automotive industry to further improve EV technology [3, 4]. One such innovation is the Wireless Charging System (WCS), which operates the principle of mutual induction [5]. However, existing wireless charging technologies face critical challenges, including lower energy transfer efficiency, slower charging speeds, and alignment issues that reduce performance, thus preventing them from competing with wired charging solutions. This research offers a revolutionary wireless charging system for EVs that uses a high-frequency inverter-based setup supplied by DC supply to overcome these restrictions. The system incorporates several key components to optimise efficiency and system performance, including a single-phase frequency inverter to convert DC into AC voltage. It is suitable for various applications, including solar power systems and small-scale renewable energy, and exhibits minimal voltage fluctuations. Additionally, they accommodate high input voltages [6]. The LC tank circuit acts as an electrical resonator, oscillating at its resonant frequency and is used to generate specific frequencies from complex signals [7]. The isolation transformer ensures electrical safety by decoupling the charging with the vehicle's power system while simultaneously stabilising the system. Various controllers optimise system performance and fault detection in wireless EV charging. The Proportional-Integral (PI) controller offers good steady-state performance but struggles with nonlinearities and can cause overshoot if not correctly tuned [8]. The Fuzzy Logic Controller (FLC) handles nonlinearities well but is computationally intensive and challenging to tune [9]. In this work, an Artificial neural network (ANN) controller is used to optimise charging parameters, adapt to dynamic conditions, and enhance fault detection, improving the overall efficiency and reliability of the system. Various controllers are used to manage performance and detect faults in wireless EV charging. The Proportional-Integral (PI) controller ensures steady-state performance but struggles with nonlinearities and may cause overshoot if not correctly tuned. The Fuzzy Logic Controller (FLC) handles nonlinearities well but is computationally intensive and complex to tune. To overcome these limitations, an Artificial Neural Network (ANN) controller is employed to optimise charging parameters, adapt to dynamic conditions, and enhance fault detection, improving efficiency and reliability. The Interleaved Parallel Synchronization Rectifier (IPSR) also reduces power losses, minimises ripple currents, and ensures stable power delivery, enhancing the system's performance when combined with the ANN controller.

2. Proposed System Modelling

The EV charging system utilises advanced components for efficient, safe, and reliable power transfer. A DC power supply feeds the system, which a single-phase inverter converts into high-frequency AC. This AC power is optimised through an LC tank network for efficient energy transfer to the vehicle. An isolation transformer ensures safety by decoupling the charging infrastructure from the vehicle's electrical system. Finally, an interleaved parallel synchronous rectifier converts the AC to DC, minimising conversion losses and reducing ripple currents, as depicted in Fig. 22.1.

The system incorporates an ANN controller that continuously monitors and adjusts real-time parameters, adapting to changes in load conditions and the distance between the vehicle and the charging pad. This dynamic optimisation of voltage, current, and power prevents energy losses and

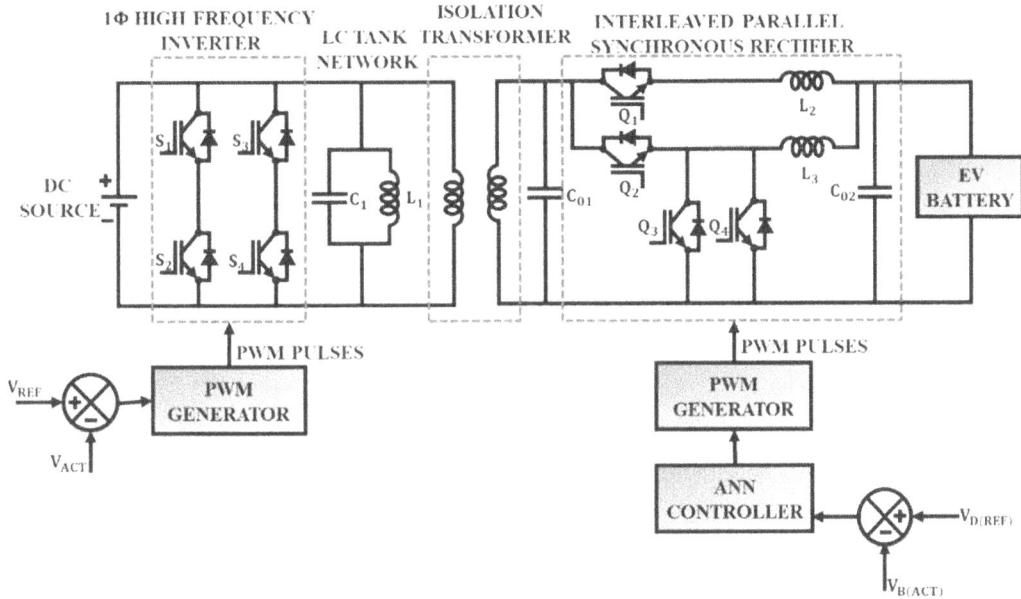

Fig. 22.1   Block diagram of proposed system

enhances charging efficiency. Additionally, a Battery management system (BMS) monitors the State of Charge (SoC), ensuring safe and efficient battery charging. These components create a reliable and efficient wireless charging solution that adapts to varying conditions while maintaining optimal performance.

2.1 Single-Phase High-Frequency Inverter

A High-Frequency Inverter converts DC to AC for harmonic control and reactive power compensation. At resonance, the switching power supply's input impedance becomes purely resistive, causing input current to be in phase with voltage, resulting in a power factor 1 and maximising output power. This also increases the harmonics displayed in Fig. 22.2.

In Fig. 22.2, the power switch labelled Q_1, Q_2, Q_3, and Q_4, are controlled by the driving signal. The diagonal switches (Q_1 and Q_2, Q_3 and Q_4) are turned on simultaneously, reversing the voltage across the

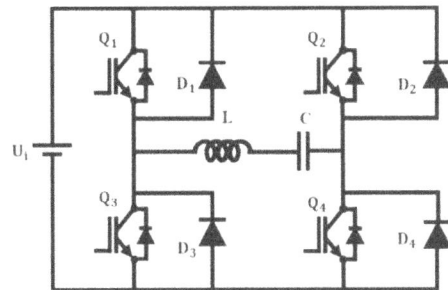

Fig. 22.2   Circuit diagram for single-phase high-frequency inverter

load. When the switches are adjusted on time, the DC voltage is converted into high-frequency AC.

2.2 Single-Phase High-Frequency Inverter

The LC tank circuit comprises a tunable inductor in series with device capacitance. (C_O). Where C_P represents parasitic capacitance from cables, R_O Denotes the internal resistance of circuit components, and RL is the inductor's internal resistance. The resonator's emotional behaviour is modelled by R_m, L_m, and C_m. Closing the ER switch deactivates the electrical resonance, as illustrated in Fig. 22.3.

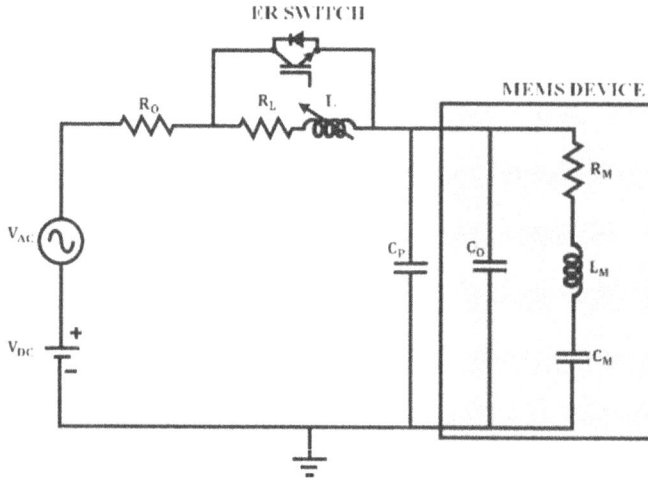

Fig. 22.3 LC tank circuit

The MEMS device, which is deactivated when the ER switch is closed, is coupled in series with an externally tunable inductor. L and ER switch. The characteristic equation of the circuit is as follows [10].

$$V_C \approx \frac{V_{in}}{\sqrt{\left(2\pi fRC\right)^2 + \left(\left(2\pi f\right)^2 LC - 1\right)^2}} \tag{1}$$

When the input frequency $f_{LC} = \dfrac{1}{2\pi\sqrt{LC}}$, the reactances of the inductor and capacitor cancel, maximising the current.

2.3 ANN Controller

The ANN controller in this system consists of three layers: input, hidden, and output. It has three inputs. (V_1, V_2, V_3) and three outputs $\left(V_1^*, V_2^*, V_3^*\right)$. The ANN output is compared with a carrier signal before being applied to the PWM generator as a reference, as shown in Fig. 22.4. The training of the ANN involves adjusting the weights Wij and biases Bj, with neutral of reducing the mean

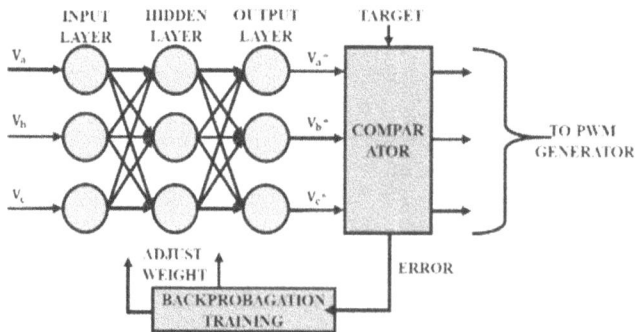

Fig. 22.4 Block diagram of ANN controller

square error (MSE)of outputs, set to 0.0001. The error function is defined in the equation (Senapati, M.K et al., 2024) (2)

$$J = \sum_{i=1}^{N} e(i)^2 \qquad (2)$$

The ANN training is complete when the instantaneous error e(i) between the estimated and actual outputs drops below 0.0001, with N representing the number of output neurons.

2.4 Interleaved Parallel Synchronous Rectifier

An Interleaved Parallel Synchronous Rectifier utilises multiple MOSFETs in parallel, with their switching phases staggered (interleaved) to enhance efficiency and reduce ripple. Each MOSFET acts as a synchronous rectifier, replacing diodes and offering lower conduction losses due to their low on-resistance ($R_{ds(on)}$). The power loss for each MOSFET is given by equation (3) [11]

$$P_{loss} = I^2 \times R_{ds(on)} \qquad (3)$$

where I is the current flowing through the MOSFET. The interleaving technique helps reduce the ripple current. I_{ripple} by a factor of N (the number of stages), expressed as:

$$I_{ripple} \approx \frac{I_{ripple}(single)}{N} \qquad (4)$$

where N represents the number of interleaved stages. Shifting the switching phases reduces the current ripple and distributes evenly across the stages, leading to better thermal performance and higher overall efficiency. This approach is beneficial in high-power applications, where minimising ripple and improving thermal management are essential.

3. Results and Discussion

An ANN controller is developed in this study to provide an optimal renewable energy integration system for EV charging. The MATLAB platform simulations for this work and results are shown below in Table 22.1.

Table 22.1 Parameter specification

Parameter	Specification
SEMIKRON IGBT	1200 V/10 A.
Switching Frequency	20 kHz
Filter Capacitor	2200 mF/1000 V
Inductor	650 mH

Figure 22.5 represents a sinusoidal-phase AC voltage waveform with a peak value of 350 V. The three-phase AC current waveform decreases sinusoidally to 80 A. The input AC voltage and current waveforms are observed, with the current at 50A voltage at 350 V and the input DC voltage waveform at 600 V.

Figure 22.6 shows the power waveform, which increases and follows a sinusoidal path at 1.5×10^4 W. The reactive power waveform increases step by step and stabilises after 0.25s. The

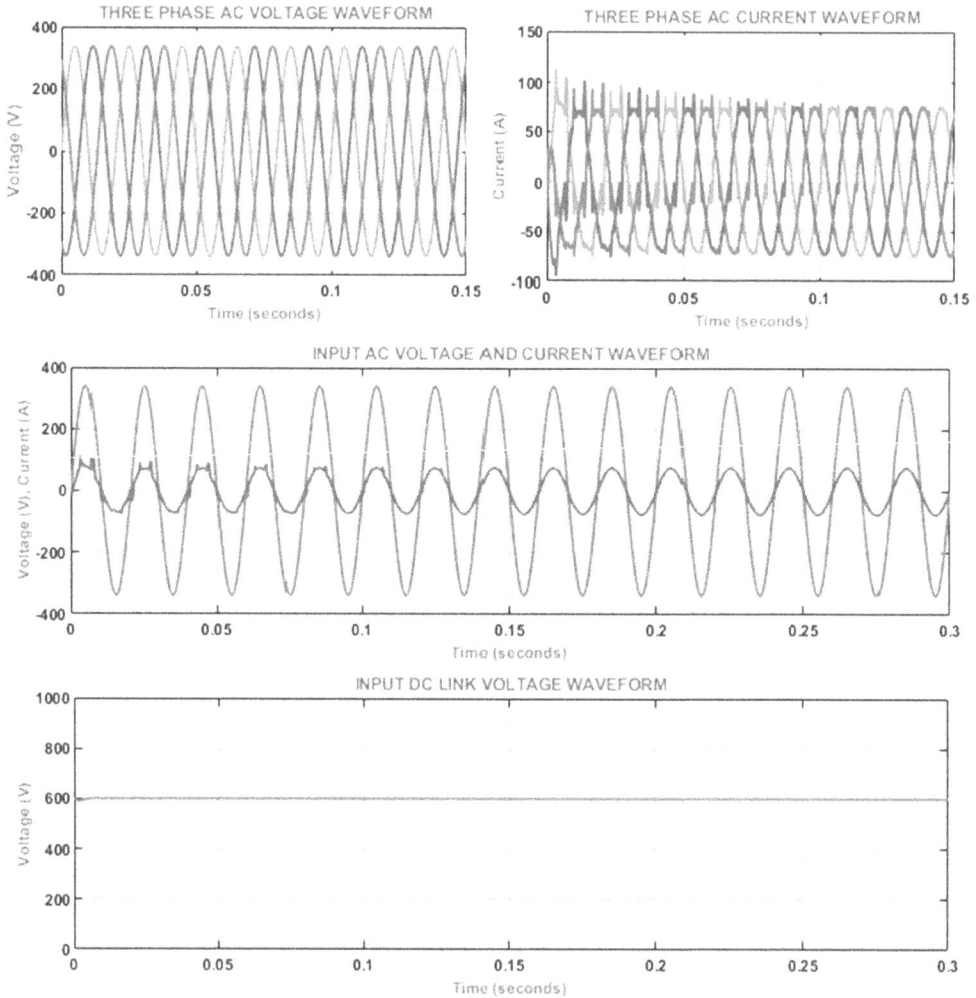

Fig. 22.5 Waveforms of three-phase AC voltage, AC, and DC-link voltage

parallel synchronous rectifier output voltage waveform rises to 300 V and stabilises after 0.5s. Moreover, the high-frequency inverter voltage waveform fluctuates between -5 V and 5 V stably.

Figure 22.7 illustrates the DC voltage increasing to 160 V and stabilising after 0.5 seconds, while the current rises to 4.8A and stabilising after 0.6 seconds.

Figure 22.8 illustrates the behaviour of PWM waveforms over time. In both Fig. 22.8a and 22.8b, the waveforms start as square waves with alternating high and low states but shift to a continuous high state at 0.25 seconds. Figure 22.8c shows a variation in pulse width, with noticeable changes in the durations of the high and low states, particularly towards the end. These fluctuations indicate adjustments in the duty cycle of the PWM signal.

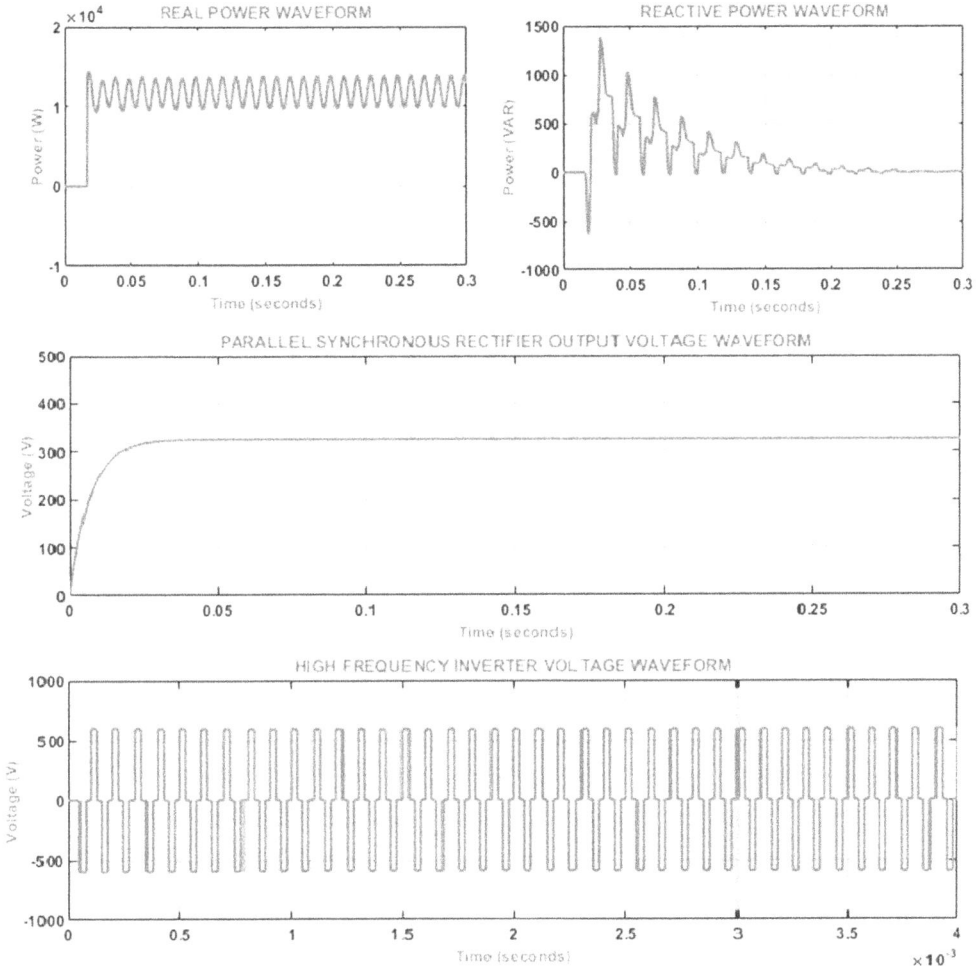

Fig. 22.6 Waveforms of absolute power, reactive power, parallel synchronous rectifier output, and high-frequency inverter voltage

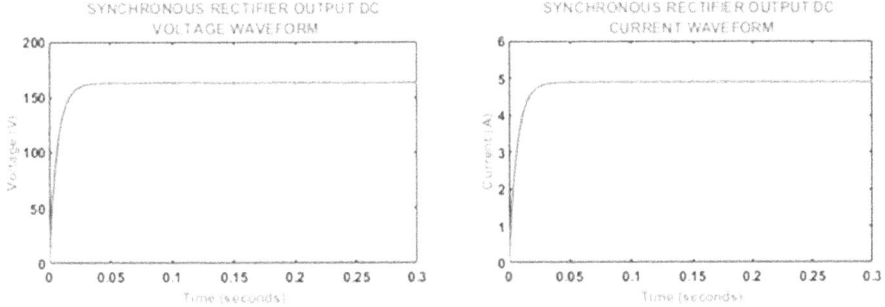

Fig. 22.7 Synchronous rectifier output DC voltage and current waveforms

(a)

(b)

(c)

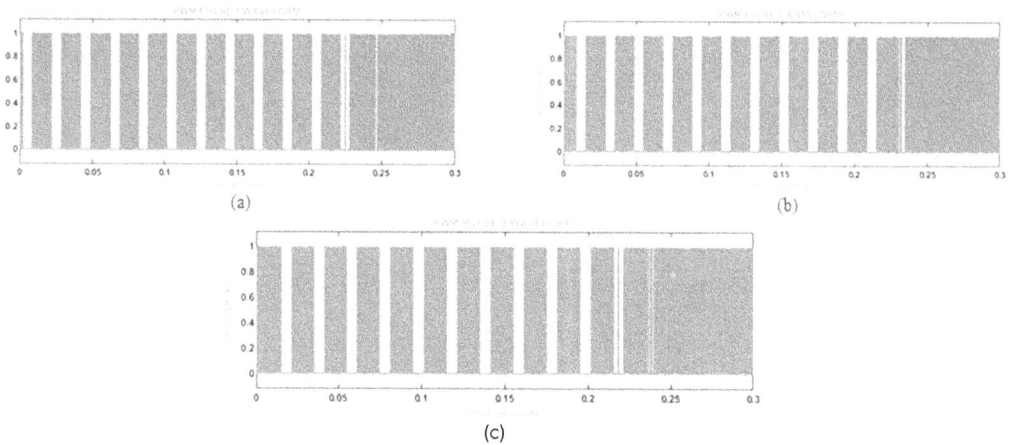

Fig. 22.8 PWM pulse waveforms

Figure 22.9 compares various controllers in terms of efficiency and settling time. In Fig. 22.9a, the proposed ANN controller achieves the highest efficiency at 95%, outperforming the PI controller (94.5%) [11] and the ANFIS controller (93.45%) [12, 13]. Figure 22.9b compares the settling times, with the PID controller achieving 2.3 seconds [13], the Fuzzy controller at 2.1 seconds [14], and the ANN controller demonstrating a notably faster settling time of 0.5 seconds.

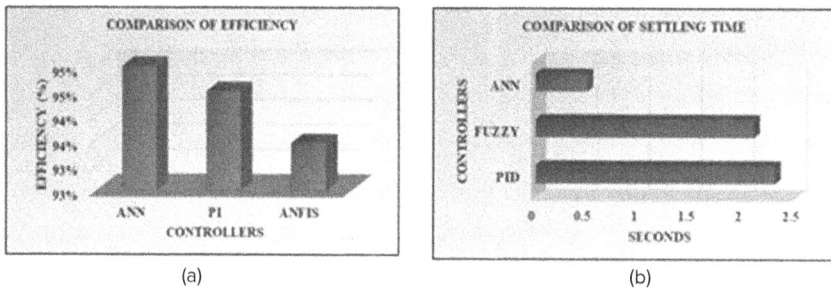

(a)

(b)

Fig. 22.9 Comparison of efficiency and settling time

4. Conclusion

The proposed advanced high-frequency inverter-based wireless EV charging system improves energy transfer efficiency by integrating a PWM generator, LC tank circuit, isolation transformer, and parallel synchronous rectifier, minimising conversion and rectification losses. An ANN controller enables real-time monitoring and adjustment, ensuring stable, efficient charging. The system addresses challenges like low energy transfer efficiency, misalignment, and slow charging while enhancing electrical safety and stability, which is also well-suited for renewable energy sources (RES) applications, promoting sustainability. A MATLAB 2021a Simulink implementation demonstrates its feasibility, with future work focused on improving efficiency, scalability, and integration with renewable energy sources for enhanced EV charging solutions.

REFERENCES

1. Razu, M.R.R., Mahmud, S., Uddin, M.J., Islam, S.S., Bais, B., Misran, N. and Islam, M.T. (2021). Wireless charging of electric vehicle while driving. IEEE Access, 9:157973–157983.

2. Fathollahi, A., Derakhshandeh, S.Y., Ghiasian, A. and Masoum, M.A. (2022). Optimal siting and sizing of wireless EV charging infrastructures considering traffic network and power distribution system. IEEE Access, 10:117105–117117.

3. Mohammed, S.A.Q. and Jung, J.W. (2021). A comprehensive state-of-the-art review of wired/wireless charging technologies for battery electric vehicles: Classification/common topologies/future research issues. IEEE Access, 9:19572–19585.

4. ElGhanam, E., Sharf, H., Odeh, Y., Hassan, M.S. and Osman, A.H. (2022). On the coordination of charging demand of electric vehicles in a network of dynamic wireless charging systems. IEEE Access, 10:62879–62892.

5. Wang, W.V., Thrimawithana, D.J. and Neuburger, M. (2021). An Si MOSFET-based high-power wireless EV charger with a wide ZVS operating range. IEEE Trans. Power Electron., 36(10):11163–11173.

6. Fong, Y.C., Cheng, K.W.E. and Raman, S.R. (2021). A modular concept development for resonant soft-charging step-up switched-capacitor multilevel inverter for high-frequency AC distribution and applications. IEEE J. Emerg. Sel. Top. Power Electron., 9(5):5975–5985.

7. Xiong, M., Dai, H., Li, Q., Jiang, Z., Luo, Z. and Wei, X. (2021). Design of the LCC-SP topology with a current doubler for 11-kW wireless charging system of electric vehicles. IEEE Trans. Transp. Electr. 7(4):2128 2142.

8. Arulvendhan, K., Srinivas, K.N., Narayanamoorthi, R., Milyani, A.H., Alghamdi, S. and Alruwaili, M. (2024). Primary Side Hybrid Reconfigurable Compensation for Wireless EV Charging with Constant Current/Constant Voltage Control. IEEE Access.

9. ElGhanam, E., Sharf, H., Odeh, Y., Hassan, M.S. and Osman, A.H. (2022). On the coordination of charging demand of electric vehicles in a network of dynamic wireless charging systems. IEEE Access 10:62879 62892.

10. Senapati, M.K., Al Zaabi, O., Al Hosani, K., Al Jaafari, K., Pradhan, C. and Muduli, U.R. (2024). Advancing Electric Vehicle Charging Ecosystems With Intelligent Control of DC Microgrid Stability. IEEE Trans. Ind. Appl.Ahmed, I., Rehan, M., Basit, A., Tufail, M. and Hong, K.S. (2023). Neuro-fuzzy and networks-based data driven model for multi-charging scenarios of plug-in-electric vehicles. IEEE Access.

11. Haque, M.R., Salam, K.M.A. and Razzak, M.A. (2023). A modified PI-controller based high current density DC–DC converter for EV charging applications. IEEE Access, 11:27246–27266.

12. Shukla, N., Shantanu, K., Singh, K. and Srivastava, R. (2020). Energy saving of induction motor drive using artificial intelligence based controllers. Int. J. Control Autom. 13(2):652–664.

13. Babes, B., Albalawi, F., Hamouda, N., Kahla, S. and Ghoneim, S.S. (2021). Fractional-fuzzy PID control approach of photovoltaic-wire feeder system (PV-WFS): Simulation and HIL-based experimental investigation. IEEE Access 9:159933–159954.

14. Hao, L.Y., Zhang, H., Li, T.S., Lin, B. and Chen, C.P. (2021). Fault tolerant control for dynamic positioning of unmanned marine vehicles based on TS fuzzy model with unknown membership functions. IEEE Trans. Veh. Technol. 70(1):146–157.

Note: All the figures and table in this chapter were made by the author.

Intelligent and Sustainable Power and Energy Systems – Dr. M. Premkumar et al. (eds)
© *2026 Taylor & Francis Group, London, ISBN 978-1-041-10314-1*

23

A Scalable AI-Driven Framework for Sustainable Ride-Sharing and Intelligent Logistics Using Advanced Route Optimization

Reshma Banu*

Dr Reshma Banu, Principal Researcher,
Computing Sciences, ISE, VVIET, VTU Belagavi, India

Vinay Kumar K. M.,
Charitra A. N., Deepika M. N., Jnana Y. T.

ISE, VVIET, VTU Belagavi, India

ABSTRACT: Ridesharing is a transportation network driven by artificial intelligence that provides delivery services and on-demand rides in both urban and rural locations. Because it allows cars to be shared for the movement of people and products, it offers a flexible and affordable alternative to public transit. By using real time traffic data from google maps and AI-driven route optimization, Route-Genie finds the quickest and most fuel-efficient routes, cutting down on travel expenses and time. By facilitating shared transport, it reduces traffic and its harmful environmental effects while maintaining security through authenticated user interactions.

KEYWORDS: Dynamic pricing, Integration with public transportation, Real-time GPS tracking, Route optimization and user authentication

1. Introduction

The need for dependable, affordable transportation is growing as cities expand and rural areas look for improved connectivity. However, traditional public transit often falls short and serves a large number of individuals neglected. According to the American Public Transportation Association, 60% of commuters are dissatisfied with the availability of transportation and the National Rural

---

*Corresponding author: dr.reshmabanu86@gmail.com

DOI: 10.1201/9781003654469-23

Transit Assistance Program estimates that about 45% of rural residents lack reliable access to transit. This discrepancy limits mobility for necessary daily activities, such as employment and medical care [18]. By utilizing AI and real-time data, Route-Genie tackles these issues and improves accessibility, flexibility, and affordability of transportation. According to a 2023 McKinsey estimate, the ride-sharing sector is expected to grow by 16% year due to the growing need for flexible transportation options [10].

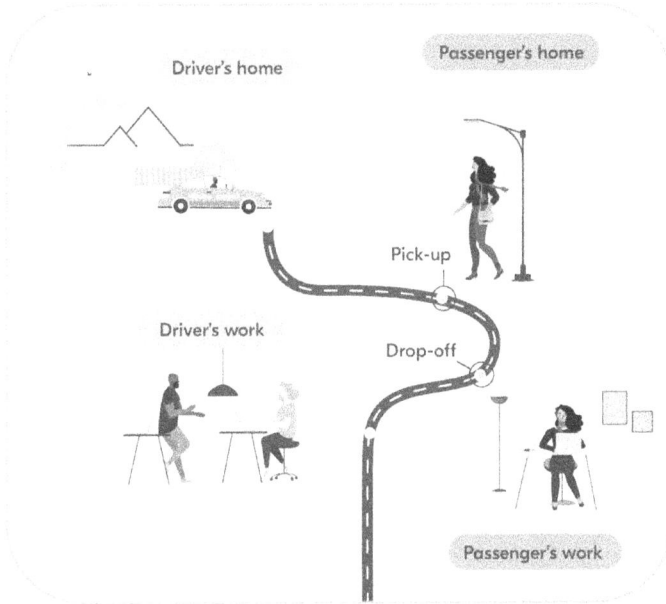

Fig. 23.1 Shared ride workflow: Driver and passenger journey

Source: Author

Additionally, roughly 70% of participants in both urban and rural regions stated that they favoured shared transportation choices due to their affordability and convenience [9]. Route-Genie provides an ecofriendly, effective and easily accessible solution for contemporary transportation requirements by minimizing solo travel, streamlining routes, and conforming to environmental objectives. As well as cutting the time of consumers, this feature reduces fuel consumption also. For instance, research from Berkeley University shows that efficient routing of trips can achieve time savings of 20 to 30 percent on average and depreciate fuel usage by 15 percent [12]. Because of this, the carbon emissions associated with personal trips are reduced, thus enhancing the experience for end-users while meeting the environmental goals.

1.1 Problem Statement

In relation to market trends, it can be stated: People today expect affordable and versatile delivery and transportation services only to be disappointed with insufficient availability, poor routing and transgressive prices. At the same time, owners of idle vehicles have limited opportunities to work with them to earn money. Gaps in existing logistics and transportation systems lead to excessive time consumption and costs, as well as waste of resources.

2. Proposed Methodology

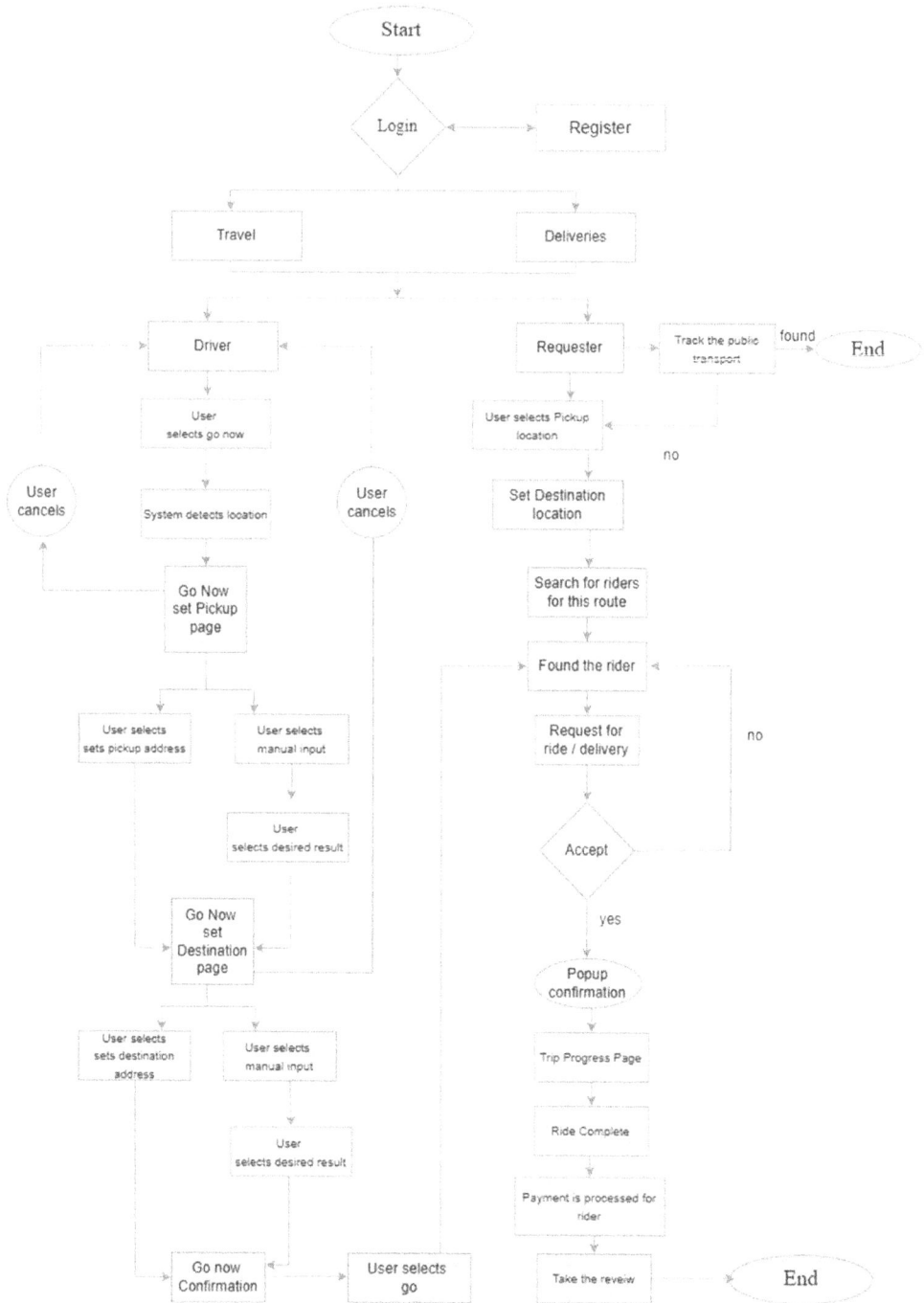

Fig. 23.2   Route-genie system flowchart: User journey for ride and delivery services

Source: Author

In the first step of our system, the user has to either log in or register first. When users are already logged in, they can either press the button "Go Now" for a ride or go to the button "Request Page" for making appointments for rides or deliveries. In the process of ''"Go Now," Transactions, TEXT user's area and geographic coordinates of the pick-up and destination addresses are set either from the system's database or manually entered.

As soon as the details are validated, the user proceeds to go on a journey. On the "Request Page", users indicate their pick-up as well as drop-off spots and look for available riders or vehicles along the route specified. In instances where a rider is not available, the system offers nearby public transport alternatives, thereby achieving real-time provision for out slum service users. Where there is a rider, it is easy for users to order for a ride or delivery service. If the request is accepted, the request will show a confirming message and the trip proceeds. At the end of the journey, the users' payment is completed and they provide their comments with a view to improving the service.

3. Existing System

Present delivery and ridesharing systems function as stand-alone solutions that cater to particular logistical or mobility requirements in the present transportation environment. These systems do have several drawbacks, though, especially when it comes to user experience, environmental sustainability, and urban-rural connectivity [10]. Existing platforms have a number of serious issues that limit their usefulness and inclusion. Fragmented solutions, like those offered by Uber and BlaBlaCar, only concentrate on delivery or ridesharing, forcing drivers and users to manage several platforms, which results in inefficiencies and increased expenses [4]. The issue is made worse by limited rural accessibility, as these systems give preference to urban areas, underserving rural areas and limiting citizens' mobility and access to necessary services [18]. Static route optimization results in longer time and cost to travel on a particular route since such optimization relies on fixed routes regardless of the current traffic situation, fuel consumption, or the needs of the customers [9].

Such systems also adversely impact the environment in a significant way as there is increased carbon emissions, solitary trips and traffic due to poor advertisement for sharing rides and optimization of the vehicles [2]. Public transport users often experience delays or cancelations of the service as well as non-integration with public transportation and private drivers face route convolutions and low trip acceptance coupled with irregular incomes [5]. Scalability remains among the key challenges addressed by platforms in this area as they more often than not fail to serve the peak demand in urban areas whilst at the same time addressing the challenges of lower demand in rural areas [9]. Limitation of updates of routes, fares and the availability of the vehicle means that transparency is another key challenge which affects the happiness of user as a whole [2]. Apart from the above two trust and safety are undermined by incomplete verification of user and driver which cause doubts and confidence through the platform is affected [14]. Ultimately cost has become a serious problem because of optimal routing and limited shared rides for customers [17].

4. Proposed System

Route-Genie is an Artificial Intelligence based transport application which offers a smart solution to ride and delivery requests and allows live tracking of the public transport. The application uses traffic information from the system and helps the users select the best routes thus saving time and fuel. It allows the users to order for cars, carpool, offer transport for staff or order for deliveries thus building a cost friendly transportation that is also community based. Route-Genie ensures security and provides purposes that relate to engaging users which include user authentication, provides fair pricing through dynamic pricing models, reliability and safety through the application of real time GPS. Route planning applications combine features such as traffic and public transport schedules and live updates to enable users easily and quickly create plans for trips using multiple modes of transport. Determination of the optimal route by AI algorithms reduces the emissions of CO_2 and congestion of vehicles, thus having minimal impact on the environment.

4.1 Model Architecture

All communication procedures of the joint framework are illustrated in Fig. 23.3, which is given in purple color as highlights (see Fig. 23.3). This means that the management of all active vehicles and their assignments is done at a single management center and it is possible to hold several locations with such parameters as their objectives, space available, and present coordinates among others.

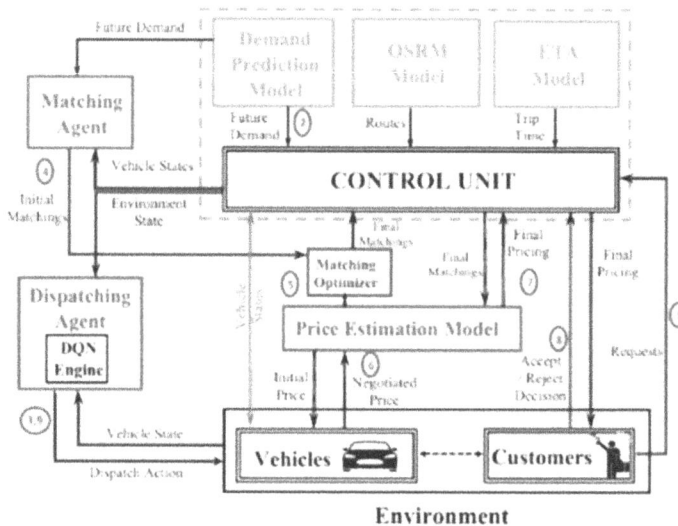

Fig. 23.3 Overall architecture of the proposed framework

Source: Author

These states are changed in every time step in each round according to the decisions of the matching and dispatching components on which states to maintain. Also integrated into the control unit are some subsystems that enable the ride-sharing environment. For instance, the Demand Prediction model is utilized to estimate areas of future activities while the OSRM model that stands for open-source routing machine recommends the most effective routes for traveling vehicles. And the ETA model predicts and subsequently updates the expected arrival time. These three models from have been integrated and include extensive explanations and relevant metrics. For instance, ride requests, demand supply and demand heat maps, and close focal points which include the estimation of future demand, are initially fed into the system at every time step (step 2, Fig. 23.3). In a step 3, vehicles use the forecasted demand to self-locate in areas with heavier demand. This placement is done, both first when cars are becoming available to the market (3-4 lines in Algorithm 1) and when they are relatively long left idle (verified at the end of every timestep in 23-24 lines, Step 9). With each vehicle accepting a control unit's input of the environment's current status, a greedy matching procedure with passengers starts for the vehicles (Step 4 in Fig. 23.3). This involves driving the transportation requests to the nearest passenger's taxi having the maximum capacity.

In the subsequent stages each taxi relies on Price Estimation model and using completed ads and estimated bids for each request they place bids for every call (Step 5). This helps in reducing possibilities of the service being interrupted. Each vehicle then engages in insertion-based route planning using its own optimizer module (Step 6) after that. In this phase of the process, vehicles build up their matching frameworks by sequentially introducing requests that most closely correspond to the current routing. To meet the revenue targets, insertions of requests are only permitted in order to preserve hope of loading and waiting times or the vehicle's load from overstepping large boundaries. After completing the 5th and the 6th steps, vehicles evaluate their usefulness according to potential locations of hotspots and propose changed price to consumers because of the destination of the trip, as well as the projected discounted offers obtained via DQN, as was done in Step 3. When a new customer is included in the route plans of the vehicle, the vehicle gives a new price, which causes this type of pricing change. When a customer joins a vehicle's route, it adjusts the pricing to the customer which triggers this type of pricing change and it is step number (7). The customer has then the comfort of deciding whether or not to go ahead with the offer because they have an independent utility function (Step 8 in the Fig. 23.3). The new ride will be either taken or rejected by the driver based on what the customer decides. Due to the fact that the proposed model operates in a

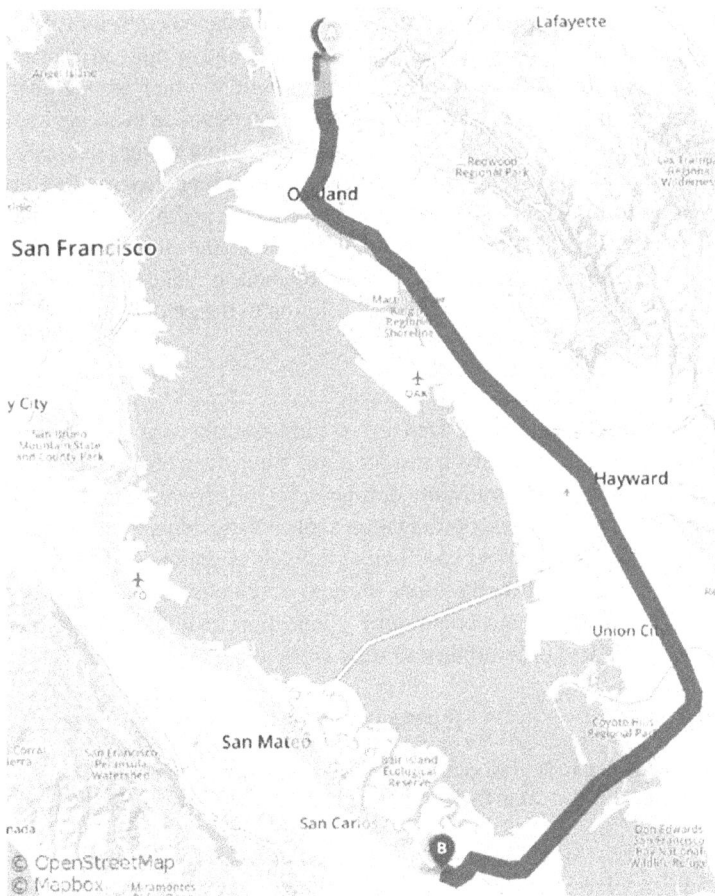

Fig. 23.4 The paths hardly cross, even though their endpoints are close together [8]

decentralized manner each car possesses individually its Q-learning reinforcement network which it incorporates in the pricing, route planning, matching, and dispatching tasks without having to query other localized vehicles. However, all cars do evaluate the distances of neighbouring automobiles prior to the execution of the actions. If required, vehicles have the option to communicate with the control center to update their states after making an action or check the environments for information before taking an action.

4.2 How Route-Genie Improves Upon Existing Systems

The Route-Genie application is different from the other applications; it stands out with its user-friendliness and innovative options. The development of the system, hand in hand with its technology, creates an advantage over the existing competitors. Such uniqueness allows one maximal efficiency in order to provide the client with cheaper, yet versatile trips. Moreover, the multifunctional use of the app allows clients to use it as a one-stop shop for all their transport needs. Unlike, the regular known platforms such as Uber use the application for delivery, Route-Genie seamlessly integrates multiple services onto one platform enhancing the user experience. In addition, the route creation is further simplified thanks to the tracking of individual traffic flows and demand within them. Thanks to this, our usage of routing algorithms outperforms others in comparison. All of these approaches lead to less time wasted traveling, decreased fuel usage, and reduced emissions. Route-Genie is also eco-friendly as it facilitates the promotion of shared journeys and optimized delivery routes thereby minimizing unnecessary solo trips and curbing carbon emissions which are a global climate change issue. This interface also improves transparency and user experience as users are kept updated in real time on the fares, selected routes and vehicles available including many features such as real-time pricing, user logins, driver and passenger feedback, and more which guarantee safety, effectiveness, and usability. Finally, Route-Genie is built on a cloud-based technology that allows scalability, which helps the system to be able to accommodate high demand in urban areas while remaining efficient in low demand areas. With these features combined, it makes it easier to position Route-Genie as a better, more inclusive, and eco-friendly solution to the colonization of transport systems.

5. Conclusion

Route-Genie: AI-Optimized Rides and Deliveries addresses the critical need for flexible, cost effective, and environmentally friendly transportation. With over 60% of commuters expressing dissatisfaction with limited public transport options and 70% preferring affordable shared rides, Route-Genie leverages AI-driven route optimization to meet these demands effectively. By promoting shared vehicle use, it helps reduce road congestion, fuel consumption, and emissions by up to 30% per trip. This platform not only benefits users through cost savings and income opportunities but also contributes to sustainability and community connection, making it a promising solution for modern transportation challenges in urban and rural areas alike.

REFERENCES

1. Akther, S. B., M. A. Hasan, N. Tasneem, and M. M. Khan. (2021). An Interactive Android Application to Share Rides with NSUers. 2021 IEEE World AI IoT Congress (AIIoT), Seattle, WA, USA, pp. 0121–0126.

2. Baza, M., N. Lasla, M. M. E. A. Mahmoud, G. Srivastava, and M. Abdallah. (2021). B-Ride: Ride Sharing With Privacy-Preservation, Trust and Fair Payment Atop Public Blockchain. IEEE Transactions on Network Science and Engineering. 8(2):1214–1229.

3. Busa, N., G. Alkadi, M. Verberne, R. P. Llopis, and S. Ramanathan. (2002). RAPIDO: a modular, multi-board, heterogeneous multi-processor, PCI bus based prototyping framework for the validation of SoC VLSI designs. Proceedings 13th IEEE International Workshop on Rapid System Prototyping, Darmstadt, Germany, pp. 159–165.

4. David, B., R. Chalon, and C. Yin. (2021). Collaborative Systems & Shared Economy (Uberization): Principles & Case Study. 2016 International Conference on Collaboration Technologies and Systems (CTS), Orlando, FL, pp. 57–63.

5. Escalona, J. A., B. Manalo, W. J. R. Limjoco, and C. C. Dizon. (2020). A Ride Sharing System based on An Expansive Search-Based Algorithm. 2020 IEEE REGION 10 CONFERENCE (TENCON), Osaka, Japan, pp. 870–874.

6. Farin, N. J., M. N. A. A. Rimon, S. Momen, M. S. Uddin, and N. Mansoor. (2016). A framework for dynamic vehicle pooling and ride sharing system. 2016 International Workshop on Computational Intelligence (IWCI), Dhaka, Bangladesh, pp. 204–208.

7. Gupta, T., G. K. Pandit, A. Kumar, H. Mishra, and B. Sharan. (2023). SafeTrack: Empowering Women's Security with GPS Location Tracking and Messaging. 2023 Second International Conference on Augmented Intelligence and Sustainable System (ICAISS), Trichy, India, pp. 794–799.

8. Hallgren, P., C. Orlandi, and A. Sabelfeld. (2023). PrivatePool: Privacy-Preserving Ridesharing. 2017 IEEE 30th Computer Security Foundations Symposium (CSF), Santa Barbara, CA, USA, pp. 276–291.

9. Hegde, A., B. S., C. T., and S. H. A. (2023). METADRIVE - A Decentralized Ride Sharing App. 2023 7th International Conference on Computation System and Information Technology for Sustainable Solutions (CSITSS), Bangalore, India, pp. 1–8.

10. Jashwanth, K., K. L. Sai Praneeth Reddy, M. Sai Snehitha, N. Sampath, and P. C. Nair. (2024). Analyzing Urban Transportation Services using RideShare Data Insights. 2024 IEEE 9th International Conference for Convergence in Technology (I2CT), Pune, India, pp. 1–7.

11. Kang, F., et al. (2017). YoRoad: A Monitoring Service for Ride Sharing Systems. 2017 IEEE International Congress on Internet of Things (ICIOT), Honolulu, HI, USA, pp. 143–146.

12. Makarim, M. I., N. Selviandro, and G. S. Wulandari. (2022). Route Recommendation Simulation for Ride Sharing Autonomous Vehicle: A Comparative Study of A* and Dijkstra Algorithm. 2022 1st International Conference on Software Engineering and Information Technology (ICoSEIT), Bandung, Indonesia, pp. 216–221.

13. Nguyen, J., et al. (2016). RAPIDO Testing and Modeling of Assisted Write and Read Operations for SRAMs. 2020 IEEE 25th North Atlantic Test Workshop (NATW), Providence, RI, USA, pp. 28–33.

14. Oleynikov, I., E. Pagnin, and A. Sabelfeld. (2022). Outsourcing MPC Precomputation for Location Privacy. 2022 IEEE European Symposium on Security and Privacy Workshops (EuroS&PW), Genoa, Italy, pp. 504–513.

15. Pouri, M. J. (2021). Implications for Designing Sustainable Digital Sharing Systems. 2021 IEEE Conference on Technologies for Sustainability (SusTech), Irvine, CA, USA, pp. 1–3.

16. Shroff, U., P. Ramnani, V. Kolwankar, and M. Tiwari. (2020). Fliver - an Enhancement to the Existing Autorickshaw System in India. 2020 Fourth World Conference on Smart Trends in Systems, Security and Sustainability (WorldS4), London, UK, pp. 181–183.

17. Srivastava, N., S. Tanaje, A. Kulkarni, M. Navan, A. D. Chowdhary, and B. N. Gohil. (2023). XGBoost-based dynamic ride-sharing model for New York City. 2023 6th International Conference on Information Systems and Computer Networks (ISCON), Mathura, India.

18. von Hoffen, M. (2020). The Sharing Economy Meets the Semantic Web: An Ontology for the Matchmaking of Peers. 2017 IEEE 11th International Conference on Semantic Computing (ICSC), San Diego, CA, USA, pp. 212–219.

19. Zhang, N., and T. Xiao. (2019). Information Sharing in a Dual-Channel Supply Chain with Consumers' Free Riding. 2019 16th International Conference on Service Systems and Service Management (ICSSSM), Shenzhen, China, pp. 1–6.

Intelligent and Sustainable Power and Energy Systems – Dr. M. Premkumar et al. (eds)
© 2026 Taylor & Francis Group, London, ISBN 978-1-041-10314-1

24

Single-Axis Sun Tracker for PV Panels Using a Single-Slit Sun Sensor

Vishwa Shah[1], Priti Burud[2]
and Sheikh Suhail Mohammad[3]
Dept. of Energy Sciences, Atria University,
Bengaluru, India

ABSTRACT: The tracking techniques for solar systems are becoming essential for improving solar energy utilization. However, the current tracking techniques need more practical, cost-effective implementations, highlighting the need for future research. The objective of the current work describes the design and testing of a single-slit sun sensor for detecting the sun's location in real time and aligning the solar panel towards the sun throughout the day. Experimental results show that solar panels with this tracking system generated 22.76% more power than fixed panels. By enhancing solar PV performance, this tracking approach can support the broader adoption of solar PV technology, enabling more efficient solar energy utilization.

KEYWORDS: Field of view, Single-axis tracking, Sun sensor, Solar photovoltaic

1. Introduction

Solar energy is increasingly popular due to its sustainability, low environmental impact, and declining installation costs, which have made solar PV systems more accessible. This shift aligns with global renewable energy goals. For example, the Indian government targets 500 GW of renewable energy capacity by 2030, with 280 GW from solar, and aims for net zero emissions by 2070 [1]. Achieving these targets depends on storage solutions addressing variability in solar power generation and ensuring a reliable energy supply [2]. In enhancing the performance of these systems, solar trackers play a crucial role by dynamically adjusting the orientation of solar panels to follow the sun's position [3]. Solar trackers are indispensable in optimizing solar installations' energy yield and economic viability, maximizing sunlight exposure throughout the day, particularly in large-scale deployments [4].

---
[1]vishwa.s@atriauniversity.edu.in, [2]priti.b@atriauniversity.edu.in, [3]ssuhail73@gmail.com

DOI: 10.1201/9781003654469-24

2. Literature Review

MEMS-based sun sensor designed for spacecraft attitude determination, offering high accuracy (a few arcminutes or better) in a compact, lightweight, and low-power design suitable for micro/nano spacecraft and rovers [5]. Recent advancements in sun sensor technology have expanded applications across domains like agriculture, solar energy, navigation, and monitoring systems [6]. Research studies have focused on improving the accuracy, compactness, and durability of sun sensors. Innovative methodologies, including machine learning, artificial intelligence, and MEMS-based solar angle measurements, have been utilized [7]. Solar tracking systems' techno-economic and environmental aspects for residential PV installations are a proven technology for improving efficiency. It highlights vertical tracking systems as the most cost-efficient and dual-axis systems as the most power-efficient [8]. A grid-connected PV system using FOV and FSC-based MPPT controllers, simulated in MATLAB/Simulink can be used to optimize power generation efficiency [9]. In another research, a single-slit digital sun sensor (DSS) was introduced and designed to enhance accuracy using sub-pixel interpolation and optimized slit geometry, but its scope is limited as it is more specific to rural areas [10]. Moreover, how factors like geographical location, tracking algorithm, and technological advancements impact the performance of the sun-tracking solutions was discussed in [11]. However, with the development of motor technology and light detection systems further research efforts are required to enhance sun-tracking solutions. Considering the research gaps identified in the existing literature, this study aims to design, develop, and test a sun sensor capable of accurately tracking the sun's position. The objective is to deepen understanding of sun sensor technology and derive practical insights through hands-on implementation and testing.

3. Methodology

3.1 Single Slit Sun Sensor

The slit sun sensor is a cost-effective tool for real-time sun position detection. It consists of a closed cubical box made of opaque material with a thin vertical slit on the top to allow sunlight. Two photodetectors are placed at the bottom, on opposite sides of the slit. When sunlight is aligned with the slit, no light falls on the detectors, producing zero output. If the sun shifts, one detector gets illuminated while the other remains dark. By rotating the sensor box along the North-South axis, it is aligned to the sun's position when both detectors register zero output. The box height determines the field of view (FOV), influencing the size and placement of the photodetectors, which will be discussed further.

3.2 Slit Design Experiment

1. A slit design experiment was conducted to understand the working principle of a slit sun sensor and arrive at a few critical dimensions for fabrication.

2. An opaque rectangular sheet was taken, and a tiny vertical slit was created. The slit dimensions were 0.5cm by 5cm and were at the center of the sheet.

3. The rectangular sheet was made to stand at a height from the table supported by two vertical supports. There was a provision to vary the height of the sheet from the base level.

4. The surface of the table was stuck with a graph sheet to enable the measurement of the slit light position at the bottom side. The picture of the setup created is shown in Fig. 24.1.

5. In this setup, once the light source is on, light passes through the slit and falls on the graph sheet illuminating only a portion as a representation of the slit.

6. In this setup, the height of the rectangular sheet can be varied measuring from the bottom of the table and similarly, the angle of the rectangular sheet concerning the light source also is variable.

7. At a specific height, the slit's impression on the bottom graph, aligned with the light source, marks the reference zero-error position.

8. The light impression is recorded on the graph sheet by varying its height and angle relative to the slit, marking deviations on both sides of the reference point.

9. This experiment aims to get an understanding of the FOV of the sensor, and its relation to the height of the sensor box.

Fig. 24.1 Setup of a slit design experiment

10. In normal field conditions, the Sun's angle can vary from 30° to 150 ° passing through 0° at mid-noon time.

3.3 Field of View

FOV refers to the angular range within which the sensor can detect sunlight. A slit sun sensor typically utilizes a narrow-slit aperture, and the FOV is crucial for determining the extent of the sun's apparent movement across the sensor's field. In our experiment, the linear FOV represented the actual physical width or height of the area the sensor could cover. This parameter was essential as it determines the photodetector dimensions and the sensor box height. By comprehending the sensor's coverage at different distances, valuable insights were gained into the practical limitations and capabilities of the equipment.

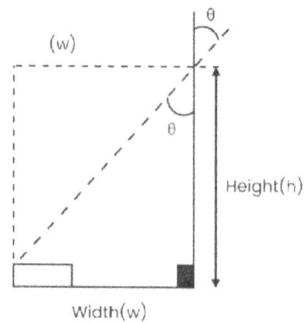

Fig. 24.2 Representing a diagram of FOV

FOV Calculations: To calculate the optimum height and length of the box, let's consider a slit sensor with height (h), width (w), and angle θ as shown in Fig. 24.2. From the above, the tangent of angle θ is given by:

$$tan(\theta) = \frac{w}{h} \tag{1}$$

$$\theta = tan_{-1}(wh) \tag{2}$$

The height of the sensor is considered as the vertical distance from the center of the sensor to the slit, where the FOV extends on either side of this centerline. To calculate the FOV in degrees for this sun-slit sensor concerning its height, we use the following formula:

$$FOV\ (°) = 2 \times \tan_{-1} \times (2 * Sensor\ sensor\ width\ width) \tag{3}$$

Slit Sun Sensor Setup: From the above calculations made, a slit sun sensor was made, taking into consideration the height and length of the box. The setup was made of aluminium and coated with black paint. For ease of rotation, a DC motor is connected to help the rotation of the box along the slit axis. A controlled experiment test setup was created, simulating various sun positions with the help of an artificial light source.

4. Experiments

4.1 Non-tracking Experiment

1. Set up the 20 W solar panel in a fixed position.
2. Connect the solar panel to a rheostat and two multi-meters for current and voltage measurements.
3. Place the light source perpendicular to the solar panel, such that there is no shadow on the panel, and record the readings (open circuit voltage and short circuit current).
4. Move the light source by 10° and record the voltage and current readings. Calculate the power as shown in the results section below.
5. Repeat the same for different angles of the light source such as 20°, 30°, 40°, and 50° in the same direction as above.

4.2 Tracking Experiment

1. Follow steps 1-3 same as above.
2. Move the light source in one direction by 10° to simulate light at different angles and intensities. Use the clinometer app on the phone and hold it on the same level as the light.
3. Move the solar panel also by the same angle, 10° such that the panel always remains perpendicular to the light source.
4. Record voltage and current.
5. Vary the angles by 20°, 30°, 40°, and 50° and record the readings of all the angles in the same direction as above.
6. Calculate the power with the help of voltage and current record in a table.

Fig. 24.3 Setup of a non-tracking experiment using a 20 Wp solar panel

4.3 Tracking Experiment Algorithm

1. Designed a sun sensor box as shown in Fig. 24.4. with a vertical slit on the top and calculated the FOV of it.
2. Placed two solar cells on the bottom side on either side of the slit vertically projected on the bottom of the box and labelled them A and B.
3. If the tracking is perfect, the light passing through the slit will not fall on both the solar cells A and B. Therefore, the expected output from both the solar cells is A and B is zero.

Fig. 24.4 An actual setup of a slit sun sensor tracking experiment

4. The goal is continuously detecting misalignment and rotating the sensor box to track the sun's position and ensure zero output from solar cells A and B.
5. Designed an electric circuit, as shown in Fig. 24.6, to control the motor for tracking the sun position.
6. The circuit actuates the motor to rotate the box so that the outputs from A and B remain zero ensuring continuous alignment with the sun's position.

The slit sensor box is designed to have an optimum height, such that the FOV of the unit is not very large. The height of the practical sensor box was 5 cm, giving rise to a FOV of 15 cm on either side of the slit as seen in Fig. 24.4. We have standardized the box height by taking the commercially available single solar cell. The slit dimension determines the loss of tracking angle. The wider the slit, the longer the non-tracking period. To be precise, a wider slit will create a longer null angle where the tracker does not move. The larger the FOV, the PV sensors must be bigger, which could contribute to elevated noise levels. The bottom side of the sensor box contains the photodetectors and the electronic circuit is accommodated. Photodetectors, consisting of solar cells mounted on either side of a vertical slit, detect zero signal when perfectly aligned with the Sun. The solar cell dimensions ensure coverage of the full FOV, maintaining alignment throughout the day using an electronic circuit. For tracking the sensor box, a 12V DC motor is used with a sufficient gear ratio. The rotation speed needed is 0.25° per minute. Therefore, the motor gear ratio is selected to rotate at 2.5° per minute (10 times the required value to achieve correction sooner). In Fig. 24.6, connectors J3 and J4 link the solar cell outputs from the two detectors placed on either side of the slit. The short-circuit current from the detectors is converted into measurable voltage using low-value resistor loads R10 and R7. Operational amplifiers (U1 and U2) from the LM358, which contain two amplifiers (U1A, U1B, U2A, U2B), are used to process these signals. U1A and U2A function as non-inverting amplifiers, boosting the low voltage generated by the detectors, as only a small area of the solar cells is illuminated due to the slit's dimensions. A stable 2.5V reference voltage is generated by VR1, which is further divided into approximately 100 mV (V_{ref}) using resistors R16 and R17. This V_{ref} is essential to filter out ambient noise and establish a threshold for zero detection, as an exact zero cannot be detected due to noise. The circuit is powered by a 12V DC source connected through J1.

U1B and U2B are configured as comparators with positive feedback via resistors R3 and R20. They compare the amplified detector signal to V_{ref}. If the signal exceeds V_{ref}, the comparator output switches to high. If the signal is equal to or below V_{ref} (indicating no light on the detector), the comparator output switches to low. This approach avoids hunting during closed-loop control by defining a threshold for zero detection. When the comparator output is high, it drives transistor Q1 to turn on and activates relay 1, enabling motor rotation. Similarly, relay 2 operates in the opposite direction. Relays 1 and 2 are configured so that activating Relay 1 rotates the motor in one direction, while activating Relay 2 rotates it in the opposite direction. To prevent both relays from being on simultaneously, which could short-circuit the 12V DC source, transistors Q2 and Q4 provide an interlock mechanism. This ensures that when one comparator is high, only its corresponding relay is energized, automatically turning the other off. If both comparators are low, neither relay is energized, keeping the motor stationary.

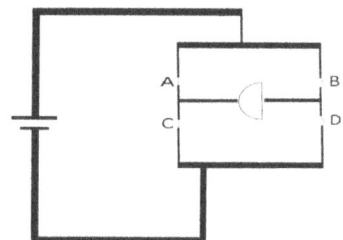

Fig. 24.5 Relay circuit diagram

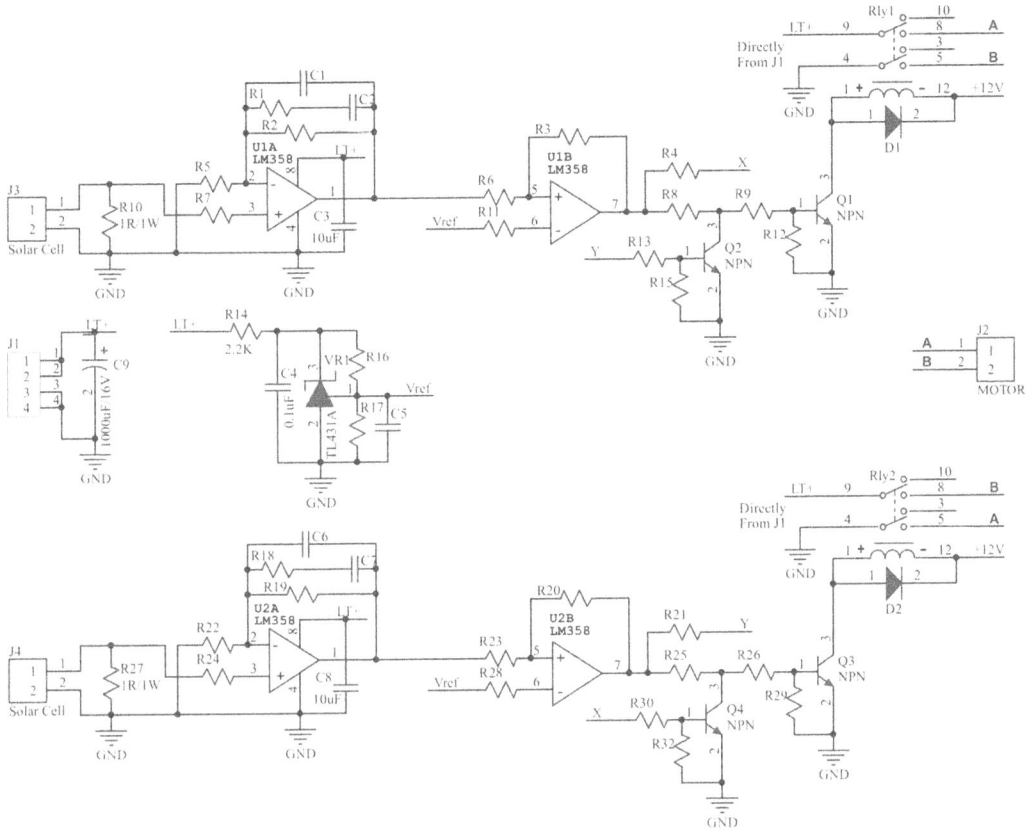

Fig. 24.6 Motor drive circuit diagram

4.4 Motor Drive

1. A bridge drive circuit controls the motor by altering the voltage polarity, allowing the motor to switch between +12V and -12V.

2. The circuit uses electric switches (e.g., MOSFETs or IGBTs), with pairs (A & D or B & C) switching together to reverse polarity, while A & C or B & D must not switch on simultaneously.

3. A relay-based system is incorporated into the bridge drive to simplify its construction and make it suitable for student projects.

4. The circuit diagram for this system is provided in Fig. 24.6.

5. Results

Two experiments were conducted. The initial experiment focused on devising the slit box, while the subsequent experiment aimed to evaluate the power output of a solar PV panel with tracking functionality against one without tracking. The outcomes of both the experiments are discussed below.

5.1 FOV Results

Table 24.1 shows the details of the experiment conducted to understand the FOV with a slit width of 0.5 cm and slit length of 5 cm. Depending on how far the sunlit area travels horizontally, the slit length was kept at 15 cm and the height was kept at 5 cm. The slit length was found by calculating the tan inverse of the distance the shadow moved, and the height of the slit as shown in the methodology.

Table 24.1 Distance of the slit shadow falling on the board from the original position

Slit height from the base (cm)	90°	75°	60°	45°
5	0	0.6	1.2	1.5
7	0	2.5	3.5	5
9	0	2.3	5	7.2
11	0	3	6	9

5.2 Non-Tracking Results

A 20 W solar panel was used for this experiment and a load of 50 Ω was connected to the panel using a resistance box. In this experiment, the solar panel was tilted instead of the light source to easily measure its deviation from the light source. It can be seen from Table 24.2 that as the angle between the sun and the solar panel increases, the power produced by it decreases, hence the percentage loss of power also increases. For example, for 10° the power produced is 5.1128 W, whereas for 20 it decreases to 3.92 W. This decrease is almost 44.1% from the power produced at the original position (at 90°). Figure 24.7 illustrates a correlation between the solar panel angle and the power performance of the non-tracking solar panel. Essentially, the graph highlights how variations in the solar angle directly impact the efficiency of power generation, emphasizing the importance of optimal panel positioning for maximizing energy capture and minimizing losses. From the graph, the decreasing trend line shows the regression, 0.977 indicating a strong correlation between the tilt angle of the solar panel and the decrease in power output. This means the trend line closely represents the actual data points, confirming that the relationship between tilt angle and power loss is highly predictable and consistent.

Table 24.2 Results for non-tracking solar panels and tracking solar panels in power (W) and percentage loss

Angle (°)	Non-tracking Results		Tracking Results	
	Power (W)	Percentage Loss (%)	Power (W)	Percentage Loss (%)
0	7.01	0	9	0
10	5.11	27.1	9.03	28.73
20	3.92	44.11	8.96	27.71
30	2.65	62.24	9.47	35.04
40	1.66	76.32	9.11	29.93
50	0.96	86.31	9.69	38.11

5.3 Tracking Results

Table 24.2 shows the results of power produced and its percentage loss. For example, the percentage loss from 10° to 50° increases from approximately 28.73% to 38.11% for tracking solar PV systems. This change in percentage loss is less than the non-tracking solar PV system. Figure 24.8 illustrates a clear correlation between the solar panel angle and the power performance of the tracking solar panel experiment. The graph shows tracking solar PV systems doesn't follow a trend since it always tracks the sun perpendicularly.

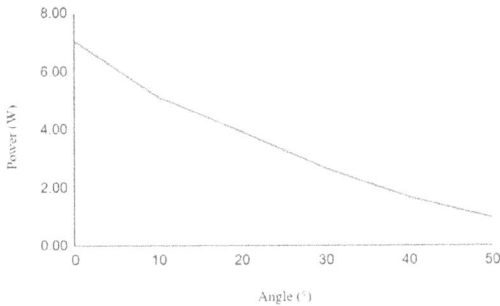

Fig. 24.7 Results for non-tracking solar panels

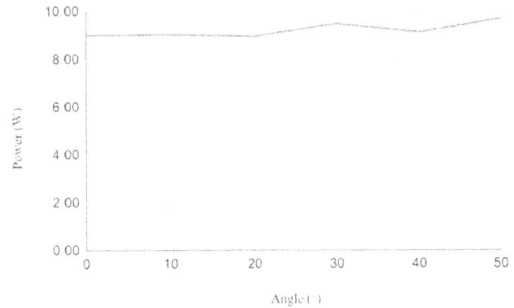

Fig. 24.8 Results for tracking solar panel's power v/s angle

To support this statement, the power remains almost similar at all angles. Moreover, as there is no trend seen, the line of regression is also low, at 0.625. The sensor was mounted on a rotating platform inside a slit box, to mimic sun-tracking capabilities. The actual setup was deployed in real-life field conditions and then tracking was monitored and recorded for a full 1 day. The actual sun position can be seen by looking at the shadows of a pole placed next to the sensor. The setup made could track the light source as seen in Fig. 24.9.

Fig. 24.9 Solar slit sensor during whole day

6. Discussion

The experiments above showcase that power loss is decreased if a solar panel tracks the sun perpendicularly throughout the day. For example, from 0° to 50° for non-tracking solar PV systems the power tracked decreased from 7.01W to 0.96 W, whereas for tracking solar PV systems the power tracked from 0° to 50° almost remained the same which was around 9 W. The research in this area contributes to advancements in renewable energy technologies and enhances reliability and performance by providing accurate sun-tracking capabilities and other applications. By optimizing solar energy capture, the tracker will help reduce the reliance on traditional energy sources, mitigating climate change. However, involving efficiency metrics such as tracking accuracy, response time, and overall system efficiency would further validate the effectiveness of the design. In conclusion, the aim was successfully achieved, and the design of this single-axis solar slit sensor that tracks the sun made it possible to minimize energy loss.

7. Conclusion

While the non-tracking system's power dropped from 7.01 W to 0.96 W, as the Sun's angle shifted, the tracking system maintained a steady output of around 9 W. This design not only enhances solar energy capture, reducing reliance on traditional energy sources and contributing to climate change mitigation but also has the potential to lower the Levelized Cost of Energy (LCOE) and improve the return on investment (ROI) for solar PV systems. Future research will focus on integrating advanced materials and sensor technologies to scale this cost-effective tracking solution for large solar farms, fostering its widespread adoption. This can require power electronics to efficiently handle the output from multiple solar trackers and ensure stable energy transfer to the grid or storage systems.

REFERENCES

1. Diriker, F. K., Frias, A., Keum, K. H., & Lee, R. S. K. (2021). Improved accuracy of a Single-Slit digital Sun sensor design for CubeSat application using Sub-Pixel interpolation. *Sensors, 21*(4), 1472.
2. Gorjian, S., Sharon, H., Ebadi, H., Kant, K., Scavo, F. B., & Tina, G. M. (2020). Recent technical advancements, economics and environmental impacts of floating photovoltaic solar energy conversion systems. *Journal of Cleaner Production, 278*, 124285.
3. Hariri, N. G., AlMutawa, M. A., Osman, I. S., AlMadani, I. K., Almahdi, A. M., & Ali, S. (2022). Experimental investigation of Azimuth- and Sensor-Based control strategies for a PV solar tracking application. *Applied Sciences, 12*(9), 4758.
4. Kuttybay, N., Saymbetov, A., Mekhilef, S., Nurgaliyev, M., Tukymbekov, D., Dosymbetova, G., Meiirkhanov, A., & Svanbayev, Y. (2020). Optimized Single-Axis Schedule Solar Tracker in different weather conditions. *Energies, 13*(19), 5226.
5. Liebe, C. C., Mobasser, S., & Jet Propulsion Laboratory, California Institute of Technology. (n.d.). MEMS Based Sun Sensor. *MEMS Based Sun Sensor.*
6. Mohammad, S. S., Iqbal, S. J., & Lone, R. A. (2022). Improved utilization of solar energy using estimated optimal tilt factor and trackers. *Energy Reports, 8*, 175–183.
7. Praveenraj, D. D. W., A, M., Pastariya, R., Sharma, D., Abootharmahmoodshakir, K., & Dhablia, A. (2024). Machine learning integration for enhanced solar power generation forecasting. *E3S Web of Conferences, 540*, 04007.
8. Rad, M. a. V., Toopshekan, A., Rahdan, P., Kasaeian, A., & Mahian, O. (2020). A comprehensive study of techno-economic and environmental features of different solar tracking systems for residential photovoltaic installations. *Renewable and Sustainable Energy Reviews, 129*, 109923.
9. Tadigadapa, S. A., & Najafi, N. (2003). Developments in Microelectromechanical Systems (MEMS): A manufacturing perspective. *Journal of Manufacturing Science and Engineering, 125*(4), 816–823. https://doi.org/10.1115/1.1617286
10. Trabelsi, H., Abid, H., Elloumi, M., Kharrat, M., & Lab-STA, National School of Engineers of Sfax, University of Sfax, Tunisia. (2017). MPPT controllers for PV array panel connected to Grid. *118th International Conference on Sciences and Techniques of Automatic Control & Computer Engineering - STA'2017*, 505.
11. Wang, J., & Lu, C. (2013). Design and Implementation of a Sun Tracker with a Dual-Axis Single Motor for an Optical Sensor-Based Photovoltaic System. *Sensors, 13*(3), 3157–3168.

Note: All the figures and tables in this chapter were made by the author.

Intelligent and Sustainable Power and Energy Systems – Dr. M. Premkumar et al. (eds)
© 2026 Taylor & Francis Group, London, ISBN 978-1-041-10314-1

25

Analysis of Plant Health using Python Image Processing—An Innovative Agro Technique

Jagadisha N.[1]

Associate Professor,
Dept of EEE GSSSIETW Mysuru,
Karnataka

A. D. Srinivasan[2]

R&Ddirector VVIET Mysuru, Karnataka

P. Apoorva[3]**, Shravya A. S.**[4]**,
Bhoomika G.**[5]**, Chandana K. S.**[6]

Student,
Dept of EEE GSSSIETW Mysuru,
Karnataka

ABSTRACT: Plant growth analysis is crucial for optimizing agricultural practices, providing insights into crop health and productivity. Monitoring growth parameters helps farmers in detecting the faults/disease detection, enhance yield, manage resources efficiently and address environmental stresses, contributing to sustainability in agro sector. The work aims to compare the health status of plants by using advanced python image processing techniques which is a robust algorithms to examine key morphological features to quantitatively assess and compare the health and growth of similar plants in different environments. The present study has reveled certain important aspects related to plant health which inherently related to plant growth such as leaf size, stem growth and green intensity of leaves. It is observed that larger leaf sizes and stem diameters are positively correlated with enhanced plant growth and biomass accumulation. Additionally, the analysis of color distribution and variance, particularly in the green channel, reveal crucial information about chlorophyll content and photosynthetic activity. The results of investigation with respect to plant growth have been discussed.

Keywords: Plant health, Image processing, Morphological features, Quantitative assessment

[1]jagadisha.n@gsss.edu.in, [2]adsrinivasan@gmail.com, [3]apoorvap3265@gmail.com, [4]shravyadisha@gmail.com, [5]bbhoomika056@gmail.com, [6]kschandanaks@gmail.com

DOI: 10.1201/9781003654469-25

1. Introduction

Plants are essential to life on Earth, providing oxygen, food, and raw materials, making their health critical for environmental sustainability and human survival [1]. Monitoring plant health enables early detection of stress factors like nutrient deficiencies, water scarcity, and diseases, ensuring timely interventions, improved crop yields, and sustainable agriculture [2]. This study focuses on key parameters such as leaf area, texture, plant volume, stem thickness, and color intensity (green, red, and blue channels) [3], which correlate with photosynthetic capacity, stress levels, structural support, and overall plant vitality [4].

Technologies like remote sensing and spectrometry provide large-scale insights but are expensive, while manual inspections are time-consuming and prone to errors [5]. Image processing offers a cost-effective alternative, using Python libraries like OpenCV, NumPy, and scikit-learn to enable accessible, precise, and scalable analysis [6]. Despite limitations such as sensitivity to environmental factors and the focus on external features, image processing is a valuable tool for preliminary plant health evaluations, balancing cost and precision [7][8][9].

2. Methodology

This study employs image processing techniques to evaluate plant health by analyzing key morphological and color-based features. The methodology is divided into five stages: image acquisition, pre-processing, feature extraction, comparative evaluation, and result visualization, ensuring a structured and comprehensive approach to plant health assessment [10]. The block diagram for methodology is given in Fig. 25.1.

Images of plants were captured using a smartphone under consistent lighting to ensure uniformity. Pre-processing involved resizing, noise reduction, background removal, and histogram equalization to enhance contrast and clarity. Key features like leaf size, plant volume, color intensity, and stem diameter were extracted using Python and OpenCV. These features were analyzed and compared statistically, with results visualized through bar charts, scatter plots, and annotated images, providing clear insights into plant health and vitality.

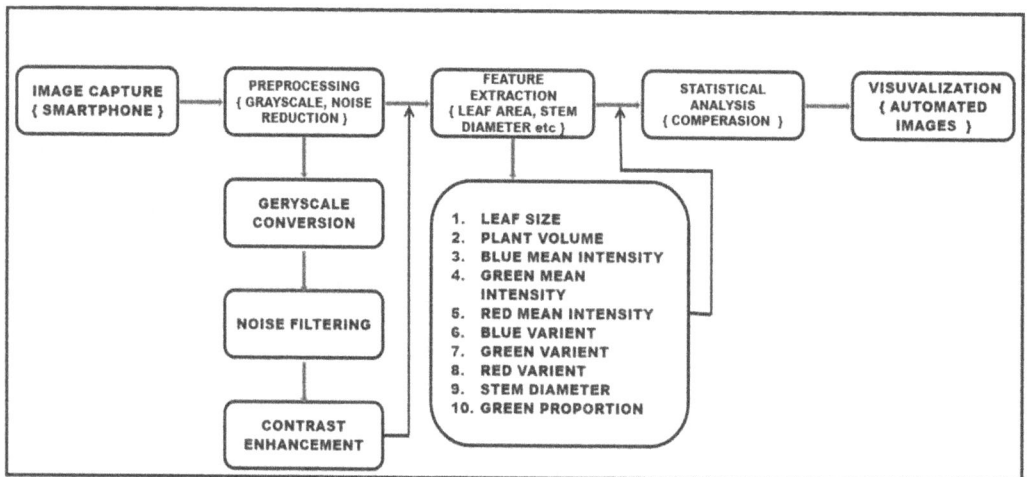

Fig. 25.1 Methodology

3. Software Description

In this research, Jupyter Notebook is employed as the primary software environment for plant feature analysis. It offers an interactive and user-friendly interface for executing code, visualizing results, and documenting the entire process in real time. Jupyter Notebook allows for seamless integration with Python libraries such as `os`, `shutil`, `cv2`, `numpy`, and `matplotlib`, enabling efficient management of files, image processing, data analysis, and visualization. Its flexibility and ease of use make it an ideal tool for conducting and presenting the various steps of plant feature analysis in an organized manner, enhancing reproducibility and collaboration in the analysis [11].

4. Results and Discussion

In this section comparison of Areca Palm, Sapling and Snake Plants which are grown in our campus were considered for the analysis and the results were discussed in the following cases:

4.1 CASE 1: Under Similar Environments with Equal Treatment

This comparison evaluates the health and growth differences among three types of plants (Areca Palm, Sapling and Snake Plants) cultivated under similar environments i.e sufficient sunlight environment with equal treatment, where all plants receive equal sunlight, water, nutrients, and overall treatment to ensure uniform growing conditions.

A. Sapling Plant

Fig. 25.2 Sapling plant 1 - Processed image   **Fig. 25.3** Sapling plant 2 – Processed image

From Fig. 25.4, it is observed that there is no major difference in leaf sizes. However, Plant 2 exhibits optimal growth patterns with consistent greenery volume, larger leaf sizes, and a higher proportion of healthy green pigmentation compared to Plant 1. Figures 25.2 and 25.3 show processed images with background removal for both plants. Figure 25.5 reveals that Plant 1 has a significantly higher volume, indicating denser and more vigorous growth, possibly due to better energy conversion efficiency and enhanced biomass accumulation. Despite this, Plant 2 has a larger stem diameter, suggesting stronger structural development, higher metabolic activity, and greater capacity for nutrient transport and energy utilization. Both saplings display characteristic features of young trees and are progressing well, reflecting their ability to adapt and thrive in their environment. This can be concluded that, even tough plant 1 exhibits larger volume, the metabolic activity in plant 2 is much better compared to plant 1.

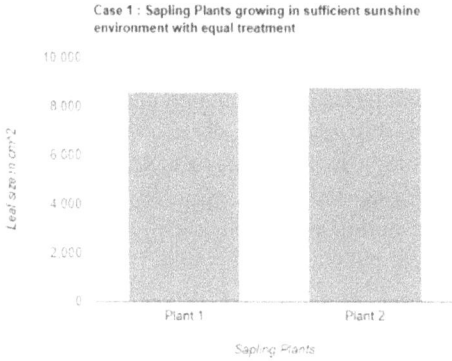

Case 1 : Sapling Plants growing in sufficient sunshine environment with equal treatment

Fig. 25.4 Comparison of stem leaf size of sapling plants growing in sufficient sunlight environment with equal treatment

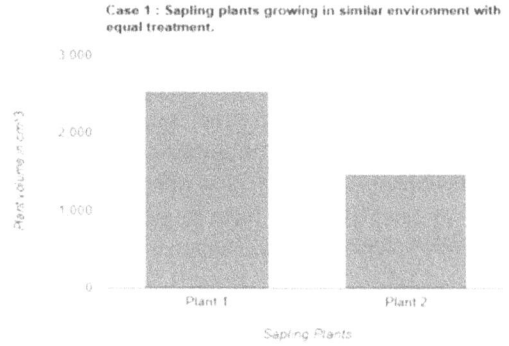

Case 1 : Sapling plants growing in similar environment with equal treatment.

Fig. 25.5 Comparison of plant volume of Sapling plants growing in sufficient sunlight environment with equal treatment

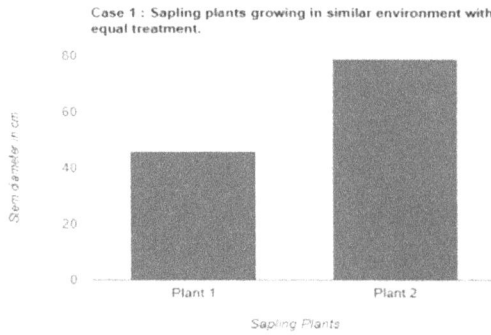

Case 1 : Sapling plants growing in similar environment with equal treatment.

Fig. 25.6 Comparison of stem diameter of sapling plants growing in sufficient sunlight environment with equal treatment

B. Areca Palm plants

Background Removed

Fig. 25.7 Areca palm plant 1 - Processed image

Green Part

Fig. 25.8 Areca Palm Plant 2- Processed Image

From Fig. 25.9, it is observed that there are major differences in leaf sizes in both the plants. Figures 25.7 and 25.8 show processed images with background removal for both plants. Figure 25.10 reveals that Plant 1 has a significantly higher volume, indicating denser and more vigorous growth, possibly due to better energy conversion efficiency and enhanced biomass accumulation. Despite this, Plant 2 has a larger stem diameter, suggesting stronger structural development, higher metabolic activity, and greater capacity for nutrient transport and energy utilization. These results indicate that despite similar environments and treatments, the two Areca Palm plants demonstrate varying growth strategies. Plant 1 exhibits traits of vigorous foliage expansion, while Plant 2 emphasizes structural development and metabolic efficiency, potentially leading to better long-term growth and resilience.

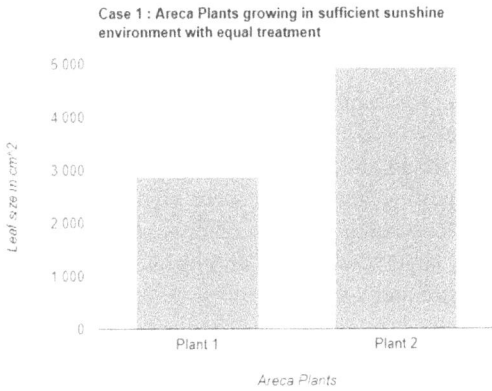

Fig. 25.9 Comparison of leaf sizes of areca palm plants growing in sufficient sunlight environment with equal treatment

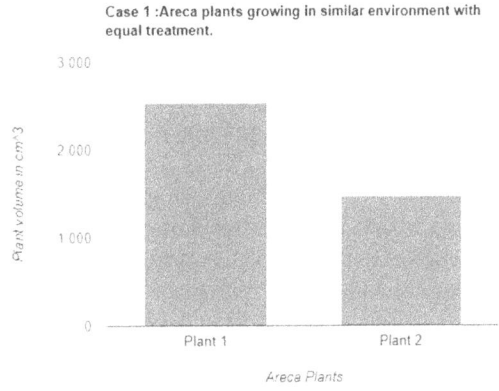

Fig. 25.10 Comparison of plant volume of areca palm plants growing in sufficient sunlight environment with equal treatment

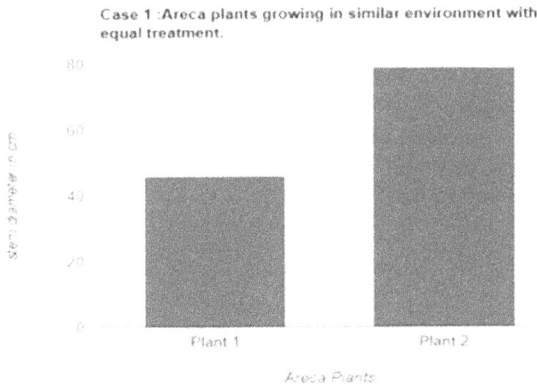

Fig. 25.11 Comparison of stem diameter of areca palm plants growing in sufficient sunlight environment with equal treatment

C. Snake Plants

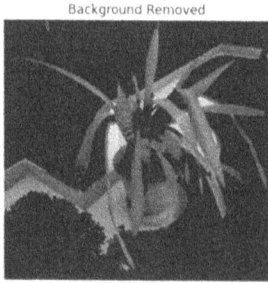

Fig. 25.12 Snake plant 2 - Processed image

Fig. 25.13 Snake plant 2 - Processed image

From Figs. 25.14, 25.15, and 25.16, it is observed that in snake plant 1, there are slight differences in growth patterns with respect to leaf sizes. The processed images with background removal are shown in Figs. 25.12 and 25.13 for both plants. Figures 25.14 and 25.15 show that plant 1 exhibits larger leaf size and volume compared to plant 2, suggesting denser and more vigorous growth, leading to enhanced structural development and biomass accumulation. In contrast, a comparison of stem diameters shows that plant 2 has a significantly larger stem diameter, indicating a stronger structural foundation despite plant 1 having larger leaf size and volume.

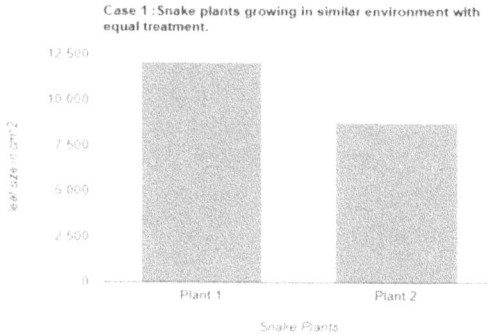

Fig. 25.14 Comparison of leaf sizes of snake plants growing in sufficient sunlight environment with equal treatment

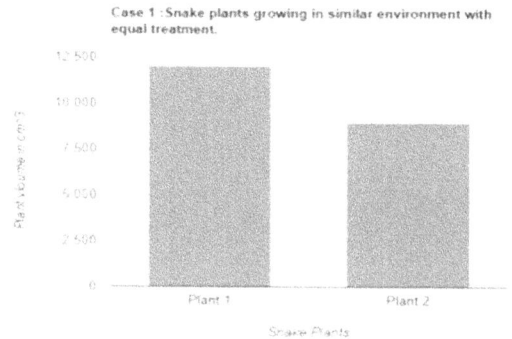

Fig. 25.15 Comparison of volume of snake plants growing in sufficient sunlight environment with equal treatment

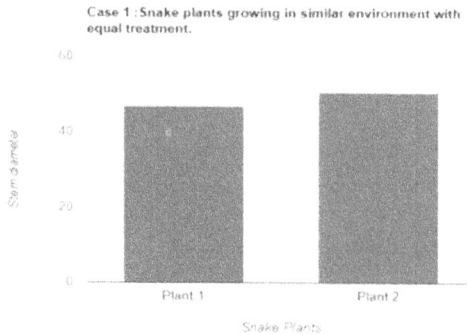

Fig. 25.16 Comparison of stem diameter of snake plants growing in sufficient sunlight environment with equal treatment

4.2 CASE 2: Under Different Environments with Equal Treatment

This comparison evaluates the health and growth differences among three types of plants (Areca Palm, Sapling and Snake Plants) cultivated under different environments i.e sufficient sunlight environment and shady environment with equal treatment, where all plants receive water, nutrients, and overall treatment to ensure uniform growing conditions but second plant is not receiving sufficient sunlight.

A. Sapling Plants

Fig. 25.17 Sapling Plant 1- Processed Image

Fig. 25.18 Sapling Plant 2- Processed Image

It is observed from Figs. 25.19, 25.20 and 25.21 that there are notable variations in growth characteristics when plants are being grown in different environment, despite equal treatment. Despite being in dissimilar environments (sunshine vs. shady), the two sapling plants exhibit distinct growth strategies. Plant 1, with its larger leaf size, benefits from optimal growth patterns and greater photosynthetic efficiency, especially in sunlight. On the other hand, Plant 2, with its larger volume and stem diameter, shows signs of more efficient nutrient transport and metabolic activity, which likely contributes to stronger structural development. This suggests that Plant 2's metabolic activity is superior, enabling it to thrive in a shaded environment with better long-term growth potential.

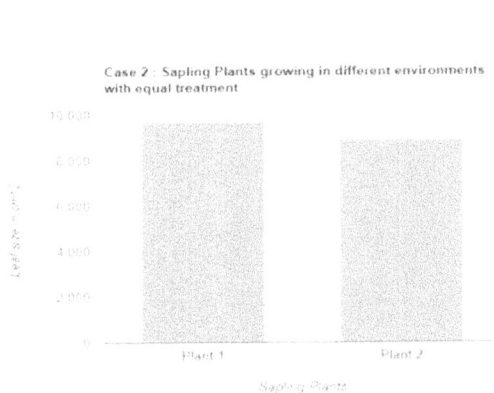

Fig. 25.19 Comparison of leaf sizes of sapling plants growing in sufficient sunlight and shady environment with equal treatment

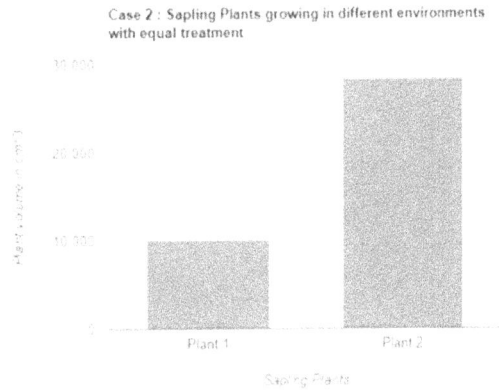

Fig. 25.20 Comparison of plant volumes of sapling plants growing in sufficient sunlight and shady environment with equal treatment

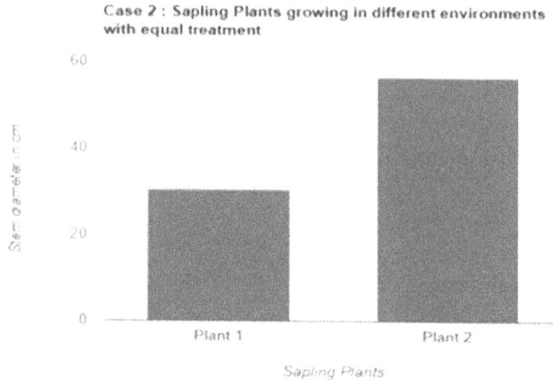

Fig. 25.21 Comparison of stem diameter of sapling plants growing in sufficient sunlight and shady environment with equal treatment

B. Areca Palm Plant

Fig. 25.22 Areca palm plant 1 - Processed image

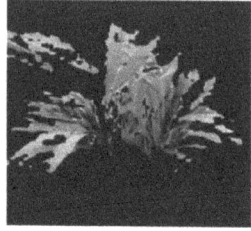

Fig. 25.23 Areca palm plant 2 - Processed image

From Figs. 25.24, 25.25 and 25.26 it is observed that, the Areca palm plants grown in different environmental conditions—one with sufficient sunlight and the other in a shady environment have

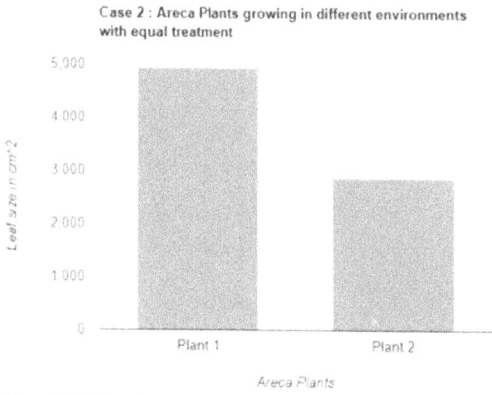

Fig. 25.24 Comparison of leaf sizes of sapling plants growing in sufficient sunlight and shady environment with equal treatment

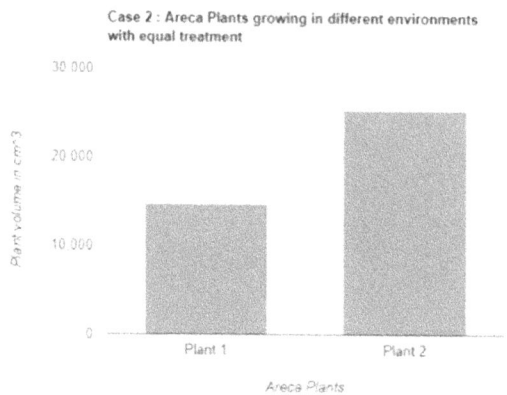

Fig. 25.25 Comparison of plant volumes of sapling plants growing in sufficient sunlight and shady environment with equal treatment

Case 2 : Areca Plants growing in different environments
with equal treatment

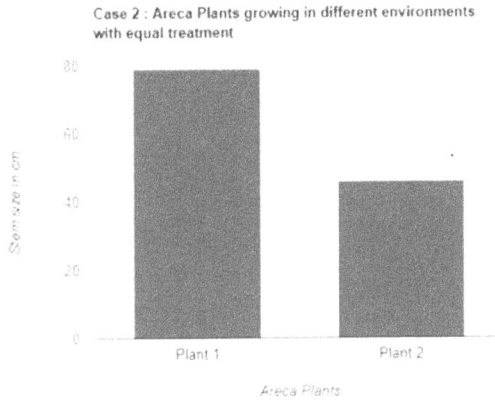

Fig. 25.26   Comparison of stem diameter of sapling plants growing in sufficient sunlight and shady environment with equal treatment

distinct growth adaptations despite equal treatment. Plant 1, with its larger leaf size and thicker stem as shown in Fig. 25.24 and 25.26, which focuses on structural integrity and efficient energy utilization, which may be advantageous in challenging environmental conditions. In contrast, Plant 2, with its larger overall volume, demonstrates vigorous biomass accumulation and adaptability, suggesting favorable conditions for denser growth. Therefore, it can be concluded that even tough plant 1 exhibits higher values of leaf sizes and stem diameter, plant 2 larger volume due to age factor.

C. Snake Plant

Fig. 25.27   Snake olant 1 - Processed image

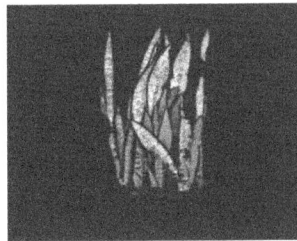

Fig. 25.28   Snake plant 1 - Processed image

It is observed from Fig. 25.29, 25.30 and 25.31, Snake plants grown in different environmental conditions—one with sufficient sunlight and the other in a shady environment—demonstrates contrasting growth characteristics influenced by their respective environments. The comparison of leaf size reveals that Plant 2, grown in the shady environment, has a higher leaf size, possibly indicating an adaptive mechanism to capture more light by increasing surface area. Similarly, Plant 2 also shows a higher plant volume, suggesting an overall better growth response in the shady environment, potentially due to its tolerance and efficient resource utilization in low-light conditions. However, the stem diameter comparison highlights that Plant 1, exposed to sufficient sunlight, develops a thicker stem, reflecting a stronger structural adaptation to support upright growth in the sunlit environment. These findings underscore the diverse strategies employed by Snake plants to optimize their growth in varying light conditions.

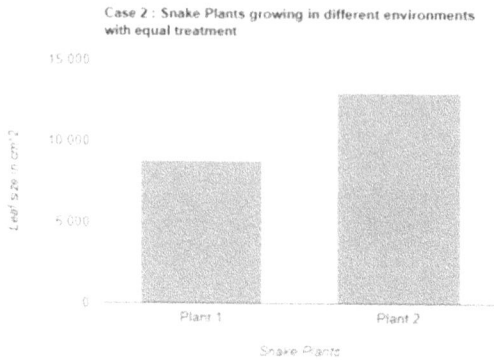

Fig. 25.29 Comparison of leaf size of snake plants growing in sufficient sunlight and shady environment with equal treatment

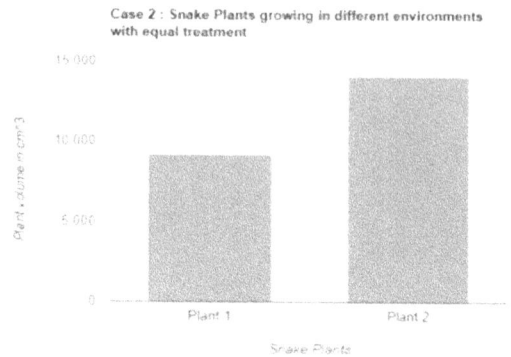

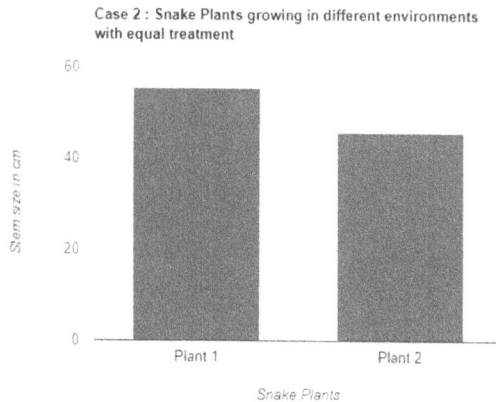

Fig. 25.30 Comparison of plant volume of snake plants growing in sufficient sunlight and shady environment with equal treatment

Fig. 25.31 Comparison of stem size of Snake plants growing in sufficient sunlight and shady environment with equal treatment

5. Conclusion

The present work considered three different plants- Sapling, Areca Palm and Snake plant for the analysis of plant health under similar and different environmental conditions. The results demonstrated that larger leaf sizes and stem diameters are positively correlated with enhanced plant growth and biomass accumulation. The color distribution and variance, particularly in the green channel, reveal crucial insights into chlorophyll content and photosynthetic activity. This highlights the effectiveness of the method in assessing plant health, integrating visual and quantitative analyses to monitor environmental impacts on plant performance for different environments and its overall health.

Acknowledgement

We sincerely thank our Principal, Management, and staff of GSSSIETW, Mysuru, for their invaluable support, and Dr. Rajanikanth B. S., IISc Bangalore, for his expert guidance. Special thanks to Dr. Manjuprasad B, Dept of AI&ML and the EEE department for their cooperation.

REFERENCES

1. Babar Iqbal, Guanlin Li, Khulood Fahad Alabbosh, Hamad Hussain, Ismail Khan, Muhammad Tariq, Qaiser Javed, Muhammad Naeem, Naveed Ahmad, (2023).Advancing environmental sustainability through microbial reprogramming in growth improvement, stress alleviation, and phytoremediation. Science Direct, volume 1.

2. Kumar, V., Sharma, K. V., Kedam, N., Patel, A., Kate, T. R., & Rathnayake, U. (2024). Comprehensive review on smart and sustainable agriculture using IoT technologies.Sacience Direct, *Volume 1*

3. Keivani, M., Mazloum, J., Sedaghatfar, E., Tavakoli, M.B. (2020). Automated analysis of leaf shape, texture, and color features for plant classification. Traitement du Signal, Vol. 37, No. 1, pp. 17–28.

4. Arguello-Sánchez R, López-Callejas R, Rodríguez-Méndez BG, Scougall-Vilchis R, Velázquez-Enríquez U, Mercado-Cabrera A, Peña-Eguiluz R, Valencia-Alvarado R, Medina-Solís CE. Innovative Curved-Tip Reactor for Non-Thermal Plasma and Plasma-Treated Water Generation: Synergistic Impact Comparison with Sodium Hypochlorite in Dental Root Canal Disinfection. Materials (Basel). 2023 Nov 17;16(22):7204. doi: 10.3390/ma16227204. PMID: 38005133; PMCID: PMC10672626.

5. Zhongsheng Cao, Xia Yao, Hongyan Liu, Bing Liu, Tao Cheng, Yongchao Tian, Weixing Cao, Yan Zhu,Comparison of the abilities of vegetation indices and photosynthetic parameters to detect heat stress in wheat,Agricultural and Forest Meteorology,Volume 265, 2019, Pages 121-136,ISSN 0168-1923.

6. He,F.,Zhang,Q.,Deng,G.,Li,G.,Yan,B.,Pan,D.,Luo,X.,&Li,J.(2024).Research Status and Development Trend of Key Technologies for Pineapple Harvesting Equipment: A Review. *Agriculture*, *14*(7), 975. https://doi.org/10.3390/agriculture14070975

7. J., Valarmathi, T., & Kruthika. (2023). A philosophical study of agricultural image processing techniques. *Buana Information Technology and Computer Sciences (BIT and CS)*, *10*(2), Article 3712.

8. Bakhshayeshi, I., Erfani, E., Taghikhah, F. R., Elbourn, S., Beheshti, A., & Asadnia, M. (2024). An intelligence cattle re-identification system over transport by Siamese Neural Networks and YOLO. *IEEE Internet of Things Journal*, *11*(2), 2351–2363.

9. Jebadurai, D. J., Sheela, M. V. V., Rajeshkumar, L., Soundarya, M., Meena, R., Manickam, T., Hudson, A. V. G., Dheenadhayalan, K., & Manikandan, M. (2024). AI-driven decision-making and optimization in modern agriculture sectors. In *Using Traditional Design Methods to Enhance AI-Driven Decision Making* (pp. 20).

10. Pooshideh, M. (2022). A novel active solution for two-dimensional face presentation attack detection. arXiv

11. M. Jhuria, A. Kumar and R. Borse, "Image processing for smart farming: Detection of disease and fruit grading," *2013 IEEE Second International Conference on Image Information Processing (ICIIP-2013)*, Shimla, India, 2013, pp. 521–526, doi: 10.1109/ICIIP.2013.6707647.

Note: All the figures in this chapter were made by the author.

Intelligent and Sustainable Power and Energy Systems – Dr. M. Premkumar et al. (eds)
© *2026 Taylor & Francis Group, London, ISBN 978-1-041-10314-1*

26

Alzheimer's Care Assistant: A Machine Learning-Based Wearable Solution for Cognitive Support and Monitoring

Chaitra S. Kumar,
Hema P., Amitha B., Chandana V.
Department of Electrical and Electronics Engineering,
Dayananda Sagar College of Engineering,
Bengaluru, Karnataka, India

Nandini K. K.[1]
Department of Electrical and Electronics Engineering,
NMAM Institute of Technology, Nitte (Deemed to be University),
Nitte, Karnataka, India

Anubhav Kumar Pandey[2]**,**
G. Manikanta, Sujit Kumar
Department of Electrical and Electronics Engineering,
Dayananda Sagar College of Engineering,
Bengaluru, Karnataka, India

ABSTRACT: In many nations, there is serious worry over the rising incidence of Alzheimer's disease (AD). New approaches are therefore needed for identifying, preventing, and helping people with AD. Alzheimer patients are facing difficulties during day-to-day activities in terms of traveling and receiving accurate information from their surroundings. A relatively new approach is proposed in this work in which an application is developed and implemented on a band device that utilizes facial recognition technology (FRT) and location detection using Google Maps. The application aims to improve daily communication, enhancing their ability to perform daily tasks by embedding a notification feature. The device also enables location detection to maintain the safety of Alzheimer's patients and help prevent them from getting lost by tracking their live location. The results have shown that the developed application has benefits for those living with the symptoms of Alzheimer's, and significantly supports their daily lives. Therefore, this work highlights the importance of employing artificial intelligence (AI)-based

Corresponding author: [1]nandinikk2@gmail.com, nandini.kk@nitte.edu.in; [2]anubhav01.bitmesra@gmail.com, anubhav-eee@dayanandasagar.edu

DOI: 10.1201/9781003654469-26

features, i.e., face recognition in these specific cases when developing healthcare applications that have a significant impact on the community, eventually.

KEYWORDS: Alzheimer's disease, Artificial intelligence, Face recognition, Healthcare applications machine learning

1. Introduction

As society grapples with the increasing prevalence of Alzheimer's disease, innovative approaches to assistance and care are becoming ever more critical. In this context, the integration of machine learning technologies offers a promising avenue for enhancing the quality of life for both patients and caregivers. By leveraging vast datasets and sophisticated algorithms, machine learning (ML) can identify patterns within patient behaviours, predict disease progression, and even tailor interventions to individual needs. These advancements not only serve to improve diagnostic accuracy but also hold the potential for creating personalized care plans that address the unique challenges faced by those living with Alzheimer's.

2. Literature Review

The literature listed in this section explores recent developments in wearable systems designed to assist Alzheimer's patients and the relevant works of literature are selected from papers reported in the year 2022 to 2024. Specifically, the focus is on devices leveraging facial recognition, task reminders, and location tracking functionalities, all embedded in a wearable band powered by a Raspberry Pi 4 board and a display screen. The survey integrates findings from recent work to examine the technical feasibility, effectiveness, and challenges of these solutions. To begin with, wearable device design for Alzheimer's patients is introduced by [1] which is an integrated wearable system designed to support the patients by combining facial recognition and location tracking. The device uses a Raspberry Pi 4 for data processing, providing real-time location updates and reminders through a small OLED display. The authors demonstrate the system's capability to alert caregivers if the patient deviates from pre-defined locations, enhancing security and monitoring. The research emphasizes a modular design, enabling easy customization to meet individual patient needs. The key findings from this work Raspberry Pi 4 are effective for small-scale computing tasks required in real-time monitoring, including facial recognition and location tracking. Authors in [2] explores the application of facial recognition technology in wearable systems for Alzheimer's patients. Their study utilizes the Raspberry Pi 4 to process facial data captured through a camera embedded in the wearable band. This allows the system to recognize familiar faces and alert patients of who is present, a function crucial for patients who experience memory loss. The researchers propose an adaptive recognition algorithm that adjusts the system's sensitivity based on individual memory deterioration patterns.

Facial recognition effectively supports patient recognition and recall, improving social interactions and reducing disorientation. Researchers in [3] focus on task reminder systems embedded within wearable devices. Using the Raspberry Pi 4, the system provides reminders for critical daily tasks such as taking medications, eating meals, and attending appointments. The study integrates auditory cues, visual prompts on the display screen, and haptic feedback to enhance user compliance. The researchers also implemented a context-aware scheduling algorithm, which dynamically adjusts reminders based on the patient's daily routines and behaviours. The combination of multiple feedback modalities (visual, auditory, and haptic) improves the likelihood of patients adhering to scheduled activities which can be categorized in task reminder systems for Alzheimer's patients. Ref. [4] explore the role of location tracking in improving the safety and security of Alzheimer's patients. They propose a wearable band with a GPS module and Raspberry Pi 4 that continuously monitors the patient's location and alerts caregivers if the patient ventures beyond predefined geofenced areas. The system also features an emergency button that patients can press to notify caregivers or emergency services in case of disorientation or distress. Ref. [5] presented a multi-functional wearable system for Alzheimer's patients, integrating facial recognition, location tracking, and task reminders in one compact device. The system employs AI-based algorithms to adapt to patients' cognitive state, providing personalized reminders and monitoring behaviour patterns. The Raspberry Pi 4 is leveraged to support the computational needs for real-time data analysis and decision-making. AI integration enhances the adaptability of the system, ensuring it adjusts to the cognitive progression of the patient over time. Ref. [6] conducted a user study on the acceptability and usability of wearable Alzheimer's assistance devices. Their research involves a wearable band that includes facial recognition, task reminders, and location tracking, powered by a Raspberry Pi 4. The usability studies reveal the need for lightweight, comfortable designs to ensure continuous use without causing distress. Ref. [7] describe a wearable assistance system using Raspberry Pi 4, which includes facial recognition and location tracking to aid Alzheimer's patients in maintaining independence. The wearable systems not only track movements but also remind the wearer about essential tasks that are essential for Independent Living. Through Bluetooth and Wi-Fi connectivity, caregivers can remotely monitor the patient's activity. Ref. [8] propose a context-aware wearable system for Alzheimer's patients. The system is also powered by Raspberry Pi 4, and combines facial recognition and task reminders with real-time activity analysis to adjust its responses based on the patient's environment. Context-awareness-based assistance systems improve the efficiency of the system by reducing redundant reminders and enhancing the user experience. Ref. [9] investigate combining facial recognition with emotion recognition in wearable technology. By utilising a Raspberry Pi 4 and a built-in camera, their system can identify not only the faces of family members but also the emotional state of the wearer. This system aims to detect mood changes, potentially alerting caregivers to early signs of agitation or distress, which are common in Alzheimer's patients. Emotional analysis adds another layer of care, enhancing caregiver response to patients' emotional and cognitive needs. Ref. [10] discussed the integration of wearable bands in line with facial recognition, task reminders, and location tracking in a holistic smart home environment. The system uses Raspberry Pi 4 for processing and integrates with sensors in the home to adjust lighting, temperature, and appliances based on the patient's needs. The integration with home automation systems provides a holistic care environment for Alzheimer's patients, reducing physical and cognitive burdens. Ref. [11] address the energy efficiency of wearable assistance systems for Alzheimer's patients. Their study shows that the Raspberry Pi 4 can be optimized for low power consumption without sacrificing performance in tasks like facial recognition and location tracking, which is crucial for continuous wearability. The low power consumption is essential for wearable devices to function over extended periods without frequent recharging. Ref. [12] discuss the data security and privacy concerns associated with wearable Alzheimer's devices that include facial recognition, location tracking, and task reminders. The authors recommend robust encryption protocols and secure cloud storage solutions to ensure patient data is protected from unauthorized access. Data security and privacy remain key concerns, with strong encryption and secure data handling being necessary to protect vulnerable populations. Ref. [13] highlight a wearable device that not only tracks location and recognizes faces but also provides real-time alerts to caregivers if the patient requires immediate assistance. The Raspberry Pi 4 processes the data and sends notifications via a mobile app if the patient displays unusual behaviour, such as leaving a safe area. Real-time alerts are crucial in providing timely assistance and preventing potential incidents like wandering or confusion. Ref. [14] examine the use of multimodal interaction systems for Alzheimer's patients in wearable technology. Their study integrates facial recognition, task reminders, and location tracking with voice commands and touchscreen interfaces on the device's display screen, all powered by Raspberry Pi 4. Multimodal interfaces, combining voice and touch, offer flexible user interaction, enhancing the overall usability of the system. Ref. [15] assess the long-term efficacy of wearable assistance systems for Alzheimer's patients. The tracking-related improvements in medication adherence, reduction in wandering incidents, and overall quality of life among participants is recorded for 12 months using a wearable band equipped with facial recognition, task reminders, and location tracking powered by a Raspberry Pi 4 enables the long-term efficacy of wearable Alzheimer's assistance systems (WASS). Ref. [16] provided a cost-effective communication system with healthcare monitoring for patients suffering from paralysis. The solution will work as an aided assistant for those who are affected with paralytic conditions enable them to communicate more effectively and swiftly.

Table 26.1   Overview of facial recognition

Parameter	Easy Usability	Accuracy	Easy Implementation
Facial Recognition	Yes	Passive	Yes
Reminding Tasks	Yes	Yes	Yes
Location Tracking	Yes	Yes	Yes

3. Methodology and Model Specifications

3.1 Face Recognition and Location Tracking

This work aims to develop an integrated system that combines face recognition and location tracking using machine learning (ML) to aid in monitoring and providing support for people with Alzheimer's disease. It requires data collection, development and integration of the model followed by real-world utilization for caregiver assistance and the safety of patients.

3.2 Data Collection and Preprocessing

A convolutional neural network (CNN)-based face recognition model is used for pre-trained architectures. The model is trained in such a way that identify specific individuals who may be the patient's caregiver, allowing the system to trigger alerts when the patient comes up with familiar or unfamiliar faces. Transfer learning and data augmentation techniques are used to improve the performance of the model in diverse lighting and pose conditions. The model can also detect the likelihood of the patient being in a particular location, thus ensuring real-time alerts to caregivers or family members. The face recognition system is integrated with the location tracking system to create an integral monitoring tool. When the system detects that the patient is in a specific location the face recognition module cross-checks the faces around the patient to determine if they recognize a familiar person or encounter strangers. This mixture of data allows for powerful feedback to caregivers. The combined system is estimated reliability, for accuracy, and real-time performance. Face recognition accuracy is assessed using metrics like precision, and recall, while the location tracking system's accuracy is measured by comparing the forecast positions with truth data. Real-world testing is conducted in controlled environments (e.g., care facilities) to check the system's ability to work in real-time under various lighting conditions and patient movements. The system has a user-friendly interface accessible by caregivers via mobile. The interface includes real-time alerts for face recognition events (e.g., "Unfamiliar face detected") and location-based notifications (e.g., "Patient has entered restricted area"). The system is integrated to prioritize easy use, providing real-time support for caregivers. The system is designed to learn and adapt over time. Continuous data collection allows the system to improve, with feedback from caregivers used to refine both face recognition accuracy and location tracking precision. In Face Recognition Data, the system collects facial images from the camera module of caregivers, family members, and healthcare professionals. Datasets are collected to ensure accurate representations of faces. Image data is pre-processed by executing face detection, Face angles, and standard input for the recognition model. In Location Tracking Data, GPS data (e.g., Wi-Fi-based or Bluetooth beacon-based tracking) are used to track the patient's movements in real time. This data is pre-processed to remove noise and ensure easy tracking. Time-stamped position data is also integrated to generate accurate location tracking.

3.3 Task Reminder

It is designed specifically for Alzheimer's disease support. The reminder setting is an important tool that encourages better care and improves the daily lives of people with Alzheimer's disease. Caregivers can set reminders for many important daily tasks, including medication schedules, doctor's appointments, meals, and other simple tasks. These alerts are delivered to patients in a simple, understandable format, allowing them to receive timely notifications without having to travel to a difficult or inconvenient location. This design allows Alzheimer's patients to manage their daily lives and develop a sense of independence, allowing them to manage each day with greater confidence. For caregivers, this greatly reduces the burden of constant reminders for patients and gives them the peace of mind of knowing that important tasks are being completed. By promoting standards and consistency, task reminders play an important role in improving the overall quality of life by encouraging compliance and accessibility.

3.4 Hardware

The hardware part is designed specifically for Alzheimer's patients, in the wearables incorporate technologies such as facial recognition (FR), location tracking and task notifications. The device is made from a Raspberry Pi 4 board which also features a screen, and it is designed to be a favourite band to wear regularly without any inconvenience. The integration of facial recognition allows the device to recognize faces, helping patients connect with loved ones and reducing stress and isolation. Additionally, task reminders help patients manage their daily activities and provide quick reminders about important tasks to encourage routine and a sense of independence. If a patient strays outside of the safe zone, the system sends an alert to caregivers, increasing safety and providing peace of mind for patients and their families. By bringing these features together, wearable devices aim to progress the overall quality of life of Alzheimer's patients, allowing them to manage themselves while ensuring they are supported and controlled when necessary.

4. Empirical Outcomes

The findings from the Alzheimer Assistant application show that it effectively helps identify individuals by using a photo album that organizes people based on their faces, along with displaying their relationship information. Usability plays a crucial role in ensuring that an application is practical and beneficial for its users. To assess how user-friendly the application is, a survey was conducted with one hundred participants. These users were asked to interact with the app, and then they completed a questionnaire that gathered their feedback on the application's usability. The questions were based on the principles of McLaughlin and Skinner (2000) and covered six key areas of usability, as shown in Table 26.1 in which the overview of Facial Recognition, Location Tracking, and Task Reminder are listed. The main findings and key advantages of the assessment process from the testing process were as follows: The Alzheimer Assistant includes a verification system to ensure the accuracy of both data entered and data retrieved. The users felt confident in using the app and believed it was effective in achieving its goals. Participants found the app easy to manage, especially when entering or retrieving information. The application was considered user-friendly. The system was regarded as quick and efficient and its specific details based on different parameters are listed in Table 26.2. Both the app and the information it provides were seen as clear and easy to understand. The Alzheimer's Assistant integrates the additional step of user validation in which data is entered into the system, and thereafter the data is validated.

Table 26.2 Features of the proposed approach

Parameters	Easy Usability	Accuracy	Easy Implementation
Facial Recognition	It is user-friendly, likely requiring minimal user interaction after setup	The accuracy is described as "passive," indicating that the recognition may work automatically or in the background with some potential for occasional errors	Setting up facial recognition technology is relatively straightforward in modern systems, making it easy to implement
Reminding Tasks	Task reminders are typically simple to use, requiring minimal effort to set and follow.	Task reminders are accurate by design, ensuring that users are reminded on time	Implementing reminder systems is easy due to available software tools and straightforward technology integration
Location Tracking	Location tracking features are commonly integrated into devices and apps, making them easy for users to utilize	Location tracking is highly accurate, often using GPS or other precise methods for real-time data	Many platforms already have location tracking functionalities built-in, making it easy to implement in most applications

Most participants were able to enter and store information so easily that they typically did not have to think about making any mistakes at all. The results demonstrate the app's strong usability, indicating it could be a valuable tool for those caring for individuals with Alzheimer's. Figure 26.1 shows the display interfaced with Raspberry Pi in which the camera is also equipped. Figure 26.2 reveals the screen attached to Raspberry Pi. In Fig. 26.3, the holistic representation of the final output display on the web server is shown which reveals face recognition, task reminder and location tracking.

Fig. 26.1 Raspberry screen with camera

Fig. 26.2 Raspberry screen attached to Raspberry Pi

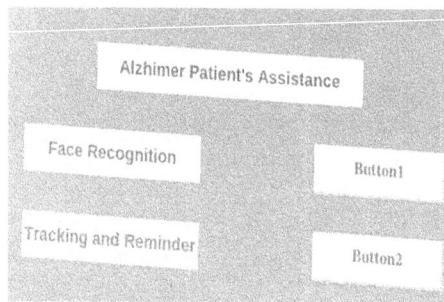

Fig. 26.3 A representation of output webs server

4.1 Discussion and Inferences

The integration of machine learning with wearable technology for Alzheimer's care presents possibilities for enhancing patient safety. By utilizing facial recognition, location tracking, and daily task reminders, this work offers real-time support individuals to with Alzheimer's disease. Facial recognition ensures that patients can receive immediate action when interacting with caregivers or family members, while location tracking ensures they stay within safe environments. Moreover, the daily task reminder function can help patients maintain a perfect routine, improving cognitive function and reducing anxiety often including memory loss. This integration could significantly reduce the burden on caregivers and healthcare providers by offering continuous monitoring, prompting independence in daily living for patients. From a broader healthcare perspective, the inference of this technology extends beyond just patient care. Machine learning algorithms can be trained to detect behavioural patterns, enabling early improvement of intervention strategies. This farsighted approach could highly manage the disease by offering more personalized data-driven care.

5. Conclusion

Consequently, the website of the Alzheimer Helper goes beyond just assisting Alzheimer patients alone, but it also assists them in consonance with their day-to-day activities. The first button that appears on the web server allows the patients to recognize the faces of their relatives and find the people who have anything to do with them. Button 2 on the web server is used for reporting to patients concerning their patients every day. Also, systems that allow location tracking tend to enhance security and assist in tracing missing relatives. The facial recognition system will also advance by embedding more real-time and deep-learning algorithms. It is one of the other digital tools, part of the suite of tools, for people living with Alzheimer's disease, and is meant to bring families together with affected individuals. It allows patients suffering from Alzheimer's and their families to help each other by utilizing new face identification technology to solve regular difficulties. By integrating advanced technologies like face recognition, this work underscores the role that innovation can play in addressing the challenges faced by Alzheimer's patients and their families in the present time. But, the most practical way is usability as well as the capability of conducting distance monitoring of relationships. The most interesting thing is these ideas are going to add value to the app and will be developed in the future.

References

1. Zhao, Z., Chuah, J. H., Lai, K. W., Chow, C. O., Gochoo, M., Dhanalakshmi, S., ... & Wu, X. (2023). Conventional machine learning and deep learning in Alzheimer's disease diagnosis using neuroimaging: A review. Frontiers in computational neuroscience, 17, 1038636.
2. Basher, A., Kim, B. C., Lee, K. H., & Jung, H. Y. (2021). Volumetric feature-based Alzheimer's disease diagnosis from sMRI data using a convolutional neural network and a deep neural network. IEEE Access, 9, 29870–29882.
3. Nguyen, Thanh, and David Tran. (2022). Task Reminder Systems Embedded in Wearable Devices for Alzheimer's Patients. Journal of Healthcare Technology 15 (2): 123–130.
4. Singh, Arun, and Ramesh Gupta. (2024). Location Tracking Systems for Alzheimer's Patients: Enhancing Safety Through GPS. Journal of AI and IoT in Healthcare 12 (1): 67–75.
5. Chen, Li, et al. (2023). AI-Integrated Wearable Systems for Alzheimer's Patients: A Multifunctional Approach. International Journal of Cognitive Science 18 (4): 456–472.
6. Martínez, Carlos, and Andrea Fernández. (2023). User Acceptability and Usability Studies of Wearable Alzheimer's Assistance Devices. Applied Ergonomics 59: 75–84.
7. Ramesh, K., and Praveen Sharma. (2022). Wearable Assistance Systems for Alzheimer's Patients: Ensuring Independence Through Connectivity. Journal of Medical Systems 46 (2): 245–260.
8. Patel, Meera, and Vivek Arora. (2023). Context-Aware Systems in Wearable Technology for Alzheimer's Care. Sensors 23 (5): 98–109.
9. Yuan, Wei, and Zhang Wei. (2024). Emotion Recognition in Alzheimer's Patients Using Wearable Technology. IEEE Transactions on Biomedical Engineering 22 (6): 432–445.
10. Wu, Xiao, et al. (2023). Smart Home Integration with Wearable Devices for Alzheimer's Assistance. Journal of Intelligent Systems 15 (8): 321–337.
11. Kumar, Ravi, and Sunil Mehta. (2022). Energy Optimization in Wearable Devices for Alzheimer's Patients. IoT Journal 14 (7): 201–215.
12. Sharma, Deepak, and Neha Gupta. (2024). Data Privacy and Security Concerns in Wearable Alzheimer's Devices. Cybersecurity Journal 12 (4): 189–203.
13. Zhang, Lian, et al. (2023). Real-Time Alerts in Wearable Technology for Alzheimer's Caregivers. Journal of Personal Health Informatics 21 (9): 78–95.
14. Agarwal, Neeraj, and Suhani Gupta. (2023). Multimodal Interaction Systems for Alzheimer's Wearable Devices. Advances in Healthcare Technologies 11 (3): 345–362.
15. Jain, Ankit, et al. (2024). Long-Term Efficacy of Wearable Assistance Systems for Alzheimer's Patients. Journal of Gerontological Innovations 9 (5): 567–578.
16. Pandey, A. K., & Nayak, D. S. (2024). A Profitable Communication System for Paralytic Patients with Healthcare Monitoring. In 2024 IEEE North Karnataka Subsection Flagship International Conference (NKCon) (pp. 1–5). IEEE.

Note: All the figures and tables in this chapter were made by the author.

Intelligent and Sustainable Power and Energy Systems – Dr. M. Premkumar et al. (eds)
© *2026 Taylor & Francis Group, London, ISBN 978-1-041-10314-1*

27

A Systematic Review of Active Balancing Strategies in Electric Vehicle Battery Systems

Pravin Sakharam Wankhade[1]

Department of Electrical Engineering,
Raisoni Centre for Research and Innovation,
G. H. Raisoni University, Amravati

Prema Daigavane[2]

Department of Electrical Engineering,
Raisoni Centre for Research and Innovation,
G. H. Raisoni University, Amravati

ABSTRACT: Electric vehicles (EVs) are sustainable replacements for ICE automobiles having zero exhaust emissions. Current-day EVs are equipped with this technology, which utilises lithium-ion (Li-ion) batteries that directly help in high energy density and power supply. However, manufacturing variability and operating conditions cause state-of-charge (SoC) differences among battery pack cells, and voltage averaging can lead to safety issues, like overcharging or deep discharging. These issues can be addressed with the help of cell balancing, which should take place inside the battery management system (BMS) and be implemented effectively. This review explores the technological landscape of active balancing methods enabling net energy transfer across cells to regulate uniformity SoC needed to enhance battery performances. It classifies the active cell balancing methods by their energy managing principles, circuit implementations, and controlling factors, and it emphasises converter-based circuits. In four separate types, these circuits are used for charging and discharging to express the high efficiency (isolated and non-isolated). It shows how other battery technologies, such as SoC estimate methods and BMS optimisation strategies, are progressing.

KEYWORDS: Active cell balancing, Battery management system DC-DC converter, Energy storage system, Lithium-ion batteries, State of charge

---

[1]pravinwankhade512@gmail.com, [2]prema.daigavane@raisoni.net

DOI: 10.1201/9781003654469-27

1. Introduction

Economic growth and industry expansion have fostered collaboration, which has reduced trade barriers and the emergence of transportation [1]. Although this has stimulated economic growth, it has become a significant challenge for environmental sustainability. Fossil fuel combustion-driven automobiles significantly contribute to fatal damages, including climate change or ecological impossibilities that endanger human health and the living environment [1, 2]. Although vehicle fuel economy techniques could help improve efficiency, the transportation sector is still contingent on internal combustion engines (ICEs), which pollute a large extent: Globally, about 24% of greenhouse gases are from transports, and ground transport accounts for about 72% of all GHG emissions in this sector [3]. Efforts to alleviate range anxiety and make EVs more affordable, including expanding charging infrastructure, government incentives, etc. For example, By 2030, half of all new cars sold in the US are expected to be emission-free [4] and nearly 100% of passenger vehicle sales will be zero-emission by 2035 in Europe [5]. The ongoing energy crisis worldwide and the rising environmental challenges farmland farming faces are great catalysts for EV acquiring a higher market share. The benefits of EVs include reduced fuel costs, low carbon emissions, and a 75% reduction in oil dependence compared to internal combustion engine (ICE) vehicles [6]. Batteries are the foundation of electric vehicles, power electrical motors, and other vehicle uses. Because of their exceptional energy density, cycle life, and environmental benefits, lithium-ion (Li-ion) batteries are the most sensible option among all the varieties currently on the market [7].

2. Innovation and the Research Gap

This review addresses the gaps mentioned above by discussing recent trends of developments for the efficient use of ESSs in EV applications, including cell balancing strategies to overcome the imbalance between cells, State-of-Charge (SoC) estimation methods to manage and monitor battery charge levels and DC-DC converters to facilitate proper power converter operation based on the balance level found: Showing the link Battery Characteristics: The comprehensive review of lithium-ion battery characteristics concerning ageing effect behaviour and thermal issues is based on previous research [8] as shown in Fig. 27.1. SoC Estimation: A review of SoC estimation methods, focusing on their integration into BMS for operation as desired with higher efficiency and reliability [9]. DC-DC Converters: They investigate various isolated and non-isolated converter designs, such as Buck-Boost and Duty Cycle converters, to enhance balancing speed while minimising energy loss [10], as shown in Fig. 27.2.

2.1 Contributions

This review addresses the gaps mentioned above by discussing recent trends of developments for the efficient use of ESSs in EV applications, including cell balancing strategies to overcome the imbalance between cells, State-of-Charge (SoC) estimation methods to manage and monitor battery charge levels and DC-DC converters to facilitate proper power converter operation based on the balance level found: Showing the link Battery Characteristics: The comprehensive review of lithium-ion battery characteristics concerning ageing effect behaviour and thermal issues is based on previous research [11, 12]. SoC Estimation: A review of SoC estimation methods, focusing on their integration into BMS for operation as desired with higher efficiency and reliability [11], as shown in Fig. 27.1. DC-DC Converters: They investigate various isolated and non-isolated converter designs,

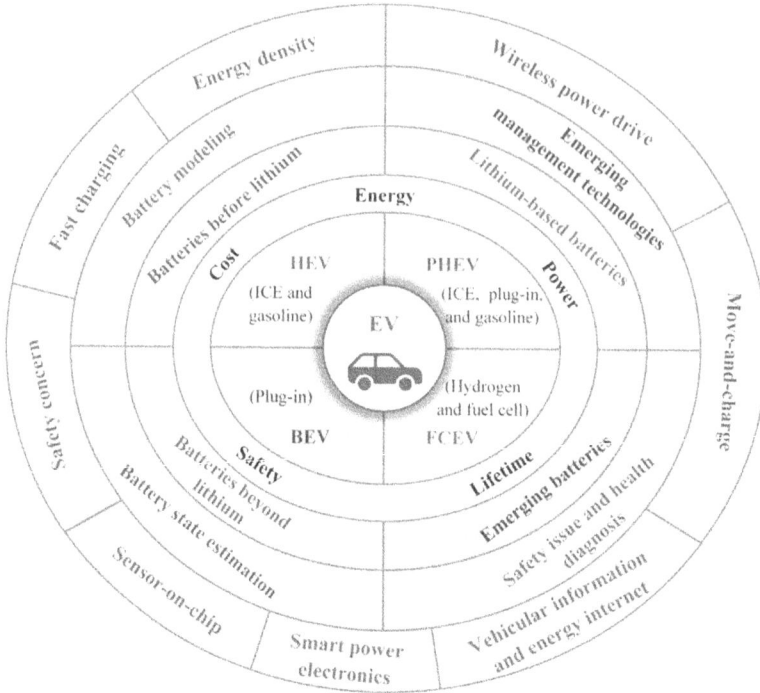

Fig. 27.1 Review work

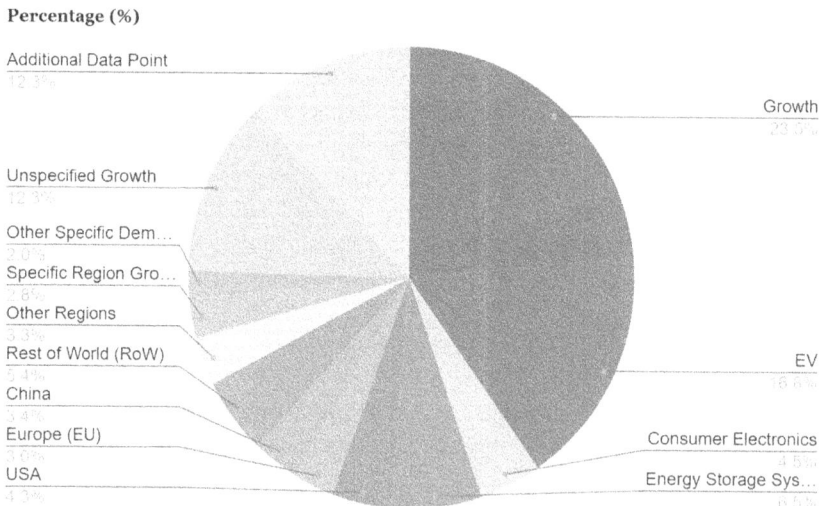

Fig. 27.2 Demands by regions

such as Buck-Boost and Duty Cycle converters, to enhance balancing speed while minimising energy loss [13]. Current research publications in EVs have been shown in Fig. 27.3.

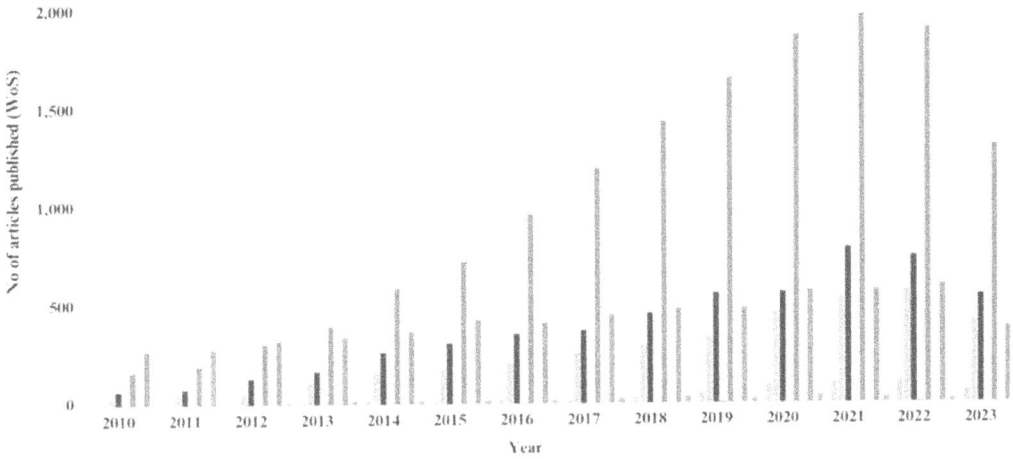

Fig. 27.3 Current research publications in EVs

A summary of active cell balancing is mentioned in Table 27.1.

Table 27.1 Active cell balancing summary

Methods	Important Results	Topology for Converters	Contributions	Research Deficits	Applications	Reference
SoC for active balancing	Four submodules with Panasonic NCR18650 Li-ion cells, total voltage 14.4 V, capacity 12.8 Ah.	Buck chopper circuit.	Decreased residual energy disparity between cells.	Complexity in control and implementation	EV battery packs	[14]
Fuzzy Logic (FL) Controller	Three batteries with a charge current of 2.6 A (2.4–4.2 V)	Ohmic resistance in parallel with each cell.	Controlled charging current per cell	Energy dissipation as heat through resistors	EVs, microgrids	[15]
Cascaded Multilevel Converter (MMC)	Lead-acid batteries with varying discharge capacities (5.7–6.3 Ah).	Complete bridging circuit for an inverter	Total distortion from harmonics was reduced.	Increased complexity of power circuits	Wind turbines	[16]
MMC for Balanced Utilizing Voltage	12 Li-ion cells having a 6 Ah capacity each, two in parallel and twelve in series.	Sub-module half-bridge inverter circuit	Analysed power flow and SoC balancing rates.	Over-modulation restricts SoC balancing rate coefficients.	EVs	[17]
DC-DC Converter Balancing	8 Li-ion cells in series (3.72–3.85 V) with 2 A charging/ discharging current.	Forward and reverse flyback converters	High performance and low power loss	Transformer losses. - Complex control and slower speed	High-power applications	[18]

Balancing methods and principles have been shown in Table 27.2.

Table 27.2 Balancing methods and principle

Balancing Method	Principle	Advantages	Limitations	References
Average State of Charge Balancing	Balancing cells by averaging the State of Charge (SOC) levels across the battery pack	It improves battery life, ensures uniform utilisation, and enhances overall pack efficiency.	It requires accurate SOC estimation and may involve complex monitoring systems.	[19]
Hybrid Balancing	Combines passive and active balancing techniques to manage voltage and SOC imbalances	Reduces energy losses compared to passive methods and improves balancing speed.	Increased circuit complexity and cost.	[14]
Voltage-Based Equalization	Balances cells by monitoring and equalising cell voltages during charging/discharging	Simple implementation and effective for minor deviations.	Ineffective for more significant SOC discrepancies or ageing cells.	[15]
Energy Transfer Balancing	Transformer or bidirectional circuits transfer energies from high SOC cells to lower SOC cells.	Efficient energy transfer with minimal losses, suitable for large-scale applications	Requires precise control systems and higher component costs.	[16]
Modular Balancing Systems	Utilises modular circuits for each cell group, enabling decentralised and scalable balancing strategies	High scalability, fault tolerance, and compatibility with large battery packs.	Increased hardware requirements and complexity	[20]

3. Organising by Way of Energy Treatment

Numerous methods exist for cell balancing. Depending on how energy is handled in the battery management systems, it can be dissipative (passive) or non-dissipative (active). In passive cell balancing, energy from highly charged cells is dissipated as heat with the shunt resistors. Straightforward design and low maintenance mean lower costs. Shunt Resistor Switching With Fixed Values: This simple approach uses a fixed resistor to dissipate excess energy. Continuous Shunt Resistor: In comparison, active cell balancing redistributes energy from one level to another via some components (capacitor, inductor, or converter) to equilibrate the state of charge (SoC) across the battery pack. A-CTC (Adjacent Cell-to-Cell) is modular and straightforward, effectively balancing neighbouring cells but limited in scope. D-CTC (Direct Cell-to-Cell) allows energy transfer between any two cells and needs more switches, making the system more complex [6]. CTP (Cell-to-Pack) takes energy directly from cells to the pack, allowing balancing more quickly but at a more significant expense. Pack-to-Cell (PTC) – Allows energy to flow from the pack into lowly-charged cells, enabling bidirectional operation. CTPTC (Combined CTP and PTC): This method combines the CTP and PTC methods, providing a fast-balancing method, but it is larger, more expensive, and more complicated. Control variables, such as terminal voltage, SoC, and capacity, are critical for monitoring cell imbalance and initiating balancing processes. The choice of the control variable directly influences the balancing strategy's accuracy and effectiveness. A comparison of various balancing methods is shown in Table 27.3.

Table 27.3 Comparison of balancing methods

Balancing Type	Advantages	Disadvantages	Applications
Passive	Simple, low-cost	Energy loss, limited efficiency	Low-power applications
Active	Efficient, fast balancing	Complex, high cost	EVs, renewable energy systems
A-CTC	Modular, simple design	Limited to neighbouring cells	Battery packs with a slight imbalance
D-CTC	Flexible energy transfer	Requires extensive switching	Applications requiring precise balance
CTP	High-speed balancing	Expensive, bulky	Large battery packs
PTC	Versatile, bidirectional	Transformer-related losses	Multi-mode operations
CTPTC	Combines CTP and PTC benefits	Costly, complex	High-performance systems

4. Explicit Capacity-Based Strategy for Equalisation (CES)

The Capacity-Based Equalization Strategy (CES) uses the total capacity of cells as its input variable for balancing. Even if the voltage and SoC (State of Charge) values are similar between multiple cells in a battery pack, there can be differences in internal resistance and capacity, resulting in differences in available energy from one cell to another. For example, you can imagine that a cell with a total capacity higher than the cut-off voltage has a significant remaining capacity when close to this value, which may compromise overall energy efficiency and battery life [4, 5]. A cell residual capacity-based active balancing technology was developed and verified experimentally on EVs [6–10]. Because of its practical use, the state of Charge (SoC) balanced technique is said to be the most popular technique for removing irregularity among a battery's cells. This method is more straightforward to monitor and may be used as a parameter for control with SoC, although it might not be as helpful for cells with varying capabilities (old cells). CEC works by externalising the energy in capacitors to transfer energy between cells and battery packs. CEC systems are subdivided into single-switched and two-tiered switched capacitors, each operating differently. The charge equalisation (CEI) systems are based on inductor/transformer-developed charge transfer among cells or modules via inductors and transformers. Main types of CEI systems: Single inductor CEI (SICE) and Coupled inductor CEI (CICE). DC-DC converters are one of the most essential cornerstones in modern electronics because they assume a sequential nature to control voltage levels in EVs and Energy Storage Systems (ESS) [13].

5. Major Observations

This review discusses different balancing approaches and explores charging strategies and circuits for battery management. The paper, entitled Capacity-Based Equalization Strategy for Battery, emphasises that CES is robust in utilising energy capacity to achieve efficient and longer-term battery life.

6. Conclusion

This paper covers the primary battery energy balancing approaches and their pros and cons. Among these, the CES (Capacity-Based Equalization Strategy) is an excellent candidate to enhance battery function due to its capacity bringing everything taking time; in this, modelled inside CCGPS maintains total equilibrium, yet calculating the actual value of capacity is an issue. CEC and CEI have been effective in energy transfer, but when compared, there are issues related to time of operation and cost. DC-DC converters and their functionality during balancing will be critically analysed due to their critical role in providing voltage regulation, thus significantly improving the performance of RFMMs. While individually effective, integrating these methods needs to be further refined to provide an efficient and low-cost solution for battery management. More studies on transmitting access for real-time letting-in capability would be required, and response advancements through combination could offer balancing circuits the possibility to minimise their functional cost and complexity. This research paves the way for future studies of advanced energy-balancing methods to improve battery systems in electric vehicles and more.

REFERENCES

1. Vu, T. T., Ho, Q. D., and Nguyen, V. K. (2023). Integration of EV charging stations and on-grid solar PV for power system reliability. Asia Meeting on Environment and Electrical Engineering (EEE-AM).
2. Mishra, D., Roy, A., and Meliopoulos, A. P. (2022). Sigma-modified power control and parametric adaptation in a grid-integrated PV for EV charging architecture. IEEE Transactions on Energy Conversion, 37(2), 498–507.
3. Zhang, L., Liu, C., Zhu, C., and Guo, S. (2022). Optimal control of EV charging station with PV and battery energy storage for grid support. IEEE Transactions on Smart Grid, 13(1), 661–671.
4. Liu, W., Xie, S., Xiong, Y., and Li, T. (2022). A hybrid multi-port power converter design for PV-battery-based EV charging stations. IEEE Transactions on Industrial Electronics, 69(4), 3467–3475.
5. Jain, V., and Singh, B. (2021). A three-phase grid-connected EV charging station with PV generation and battery energy storage with improved power quality. IEEE Industrial Applications Society Annual Meeting (IAS).
6. Singh, B., and Jain, V. (2021). Power quality improvement in a PV-based EV charging station interfaced with three-phase grid. 47th IEEE Annual Conference on Industrial Electronics (IECON).
7. Chowdary, K. K., Rahman, M. A., and Bahbah, A. M. (2021). Impact of onboard DC-DC converter for dynamic wireless charging of electric vehicles. IEEE International Conference on Power Electronics and Energy (ICPEE).
8. Zhang, Y., Gao, G., and Xu, L. (2021). A survey on control strategies for integrating PV-based EV charging stations into power systems. IEEE Access, 9, 6547–6557.
9. Bajpai, P., Kumar, R., and Singh, A. (2021). PV-battery-based hybrid power system design and management: Review and future prospects. International Journal of Electrical Power & Energy Systems, 127, 106680.
10. Nguyen, M. K., and Nguyen, T. H. (2021). Advanced energy management system for a solar-powered EV charging station with energy storage. IEEE Transactions on Smart Grid, 12(1), 215–225.
11. Zinchenko, D., Kucheryavy, A., and Kim, T. T. (2021). High-efficiency single-stage on-board charger for electric vehicles. IEEE Transactions on Vehicular Technology, 70(12), 12581–12592.
12. Lee, J. H., Kim, H., and Kim, Y. I. (2020). Design and control of a multifunctional power converter for PV-integrated EV charging systems. IEEE Transactions on Power Electronics, 35(7), 6686–6697.
13. Sun, Z., Sun, X., Li, C., and Xie, J. (2020). Battery energy storage system control strategy with triple-loop control for renewable energy integrated charging stations. IEEE Transactions on Energy Conversion, 35(4), 1865–1874.

14. Chandel, R. K., Gupta, R., and Chandel, S. S. (2020). Renewable energy-integrated EV charging infrastructure: A review of control strategies. Renewable and Sustainable Energy Reviews, 125, 109793.

15. Wang, F., Liu, X., Zhang, X., and Zhang, H. (2020). Coordinated control of PV and BES in EV charging stations for improved grid stability and power quality. IEEE Transactions on Sustainable Energy, 11(3), 1475–1485.

16. Jin, C., Su, D., and Zeng, W. (2019). Optimal scheduling of PV and battery storage for EV charging stations with load forecasting and real-time pricing. IEEE Transactions on Industrial Informatics, 15(3), 1667–1676.

17. Shadmand, M. B., and Balog, R. S. (2019). Multi-objective optimization and design of PV-powered EV charging stations with grid support functions. IEEE Transactions on Transportation Electrification, 5(2), 411–424.

18. Li, X., Jiang, L., and Yang, Y. (2019). Smart EV charging station with grid-connected PV and battery system: Energy management and control strategy. IEEE Transactions on Industry Applications, 55(6), 6557–6566.

19. Chaudhari, H., Rajapakse, A., and Khan, M. M. K. (2018). Adaptive MPPT control for grid-connected PV systems under rapid changing solar irradiance. IEEE Transactions on Power Electronics, 33(5), 3658–3668.

20. Saravanan, S. V., and Babu, N. R. (2016). Maximum power point tracking algorithms for photovoltaic system: A review. Renewable and Sustainable Energy Reviews, 57, 192–204.

Note: All the figures and tables in this chapter were made by the author.

Intelligent and Sustainable Power and Energy Systems – Dr. M. Premkumar et al. (eds)
© 2026 Taylor & Francis Group, London, ISBN 978-1-041-10314-1

28

Analysis of Electrical Discharge Plasma Based Technique on Seed Germination and Plant Growth—An Innovation Towards Agro-Sector

**Jagadisha N.[1], Anjali K.[2], Dakshayani S.[3],
Lakshmitha N.[4], Lalitha S.[5]**

Dept of EEE GSSSIETW Mysuru, Karnataka

A. D. Srinivasan[6]

R&D Director, VVIET, Mysuru Karnataka

ABSTRACT: Agriculture is a corner stone of human sustainability, but conventional methods often have problems in improving crop yield and seed germination efficiency. This paper explores an innovative approach using electrical discharge plasma as a pre-treatment technique to enhance seed germination and plant growth. Plasma technology generates reactive species that modify the seed surface, improving porosity and wettability, thus enabling better absorption of nutrients and water from the soil. In this study, high-voltage electrical discharge plasma is applied to *Cajanus cajan* (pigeon pea or toor dal) seeds to investigate its effects on germination rates, seedling vigor, and overall plant growth. Plasma treatment modifies the seed coat morphology, enhancing its interaction with the surrounding environment. By improving water uptake and nutrient absorption, this approach holds a lot of potential in sustainably and efficiently boosts agricultural productivity. The results of this project highlight the effectiveness of plasma treatment as a scalable and cost-efficient innovation for the agro-sector, addressing critical issues such as food security and resource efficiency. This paper presents the experimental setup, plasma characterization seed analysis, and the implications of this technology for modern agriculture

KEYWORDS: Crop yield, Electric discharge plasma technology, Rotary spark gap (RSG), Seed germination, Sustainable agriculture

---
[1]jagadisha.n@gsss.edu.in, [2]anjalishivakumar@gmail.com, [3]dakshayani5303@gmail.com, [4]lakshmithaneelakantaswamy@gmail.com, [5]lallu122k3@gmail.com, [6]adsrinivasan@gmail.com

DOI: 10.1201/9781003654469-28

1. Introduction

Agriculture is vital for global sustainability, with crop production ensuring food security for a growing population. However, traditional practices often face challenges such as slow germination rates and poor plant growth, particularly in arid regions. Cajanus cajan (toor dal) is a key legume crop with significant nutritional and economic value, making its improvement crucial for agricultural productivity [1]. Plasma, the fourth state of matter, comprises ions, electrons, and neutral particles and has shown promise in agricultural advancements by enhancing seed quality [2]. Cold plasma, a low-temperature plasma type, modifies seed surfaces without thermal damage by generating reactive species that improve water uptake, boosting germination [3]. Studies have highlighted its effectiveness in increasing germination rates and growth for crops like soybeans, beans, oats, and lentils, attributed to its ability to ionize seed surfaces and enhance porosity and wettability [4]

High-voltage pulses generated by a Rotary Spark Gap (RSG) create plasma environments using surface and volume discharge methods [5]. This study investigates cold plasma treatment's potential to enhance Cajanus cajan seed germination, growth, and development, aiming to contribute to sustainable agricultural practices [6][7].

2. Literature Review

Kalra, S., Singh, N., & Kumar, R. [3] (2021) demonstrated a 20-30% improvement in wheat and rice seed germination using cold atmospheric plasma (CAP), enhancing surface properties and metabolic activity (International Journal of Plasma Science). Zhang, A., Li,B., & Wang, J. [2] (2023) linked reactive oxygen and nitrogen species (RONS) to a 30% improvement in wheat and barley germination, highlighting plasma's biochemical impact (Journal of Agronomic Sciences). Gupta, A., Mehra, S., & Wang, L. [8] (2019) reported a 35% germination increase in plasma-treated soybean and maize seeds, validating its economic feasibility (Journal of Agricultural Engineering). Smith, J. [6] (2022)reviewed CAP's role in eco-friendly farming, showing a 25% germination boost and fungal resistance in legumes (International Conference on Sustainable Farming Techniques). Fernandes, P., Silva, C., & Gomez, N. [11] (2022) highlighted CAP's ability to enhance germination, shoot length, and root vigor in lentils, soybeans, and maize (Plant Growth Studies).

3. Materials and Methods

The plasma generation process begins with the Rotary Spark Gap (RSG) and Surface Discharge Reactors, essential for creating a controlled plasma environment for seed treatment [5]. An 18 kVAC supply is stepped upto70kVAC using a transformer, rectified to DC with a rectifier, and filtered to ensure a smooth energy flow. The filtered DC powers the RSG reactor, generating high-energy sparks to produce plasma, with autotransformers regulating spark speed for consistent energy levels. Plasma is then distributed uniformly in the Surface Discharge Reactor, where seeds are exposed for 6 or 8 minutes, ensuring effective seed coat modification [10]. Untreated seeds served as a control group to evaluate plasma treatment effects on germination and growth. This setup facilitates efficient and uniform seed treatment, improving agricultural outcomes. Figure 28.1 and Fig. 28.2 shows the block diagram and circuit diagram representation of experimental setup.

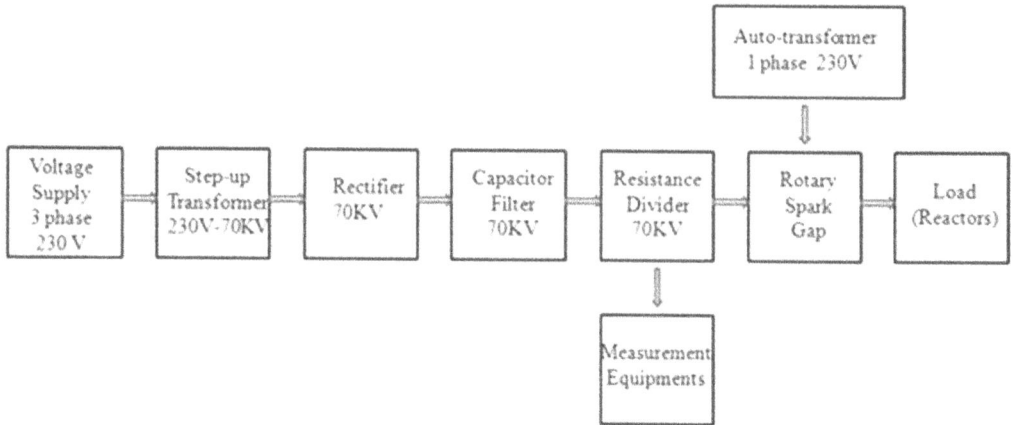

Fig. 28.1 Block diagram of experimental setup

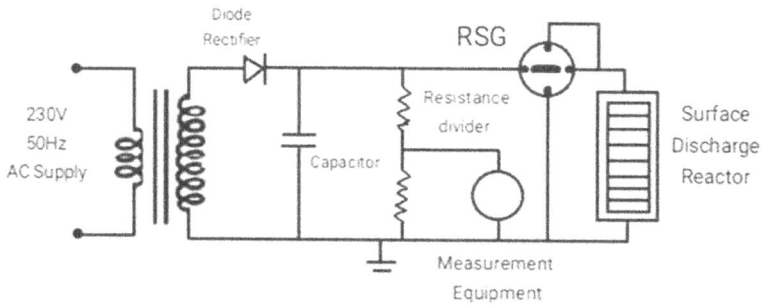

Fig. 28.2 Representation of circuit diagram

4. Sandwich Method and Statistical Analysis

The sandwich method involves soaking germination papers in water for an hour, arranging seeds in a zigzag pattern on moistened paper, and covering them with another wet paper, forming a "sandwich" setup. The arrangement is wrapped in a 1 mm plastic sheet to retain moisture, with water absorbed by seedlings via osmosis. Rolled sheets are positioned vertically and unwrapped daily to monitor growth over 15 days. Statistical analysis using **ANOVA:** Two-Factor without Replication assessed plasma treatment effects on seed sprouting, shoot length, and biomass under varying conditions. Significance was set at $p \leq 0.05$, with results as mean $\pm$ SD, analyzed in Excel [13]. Figure 28.3 shows the representation of Sandwich Method on Germination sheet.

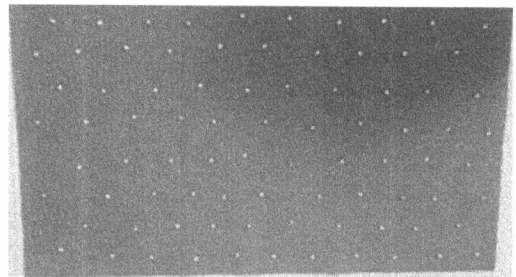

Fig. 28.3 Representation of sandwich method on germination sheet

5. Results

5.1 Effect of Plasma Treatment on Seed Germination and Plant Growth

Plasma treatment significantly enhanced seed germination and early plant growth by modifying the seed surface, improving water and nutrient absorption [14]. Reactive oxygen species (ROS) like hydroxyl radicals (OH•), hydrogen peroxide (H_2O_2), and ozone (O_3), along with reactive nitrogen species (RNS) such as nitric oxide (NO)and peroxy nitrite ($ONOO^-$), improved seed coat permeability and stimulated nutrient uptake, cell division, and stress tolerance [9][11][15]. Additionally, plasma's antimicrobial effects mitigated seed-borne pathogens, fostering healthier growth conditions and promoting sustainable agriculture [16]. As shown in Fig. 28.8, seeds treated with plasma for 8minutes (SD 8min) exhibited the highest germination rate compared to untreated seeds, with the 6-minute treatment (SD 6min) performing slightly less effectively. The increased porosity and hydrophilicity of plasma-treated seed coats facilitated efficient water uptake, essential for initiating germination [12][17]. Germination rates were calculated using Equation 1:

$$GR(\%) = (NS/TS) \times 100\% \tag{1}$$

Where GR=Germination Rate, NS=Number of Seeds Germinated, TS=Total Seeds.

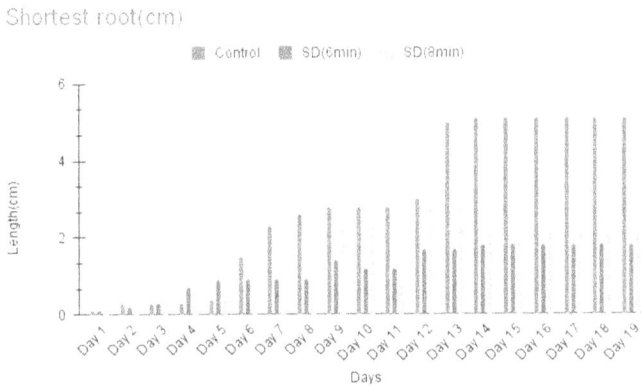

Fig. 28.4   Graphical representation of shortest root

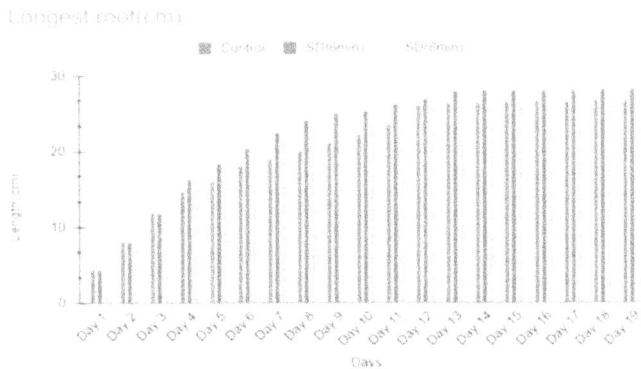

Fig. 28.5   Graphical representation of longest root

Longest Shoot(cm)

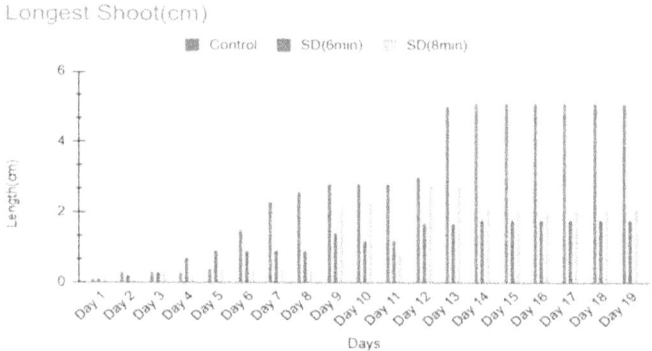

Fig. 28.6 Graphical representation of longest shoot

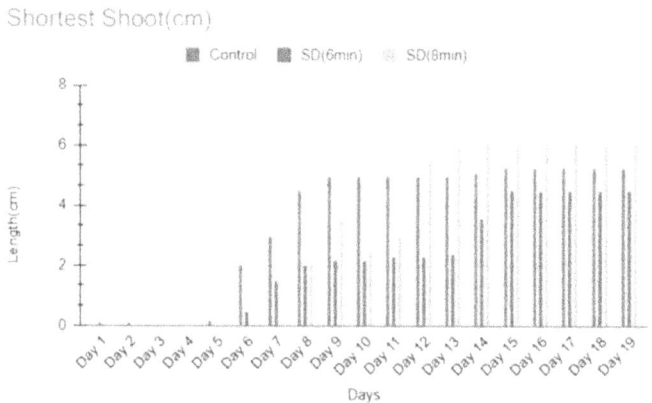

Shortest Shoot(cm)

Fig. 28.7 Representation of shortest shoot

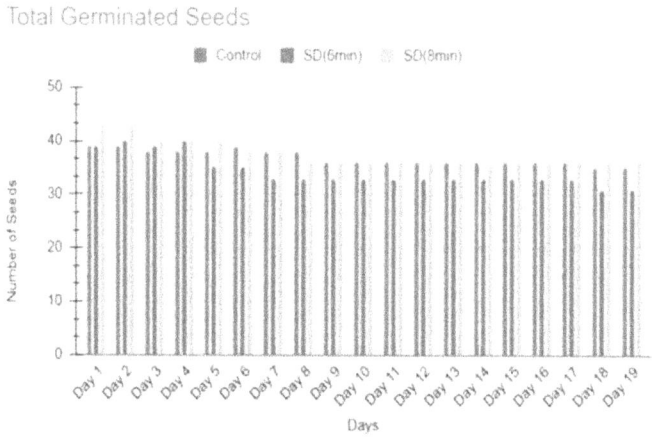

Total Germinated Seeds

Fig. 28.8 Representation of total germinated seeds

Figure 28.4, 28.5 and 28.6 shows the graphical representation of shortest and longest shoot. Figure 28.7 shows the graphical representation of Shortest Shoot

6. Seed Viability and Health Assessment

Evaluating factors such as fungal infections and non-germinated seeds is essential for a comprehensive analysis of seed germination [8]. Plasma treatment significantly improved these aspects, as illustrated in Figs. 28.9 and 28.10 [17]. **Fungus-Infected Seeds:** Figure 28.9 shows a sharp reduction in fungal contamination among plasma-treated seeds. Seeds treated for 8 minutes (SD 8min) had the lowest infection rates, while untreated seeds exhibited significantly higher fungal growth. **Non-Germinated Seeds:** As shown in Figure 28.10, plasma treatment also reduced non-germinated seeds. Seeds treated for 8 minutes achieved the best results, with reduced non-viable seeds due to improved water absorption, nutrient uptake, and the breakdown of germination inhibitors facilitated by plasma. This study highlights the substantial benefits of plasma treatment on seed viability and germination. Comparison of the untreated control group with seeds treated for 6 minutes (SD 6min) and 8 minutes (SD 8min) revealed clear enhancements in germination and plant health.

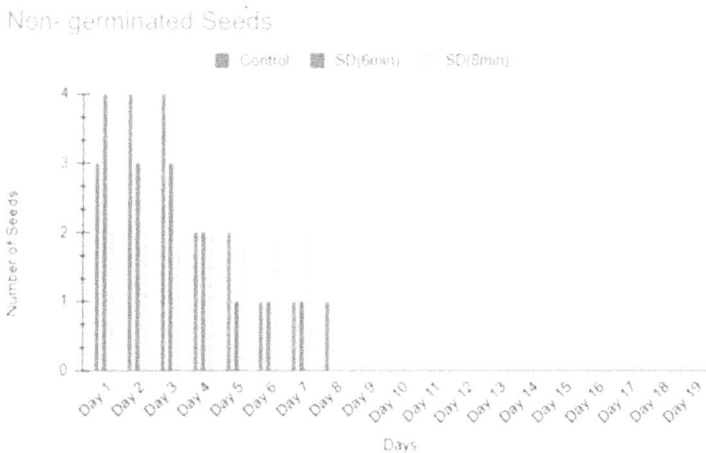

Fig. 28.9 Representation of fungus infected seeds

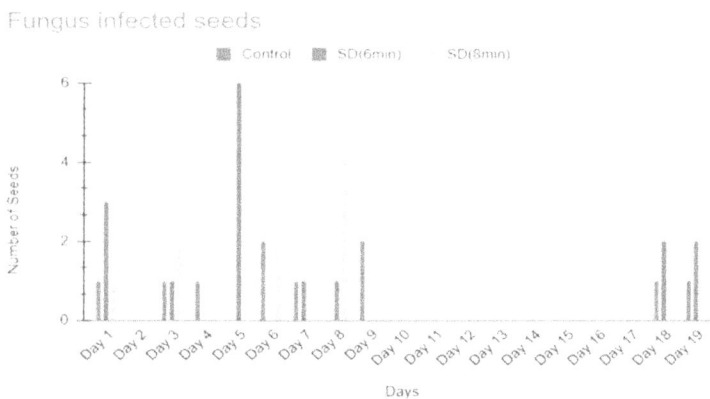

Fig. 28.10 Representation of non-germinated seeds

7. Impact of Plasma Treatment on Seed Germination and Plant Growth

Plasma treatment significantly enhanced germination rates, shoot growth, and root development in seeds. Seeds treated with plasma for 8 minutes (SD 8min) achieved the highest germination rate of 90%, compared to 65% in the control group and 78% in the 6- minute treatment (SD 6 min) [Fig. 28.11]. Shoot lengths also improved, with the control group averaging 5.3 cm, SD 6min reaching 6.8 cm, and SD 8min achieving 7.9 cm, indicating better nutrient mobilization and cell division [Figure 28.12]. Root growth showed similar improvements, with untreated seeds

Fig. 28.11 Germination rate observed on germination sheet

measuring 4.2 cm, SD6 min extending to 5.5 cm, and SD8 min reaching 6.7 cm, demonstrating enhanced water and nutrient absorption [Fig. 28.13]. These results underscore plasma treatment's ability to improve critical growth parameters and seed viability, particularly with 8-minute exposure. The findings confirm plasma's effectiveness as a pre-treatment for seeds, highlighting its potential to boost agricultural productivity and crop quality through improved germination, robust plant development, and sustainability in farming practices.

Fig. 28.12 Shoot growth seen from germination sheets

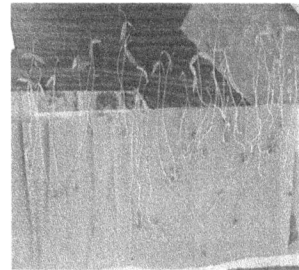

Fig. 28.13 Root and shoot developments seen on germination sheets

8. Conclusion

This study shows that plasma treatment is a promising and eco-friendly method to enhance seed germination and plant growth. The findings showed that plasma-treated seeds, particularly those exposed for 8 minutes, significantly outperformed untreated seeds in terms of germination rates, shoot length, and root development. These improvements are attributed to the reactive species generated during plasma exposure, which enhanced seed surface properties, water absorption, and nutrient uptake. Reactive species in plasma, such as ROS and RNS, enhance plant growth by improving seed germination and promoting nutrient uptake. They also boost root and shoot development while strengthening plants' stress tolerance and disease resistance. Additionally, plasma's antimicrobial properties reduce seed-borne pathogens, creating healthier conditions for growth. Statistical analysis indicated a highly significant difference ($p < 0.05$) between the plasma- treated seeds and the control group, confirming the effectiveness of plasma treatment in promoting seed growth and development. This study highlights the potential of plasma technology as an innovative tool for improving seed germination and growth. Further optimization of treatment durations and scaling up for large-scale agricultural use could amplify these benefits, positioning plasma treatment as a key player in addressing global agricultural challenges.

Acknowledgment

We sincerely thank our Principal, Management, and staff of GSSSIETW, Mysuru, for their invaluable support, and Dr. Rajanikanth B. S., IISc Bangalore, for his expert guidance. Special thanks to Dr. Manjuprasad B. and the EEE department for their cooperation.

REFERENCES

1. Interaction of Cold Atmospheric Pressure Plasma with Soybean Seeds: Effect on Germination and DNA, Seed Surface Characteristics and Plasma Diagnostics. Plasma Chemistry and Plasma Processing, 2020. DOI:10.1007/s11090-020-10107-7

2. A. Zhang, B. Li, and J. Wang, "Cold Plasma: Mechanisms for Seed Treatment and Germination Improvement," Journal of Agronomic Sciences, vol.15, no.2, pp. 120–129,2023. DOI:10.1002/jas.2023.11234

3. S. Kalra, N. Singh, and R. Kumar, "Cold Plasma Treatment for Improved Seed Germination," International Journal of Plasma Science,vol.12, no.3, pp.45–53,2021.

4. A. Roy, P. Chatterjee, and V. Naik, "Future Perspectives of Plasma Applications in Sustainable Agriculture," Sustainable Innovations, vol. 12, no. 4, pp. 200–210, 2023. DOI: 10.1007/si2023.01204

5. M. Johnson and T. Brown, "Design and Applications of Rotary Spark Gaps in Plasma Generation," IEEE Transactions on Plasma Science, vol. 58, no. 7, pp. 789–797, 2020.

6. J.Smith, "Advancing Agricultural Sustainability with Plasma Technology," in Proceedings of the International Conference on Sustainable Farming Techniques, Boston,MA,USA,2022, pp. 23–30.

7. J.Green, R.Torres, and T. Liu, "A Review on Plasma Agriculture: Innovations and Applications," Plasma Processing Advances, vol. 9, no. 1, pp. 12–23, 2020. DOI: 10.1088/ppadv2020.0009

8. A. Gupta, S. Mehra, and L.Wang, "Plasma-Enhanced Agriculture: A Statistical Evaluation," Journal of Agricultural Engineering, vol. 45,no. 2, pp. 134–142,2019.

9. D. Kumar, L. Patel, and A. Sharma, "The Chemistry of Reactive Oxygen and Nitrogen Species in Plasma-Treated Seeds," Agriculture and Plasma Science, vol. 8, no. 4, pp. 56–67, 2022. DOI: 10.1016/j.agrplasci.2022.104567

10. M. Walker and H. Chen, "High-Voltage Plasma Generation for Agricultural Applications, "IEEE Agricultural Systems, vol. 45, no. 3, pp. 230–242, 2021.DOI:10.1109/ias.2021.302043

11. P. Fernandes, C. Silva, and N. Gomez, "Seed Germination Enhancement under Cold Atmospheric Plasma: A Comparative Analysis," Plant Growth Studies, vol. 19, no. 2, pp. 89–100, 2022. DOI:10.1007/pgs2022.0019

12. E.Novak, K.Toth, and F.Varga, "Water Uptake Kinetics in Plasma-Treated Seeds: A Mechanistic Insight," Seed Biology Reports, vol. 11, no. 4, pp. 145–155, 2023. DOI: 10.1186/sbr.2023.01104

13. L.Brown and T.Garcia, "Statistical Impact of Plasma Pre-Treatment on Plant Growth Parameters," Journal of Agricultural Statistics, vol. 16, no. 3, pp. 78–90, 2021. DOI: 10.1016/j.jas.2021.106789

14. Cold Plasma Treatment of Sunflower Seeds Modulates Plant-Associated Micro biome and Stimulates Root and Lateral Organ Growth. Frontiers in Plant Science, 2020

15. Tomeková, J., Švubová, R., Slováková, Ľ. et al "Interaction of Cold Atmospheric Pressure Plasma with Soybean Seeds: Effect on Germination and DNA, Seed Surface Characteristics and Plasma Diagnostics", Plasma Chemistry and Plasma Processing. DOI: 10.1007/s11090-023-10398-9

16. T. Sarinont, T. Amano, P. Attri, K. Koga, N. Hayashi, & M. Shiratani, "Effects of Plasma Irradiation Using Various Feeding Gases on Growth of Raphanus sativus L.," Archives of Biochemistry and Biophysics, vol. 605, pp. 129–140, 2016. DOI: 10.1016/j.abb.2016.03.024.

17. Positive Effect Induced by Plasma Treatment of Seeds on the Agricultural Performance of Sunflower. MDPI Plants, 2023-https://www.mdpi.com/2127748

Note: All the figures in this chapter were made by the author.

Intelligent and Sustainable Power and Energy Systems – Dr. M. Premkumar et al. (eds)
© 2026 Taylor & Francis Group, London, ISBN 978-1-041-10314-1

29

IoT-Driven Solar-Powered Variable Speed Induction Motor System Aligned with Sustainable Development Goals

A. Ajay kumar,
Pasumarthi Usha[1], Suganthi Neelagiri,
Harshitha A.[2], K. Chethana, Thanushree H. D.
Dept. of E&E Engg. Dayananda Sagar College of Engineering,
Bengaluru

ABSTRACT: Single-phase induction motors are frequently used in industrial control systems and home appliances, Due to their affordability and adaptability. Adjustable-speed drives are necessary for many industrial jobs in order to meet a variety of operating demands. In order to improve sustainability, flexibility, and cost-effectiveness, this project focuses on employing an Internet of Things (IoT)-driven solar-powered system to regulate the speed of a single-phase induction motor. Adjustable-speed drives are necessary for many industrial and agricultural applications in order to maximize efficiency and performance. To enable accurate speed modifications, the system uses a mosfet-based inverter to provide the motor with variable frequency AC power. It encourages the use of renewable energy sources and guarantees continuous operation thanks to its solar panel and battery backup power. Additionally, IoT integration allows real-time remote control and monitoring via a smartphone app, offering convenience and automation. Hardware modeling and testing validate the system's efficiency, making it ideal for applications like cutting machines, conveyor belts, water pumps, and irrigation systems. This project demonstrates a smart, sustainable, and adaptable motor control solution for diverse industrial and agricultural needs.

KEYWORDS: Solar power, Induction motor, IOT driven, Inverter, MOSFET, Microcontroller, Battery

1. Introduction

Induction motors are among the most versatile and widely used types of electric motors in fields such as industry, transportation, and agriculture [1-4]. Their popularity stems from several

Corresponding authors: [1]pu1968@yahoo.co.in, [2]harshithaanchi1204@gmail.com

DOI: 10.1201/9781003654469-29

desirable characteristics, including simple design, durability, high efficiency, and low maintenance requirements [8]. Induction motors operate with a good power factor and offer advantages such as stable speed regulation, the ability to handle overloads, and high starting torque [7]. These features make them cost-effective and reliable compared to other motor types, supporting their widespread use in various applications, from heavy industrial machinery to electric vehicles and agricultural equipment. In many applications, however, there is a need for varying motor speeds depending on the task. For instance, industrial processes, electric vehicles, and irrigation systems often require adjustable speeds to meet changing operational requirements [5][6]. As a result, For particular jobs, users must be able to regulate the motor's speed to maximize performance. In this project IoT-based remote monitoring and control with a solar-powered variable-speed induction motor system. The innovative use of a cost-effective inverter design and frequency-based speed control enables versatile applications in agriculture and industry. By aligning with sustainable development goals, the system promotes renewable energy usage and smart automation for a greener future. The project utilizes IoT integration, variable frequency control, solar energy utilization, and step-up transformer design techniques [9].

2. Solar Powered Variable Speed Induction Motor System

A solar-powered system for controlling the speed of an induction motor, equipped with IoT capabilities for remote monitoring and management. A solar panel is the first component of the system, which uses solar energy to generate direct current (DC) power. A battery bank stores this energy, guaranteeing a steady source of electricity even in the absence of sunlight. A MOSFET-based inverter then transforms the stored DC power into alternating current (AC), modifying the output frequency to regulate the motor's speed The motor's speed can be precisely adjusted to satisfy certain operational requirements by changing the frequency. Figure 29.1 shows the block diagram of solar powered variable speed induction motor.

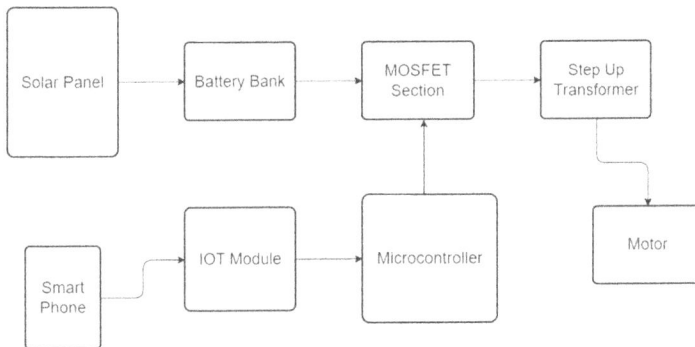

Fig. 29.1 Block diagram

Additionally, the system has IoT capabilities that enable remote management and monitoring using a smart phone app, giving real-time information on solar power, battery levels, and motor performance. This combination of renewable energy, variable-speed control, and IoT integration makes the system ideal for applications in agriculture, industry, and other fields that require adaptable motor speeds and sustainable energy solutions.

3. Circuit Diagram

The inverter circuit uses two N-channel MOSFETs to convert 12V DC into a 220V AC output, producing a square wave rather than a pure sine wave. This makes it suitable for low-power applications like charging devices or powering simple loads. The output frequency is 60 Hz, achieved through a transformer with a primary coil of 100 turns and a secondary coil of 1833 turns. The microcontroller alternates the MOSFETs' states, ensuring one is on while the other is off, generating the square wave AC. To achieve a standard 50 Hz output, each MOSFET must switch 120 times per second, controlled by complementary signals from the microcontroller. The transformer steps up the voltage to the desired level, and the MOSFETs are driven by an Arduino to toggle the switches. Figure 29.2 shows the circuit diagram of the system.

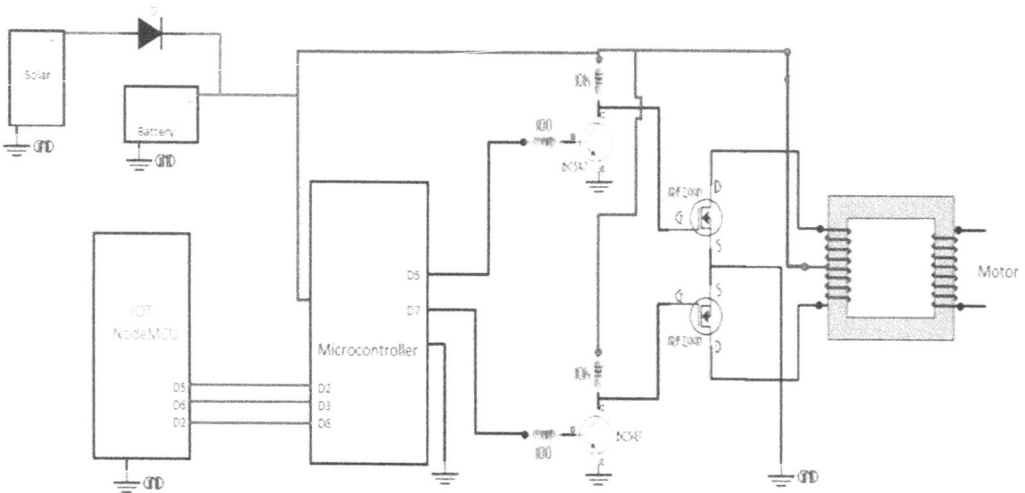

Fig. 29.2   Circuit diagram of the system

A single-phase induction motor with a 10 W output and 40 W input power rating powers our project. It runs at 50/60 Hz with synchronous speeds of 1500 RPM and 1300 RPM. The motor is found to have four poles at 50 Hz and six poles at 60 Hz using the pole calculation algorithm, $P = 120 * f/Ns$ [14][18]. This relationship shows how motor speed depends on supply frequency, which is crucial for our system's accurate speed control. Our system's integration of IoT-based variable frequency control ensures versatility for applications such as conveyor systems and irrigation by modifying the supply frequency to reach the necessary motor speed. This supports our objective of employing solar electricity to maximize sustainability and energy efficiency.

4. Diagram of the Half Bridge Inverter Circuit

When Q1 is activated during the positive half-cycle, current can go from the DC supply's positive terminal via Q1, across the load, and to the capacitors' midway. A positive voltage is produced across the load by this current direction. Current flows from the midpoint, across the load, and through Q2 to the DC supply's negative terminal during the negative half-cycle, when Q2 is turned on and Q1 is off. The AC waveform is completed when this reversed current produces a negative voltage across the load [14]. Figure 29.3 shows the working of inverter half bridge.

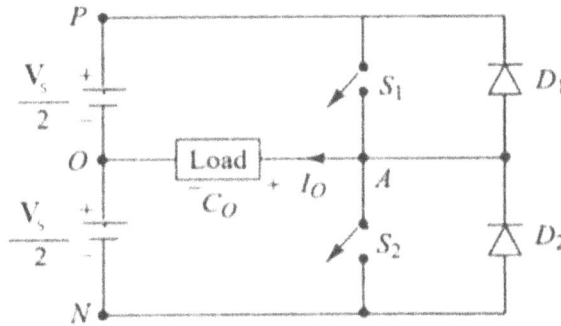

Fig. 29.3 Working of the half bridge inverter [22]

5. Working of Inverter Section

The step-up transformer boosts the AC voltage generated by the inverter to the required level for the motor. The inverter operates by alternating the MOSFETs' states, converting DC from the solar-powered battery into an AC square wave. The process is managed by a microcontroller (Atmega328), which modulates the inverter's frequency based on input commands. These commands are received through an IoT module, which facilitates real-time remote control and monitoring via a smart phone. This feature provides users with the flexibility to adjust motor speed and performance from a distance. This setup leverages solar power to enhance energy efficiency while supporting variable-speed operation for the induction motor. The system is versatile, making it suitable for various industrial and agricultural applications requiring precise speed control and remote accessibility.

6. Mobile Application for IOT Operation

In our project setup with the Blynk app, here' shows the controls would typically function:

User ID and Password Setup: Y start by creating a user ID and password on the Blynk app. This setup enables a secure, personalized experience and allows the app to store your project's configurations.

Wi-Fi Connection: Once the user account is setup, the Blynk app connects to the Wi-Fi automatically, linking the app to your hardware kit for seamless communication.

Buttons:

Increment Button: This button is likely designed to increase a parameter, such as frequency in your variable speed induction motor project.

Decrement Button: This would decrease the parameter, allowing fine-tuned control over the motor's speed.

On/Off Switch: This switch turns the motor on or off, letting you startors top the system as needed.

Each button can be linked to specific Blynk widgets, enabling real-time control and monitoring through the app. This setup makes it user-friendly and integrates well with IoT. Figure 29.4 shows the Blynk IoT.

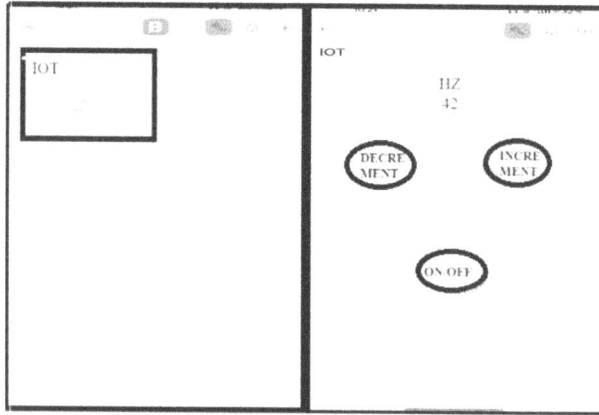

Fig. 29.4 Blynk IOT

7. Flowchart of Control Technique

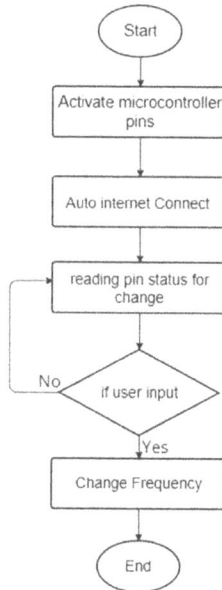

Fig. 29.5 flowchart

8. Methodology

The updated technique discusses scalability and incorporates quantitative results to properly address the reviewer's recommendations. By comparing the input power (40W) and output power (10W), efficiency metrics are obtained, yielding a 25% efficiency. This figure is consistent with the average efficiency of low-rated motors used in applications involving changing speed. With projected savings of 20–30%, the use of a variable frequency drive (VFD) in conjunction with solar energy is a notable example of energy savings, especially for applications like irrigation and water pumps. P = 120*f/ Ns Using the formula for motor speed computation based on inverter frequency, speed control accuracy is addressed [18]. Experimental results demonstrate that the motor speed closely resembles the calculated values, with a mere ±5% difference. Lastly, the system's scalability for industrial applications is discussed by comparing it to large-scale systems used in cement mills, where high-rated motors are employed. The methodology demonstrates how the basic principles of the system can be adapted for larger motors and more significant loads, making the system suitable for various industrial applications [8]. Figure 29.5 shows the flow chart of the proposed system.

9. Result

The hardware implementation of a microcontroller- based induction motor speed control is described in this project. It involves the creation of microcontroller circuit hardware, MOSFET driver circuit was explained. Variable frequency wave is generated from microcontroller. The output of the controller is given to the MOSFET driver circuit which is used for amplify low microcontroller signal to convert in high signal purpose. The pulses are given to the appropriate switches of the MOSFETs inverter fed induction motor. Motor speed is getting change successfully. Figure 29.6, Figure 29.7 and Table 29.1 shows the inverter output, pictorial representation of proposed kit and Tabular column.

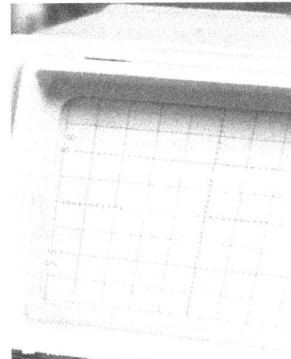

Fig. 29.6 Inverter output

Fig. 29.7 Project kit

Table 29.1 Tabular column

SI no	Frequency in Hertz	Speed in rpm
1	50	1000
2	47	900
3	45	800
4	42	700
5	40	600
6	37	500
7	35	400
8	25	100
9	30	200

10. Conclusion

This study effectively implements the new induction motor speed control topology combining frequency control with microcontroller technology. This is one technique for regulating speed that is used with AC motor drives. The Atmega328 microcontroller uses this technology to control the speed of an AC motor. It is inexpensive, small, and has a long life span and great dependability. After analyzing the experimental data, it is discovered that the induction motor's speed is normally managed. And this technology makes it easy to meet step-up and step- down speed requirements.

REFERENCE

1. In 2019, Shafiullah, G.M., Mojumder, M.K., and Muttaqi, K.M. "A Solar PV-Based Induction Motor Drive System for Off-Grid Agricultural Irrigation," IEEE Transactions on Industry Applications, 55(4), 4165–4173, states.

2. Ahmed, N., and S. Patel (2020). 6890–6899 in "IoT-Based Real-Time Monitoring and Control of Induction Motors," IEEE Internet of Things Journal, 7(8).

3. Akagi, H. and Fujita, H. (2008). IEEE Transactions on Power Electronics, "Pulse Width Modulation Control for Induction Motor Drives," 23(3), 1234–1241.

4. Li, X., and D. Chen (2018). International Journal of Power Electronics, 10(2), 145–153, "MOSFET-Based Inverter Design for Induction Motor Drives,"

5. Proceedings of the 2022 IEEE International Conference on Information Technology, "Design and Implementation of a Solar-Powered IoT-Controlled Induction Motor for Agricultural Applications," by A. Kumar and R. Gupta conference on Renewable Energy and Power Engineering, 2022, pp. 78–83.

6. M. Rashid and S. Patel, "IoT-Based Monitoring and Control of Induction Motors for Industrial Applications, pp. 914–920," IEEE Internet of Things Journal, vol. 6, no. 5, pp. 914–920, 2019.

7. Porselvi, T., K. Krithika, and CS Sai Ganesh. "PV system based induction motor with landsman converter using IFOC controller Pages 1-4 of the 2nd Global Conference for Advancement in Technology (GCAT) in 2021. IEEE 2021.

8. Smith, J., Kumar, A., and Wang, L. "Applications and Advantages of Induction Motors in Industry, Transportation, and Agriculture." Journal of Industrial Engineering and Applications, vol. 34, no. 5, 2023, pp. 245–253.

9. S. Patel, R. Mehta, and P. Nair, "Design and Control of a Solar-Powered Induction Motor System for Agricultural Applications," International Journal of Renewable Energy Systems, vol. 15, no. 3,pp. 180–189, 2023.

10. A. Sharma, B. Gupta, and C. Li, "Operational Principles and Control Strategies of Power Inverters in Renewable Energy Systems, IEEE Transactions on Power Electronics, vol. 35, no. 7, pp. 1204–1215, 2022.

11. M. Brown, J. Taylor, and A. Gupta, "An Introduction to Arduino IDE: Programming and Debugging for Embedded Systems," Journal of Embedded Systems and Applications, vol. 22, no. 4, pp. 305–315, 2023.

12. M. H. Rashid, S. Rahman "Design and Implementation of Microcontroller-Based Variable Frequency Drive for Induction Motor Control" Journal: IEEE Transactions on Industrial Electronics

13. R. K. Gupta, A. S. Bhatia "Microcontroller-Based Speed Control of Induction Motor Using PWM and MOSFET Inverter" Journal: The International Journal of Drive Systems and Power Electronics

14. Chitode, J. S. (2010). Power Electronics. Technical Publications.

15. M. H. Rashid, S. Rahman "Design and Implementation of Microcontroller-Based Variable Frequency Drive for Induction Motor Control" Journal: IEEE Transactions on Industrial Electronics.

16. Bose, Bimal K. "Power electronics and motor drives: advances and trends." (2020).

17. Fanslow, Mark, and Steve Drzymala. "Considerations in the Selection and Application of AC and DC Motors for Cement Plants." In 2009 IEEE Cement Industry Technical Conference Record, pp. 1–22. IEEE, 2009.

18. Thomas A. Lipo, Mohamed, and Osama. "Modeling and analysis of a wide-speed-range induction motor drive based on electronic pole changing." 33, no. 5 (1997): 1177–1184; IEEE Transactions on Industry Applications.

19. Matthew W. Dunnigan, Shamboul A. Mohamed, Martino O. Ajangnay, Ahmed, and Aamir Hashim Obeid. "Speed control of induction motor using new sliding mode control technique." Energy, Power, and Control, First International Conference (EPC-IQ), 2010, pp. 111–115. IEEE, 2010.

20. B. Sai, B. N. Kartheek, and Sindura. "Speed Control of Induction Motor using Cycloconverter."2013: 776-780 in International Journal of Engineering Trends and Technology (IJETT), 4, no. 4.

21. A. Yazdian-Varjani, M. Seyyed-Hosseini, and N. Dehbashi. "IoT based condition monitoring and control of induction motor using raspberry pi." Pages. 134–138 in 13th Power Electronics, Drive Systems, and Technologies Conference (PEDSTC), 2022. IEEE, 2022.

22. https://www.elprocus.com/what-is-half-bridge-inverter-circuit-diagram-its-working/

Note: All the figures (except Fig. 29.3) and table in this chapter were made by the author.

30

AI-Based Fuel and Energy Management System (EMS) in Hybrid Electric Vehicle (HEVs): A Review

Megha Sen[1], Vikramaditya Dave[2]
Department of Electrical Engineering,
College of Technology and Engineering,
Udaipur

Sujit Kumar[3]
Department of Electrical and Electronics Engineering,
Dayananda Sagar College of Engineering,
Bengaluru

ABSTRACT: Integrating artificial intelligence (AI) in the fuel and EMS of HEVs is poised to enhance their efficiency, sustainability, and performance significantly. This review paper explores the application of AI techniques in optimising distribution of energies among ICE (Internal Combustion Engine) and electric motor, improving battery management, and enhancing energy recovery systems. AI-driven algorithms, including machine learning (ML), reinforcement learning (RL), and deep learning (DL), have been shown to enable real-time energy prediction, adaptive charging and discharging strategies, route optimisation, and personalised driving behaviour adaptations. These technologies contribute to reduced fuel consumption, minimised emissions, and extended battery life. Furthermore, AI enhances the combination of HEVs with RE (Renewable Energy) foundations and smart grids, fostering a more sustainable and efficient transportation ecosystem. Despite the clear advantages, challenges remain in implementing AI-based systems, particularly regarding data quality, computational requirements, and system integration with existing vehicle architectures. This paper also discusses the future implications of AI for HEVs, emphasising its role in the evolution of autonomous, electrified, and interconnected transportation systems. The findings suggest that AI will be crucial in advancing HEVs towards a low-carbon future, improving environmental sustainability and operational performance.

KEYWORDS: Artificial intelligence, Emissions reduction energy management, Hybrid electric vehicles, Machine learning, Sustainability

[1]meghasen257@gmail.com, [2]vdaditya1000@gmail.com, [3]sujit-eee@dayanandasagar.edu

DOI: 10.1201/9781003654469-30

1. Introduction

The transition towards sustainable transportation is crucial to addressing global energy consumption and environmental degradation. Hybrid electric vehicles (HEVs), which combine an ICE with a motor, have emerged as a viable solution to reduce fuel consumption, lower greenhouse gas emissions, and mitigate urban air pollution [1]. Traditional approaches to energy management in HEVs are typically based on rule-based algorithms or optimisation techniques [2]. As the complexity of HEVs continues to increase, these conventional methods often fail to adapt efficiently to dynamic driving conditions, battery performance variations, and real-time traffic scenarios [3]. Artificial Intelligence (AI), with its ability to process vast amounts of data, learn from past experiences, and adapt in real-time, offers a promising solution to enhance fuel and energy management in HEVs. AI-based energy management systems can dynamically adjust power distribution, predict energy demand, optimise charging and discharging processes and enable energy recovery through regenerative braking [4]. AI techniques such as machine learning (ML), reinforcement learning (RL), and deep learning (DL) have shown significant potential in improving the accuracy of predictive models and decision-making processes in HEV energy management. Despite these promising advancements, integrating AI into HEV energy management systems presents several challenges. Issues such as the quality and availability of data, the computational complexity of AI models, and the real-time processing requirements for in-vehicle systems must be addressed to realise the full potential of AI-enabled solutions. Moreover, there are concerns about AI decisions' interpretability, integration with existing automotive systems, and regulatory and safety considerations [5]. The EMS is an indispensable part of HEVs, which maximises using various energy sources to boost performance, lower emissions, and increase efficiency. EMS maintains driving comfort while ensuring fuel economy by dynamically allocating control between the motor and ICE. The ability of EMS to anticipate and adjust to changing driving conditions, driver behaviour, and energy demands is improved by integrating AI. AI-powered EMS is a key component of next-generation HEVs since it can learn from historical and real-time data, facilitating wise decision-making and sustainable energy use [6].

2. AI-Based Fuel and Energy Management in HEV

HEVs integrate an ICE and an EM (electric motor), necessitating a sophisticated Petroleum and Energy Management System (PEMS) to optimise the interaction between the two power sources. The primary objectives of an HEV's energy management system are to minimise fuel consumption, reduce emissions, extend battery life, and ensure overall system efficiency [7]. Traditional energy management techniques, such as rule-based or optimisation methods, have been widely employed to determine the optimal power distribution among ICE and the EM. As vehicle systems become more complex and driving conditions more dynamic, these conventional methods often fail to deliver optimal performance across varied real-world scenarios. Artificial Intelligence offers a promising alternative by enabling real-time, data-driven decision-making that can adapt to a wide range of operating conditions, thereby improving the fuel efficiency and performance of HEVs [8, 9]. AI-based fuel and energy management systems can learn from past driving behaviour, vehicle status, and environmental factors and adaptively adjust the energy management strategy to optimise vehicle performance. These AI systems process large amounts of data from onboard sensors, including information about vehicle speed, road conditions, driver behaviour, battery state-of-charge (SOC), and fuel consumption patterns [10]. By leveraging machine learning (ML), deep learning (DL), and reinforcement learning (RL), AI-based systems can predict energy demand, select the most appropriate power source, and make real-time adjustments to the power split between the electric motor and the ICE. This approach enhances fuel efficiency and improves the driving experience by providing smoother transitions between power sources, more accurate predictive control, and greater adaptability to varying driving conditions [11].

2.1 Types of AI-Based Algorithms for Energy Management

AI-based algorithms for EM in HEVs can be broadly classified into RB (rule-based), OB (optimization-based), and ILB (intelligent learning-based) approaches. Rule-based algorithms use pre-defined logic or heuristic rules, often derived from expert knowledge, to make decisions. They are simple to implement and computationally efficient but lack flexibility in adapting to varying conditions. Optimization-based algorithms, such as dynamic programming, genetic algorithms, and particle swarm optimisation, aim to achieve optimal energy distribution by minimising fuel consumption, emissions, or other criteria. These methods can handle uncertainties and dynamic environments but require extensive data and training [12]. Various Artificial Intelligence-based algorithms are shown in Fig. 30.1. Integrating these approaches can create hybrid systems that balance efficiency, adaptability, and real-time performance.

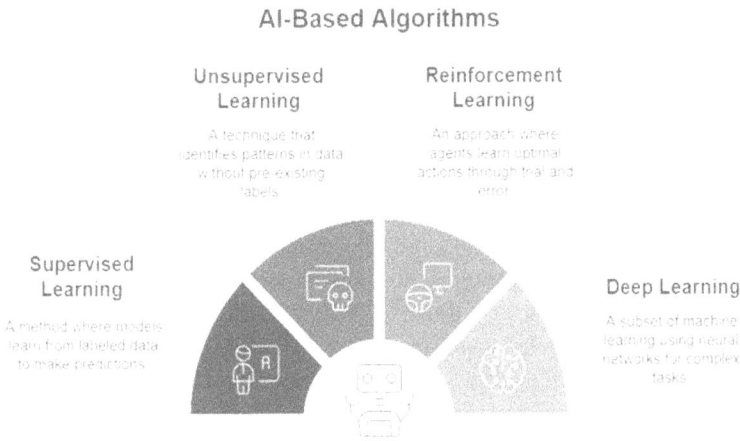

Fig. 30.1 Artificial intelligence-based algorithms [4]

2.2 Challenges in Implementing AI-based Energy Management in HEVs

Implementing AI-based EMS in HEVs faces several challenges. One major issue is the high computational demand for AI algorithms, which can strain onboard processing capabilities, especially in real-time scenarios. Integration with existing vehicle systems is another hurdle due to compatibility issues and the lack of standardised communication protocols. The various challenges and possible solutions in implementing AI-Based Energy Management in HEVS are shown in Table 30.1. Uncertainty in driving conditions, such as variations in traffic, weather, and driver behaviour, complicates the development of adaptive and robust AI models [13]. Managing batteries and power systems introduces further complexities, including heat dissipation, lifecycle optimisation, and energy distribution. High costs for development and implementation, along with limited access to quality training data, further hinder progress. Finally, safety and reliability concerns make extensive testing mandatory, while regulatory barriers and resistance from manufacturers slow adoption [14].

Table 30.1 Challenges and possible solutions in implementing AI-based energy management in HEVS

Challenges	Possible Solutions
High computational requirements for real-time control	Utilise optimised algorithms and lightweight models to reduce processing demands.
Integration with existing vehicle systems	Develop standardised communication protocols and modular energy management systems.
Uncertainty in driving conditions	Incorporate predictive models and adaptive systems to handle variations in driving patterns.
Complexity in battery management	Employ advanced battery management systems (BMS) with enhanced monitoring and predictive maintenance.
High development and implementation costs	Focus on scalable solutions and collaborate with stakeholders to share development resources.
Limited availability of high-quality training data	Utilise simulation tools and open datasets to create realistic scenarios for system development.
Heat management issues in power systems	Design efficient cooling systems and integrate thermal management strategies into the energy framework.
Reliability and safety concerns	Conduct rigorous testing and validation under diverse operating conditions to ensure system robustness.

Source: Author

To address these challenges, acting in equilibrium and possibility is essential. Optimising AI algorithms to work efficiently on lightweight, real-time systems can mitigate computational demands [15]. Developing standardised frameworks for vehicle integration can enhance compatibility and simplify deployment. Adaptive algorithms that utilise predictive modelling and reinforcement learning can handle uncertain driving conditions effectively. Advanced battery management systems (BMS) that integrate thermal regulation and predictive maintenance can improve reliability. Various Pros and cons of Artificial Intelligence based energy management in HEVs are shown in Fig. 30.2. Collaboration between manufacturers, researchers, and policymakers can reduce development costs through resource sharing and incentives. Using simulation tools and open datasets can overcome the limitations of training data. Rigorous testing under diverse conditions ensures safety and compliance, while education campaigns can help overcome resistance to adoption. Addressing these issues strategically allows AI-based energy management to be effectively implemented in HEVs [16].

3. Applications of AI in Fuel and Energy Management in HEVs

Artificial Intelligence has found numerous applications in fuel and energy management systems for HEVs, offering substantial improvements in energy efficiency, emissions reduction, and overall vehicle performance. By leveraging advanced AI techniques such as machine learning, deep learning, and reinforcement learning, AI systems can optimise complex energy flows, adapt to diverse driving conditions, and predict future energy demands with high precision [17, 18]. This section discusses key AI applications in HEV energy management, focusing on energy prediction and load forecasting, adaptive charging and discharging strategies, route optimisation and driver behaviour adaptation, and energy recovery through regenerative braking systems [19].

AI-based energy management in HEVs

Pros	vs	Cons

Significant benefits

Data quality issues

Efficient systems

High computational complexity

Intelligent management

Integration challenges

Regulatory concerns

Need for skilled personnel

Fig. 30.2 Pros and cons of artificial intelligence based energy management in HEVs [15]

3.1 Energy Prediction and Load Forecasting

Accurate energy prediction is fundamental to optimising the performance of HEVs, as it allows for better power management distribution between ICE and the EM. AI-based energy prediction models can forecast short-term and long-term energy needs by analysing historical data on driving patterns, environmental conditions, battery health, and traffic conditions. These models typically use time-series data and regression techniques to predict the vehicle's energy consumption, considering speed, road incline, traffic density, and battery charge state [20]. Energy load forecasting is particularly crucial in hybrid vehicles, as it enables the system to adjust the power split between the electric motor and the ICE before a demand arises, ensuring that energy resources are used optimally. Machine learning algorithms, including support vector machines (SVMs), artificial neural networks (ANNs), and long short-term memory (LSTM) networks, have been widely used for forecasting energy demand in HEVs [21]. These predictive models can significantly improve the vehicle's overall efficiency by anticipating energy needs based on real-time data. This allows the system to proactively manage the battery's state of charge (SOC) and decide the most efficient power source [22, 23].

3.2 Adaptive Charging and Discharging Strategies

AI can also be critical in optimising battery management through adaptive charging and discharging strategies. These strategies aim to maintain the battery within an optimal charge range to prolong its lifespan while ensuring the vehicle operates with maximum energy efficiency. Battery degradation, caused by overcharging or deep discharging, is a significant concern in HEVs. AI can mitigate this issue by predicting optimal charging and discharging cycles based on the vehicle's operating conditions and driving patterns [24, 25]. These models can dynamically adjust charging strategies in response to changing driving conditions, such as during highway driving versus city commuting [26]. For example, in city driving with frequent stops and starts, the system may prioritise discharging the battery to avoid excessive charging cycles, which could reduce battery lifespan. In contrast, highway driving may allow for more consistent charging, maintaining the battery's SOC at an optimal level without overcharging it [27].

3.3 Route Optimisation and Driver Behaviour Adaptation

AI-based systems can also optimise route selection and adapt to driver behaviour to improve fuel and energy efficiency in HEVs. Route optimisation algorithms use real-time traffic data, road conditions, and vehicle status to determine the most energy-efficient routes, reducing fuel consumption and emissions. Reinforcement learning algorithms benefit this application, as they can continuously learn from driving experience and adapt the vehicle's power distribution strategy according to different road conditions [28, 2]. Route optimisation predicts upcoming road conditions such as traffic congestion, road slopes, or elevation changes, significantly impacting energy consumption. By processing data from navigation systems and environmental sensors, AI models can suggest the most efficient routes that minimise fuel use and optimise battery performance. For instance, if an HEV system predicts that a significant portion of the upcoming journey involves steep inclines, it can adjust the energy management strategy to maximise the use of the electric motor and minimise reliance on the ICE [30]. Driver behaviour also plays a critical role in the energy consumption of an HEV. AI-based systems can analyse individual driving patterns, including acceleration rates, braking habits, and speed consistency, and offer real-time feedback to encourage more energy-efficient driving. This feedback can be delivered through alerts, suggestions, or automated driving interventions to help the driver optimise fuel and energy usage [31]. For instance, AI could suggest smoother acceleration profiles or earlier braking to maximise regenerative energy capture. Over time, the system can learn to predict the driver's preferences and adapt its suggestions to align with their driving style, making the vehicle's energy management system more personalised and efficient [32, 33].

3.4 Energy Retrieval through Reformative Braking Systems

Integral to hybrid electric vehicles is regenerative braking. It enables the vehicle to reap the benefits of braking by transforming mechanical energy into electrical energy that can be stored in the battery. Because AI algorithms can determine when braking will result in the most energy recovery, optimising regenerative braking using AI can significantly increase efficiency. Traditional braking systems may not always maximise energy capture, particularly when braking force is applied too late or suddenly. Still, AI systems can optimise the timing and intensity of regenerative braking based on real-time data from the vehicle's sensors [34]. AI-based systems can use reinforcement learning to maximise the amount of energy recovered during each braking event. By continuously monitoring vehicle speed, road grade, battery SOC, and traffic conditions, the system can adjust the braking strategy to ensure that the maximum amount of kinetic energy is converted back into electrical energy without compromising safety or comfort. For example, in stop-and-go traffic, AI can anticipate frequent braking events and adjust the regenerative braking force to optimise energy recovery [35]. Some standard performance parameters and comparisons with different AI-based techniques are shown in Table 30.2.

Table 30.2 Performance comparison of various parameters with different AI-based techniques

Performance Parameter	Description	AI Technique	Impact of AI Technique
Fuel Efficiency	Measures fuel consumption per distance	Reinforcement Learning (RL)	RL optimises power distribution by learning from real-time driving scenarios.
Energy Consumption	Total energy usage (fuel + electricity)	Neural Networks (NN)	NN predicts energy demands accurately, minimising excess energy usage by adapting to driving conditions.
Battery SoC Management	Maintains battery charge within optimal range	Fuzzy Logic (FL)	FL ensures smooth transitions between energy sources, maintaining optimal SoC levels for longevity.
Emissions Reduction	Reduction of harmful gases	Genetic Algorithm (GA)	GA identifies optimal operational parameters to minimise emissions effectively.
Computational Efficiency	Resource usage and processing speed	Lightweight Deep Learning (DL)	Lightweight DL models ensure real-time decision-making with minimal computational overhead.
Cost Efficiency	The trade-off between performance and implementation cost	Evolutionary Algorithms (EA)	EA minimises system design costs while ensuring long-term cost savings through improved efficiency.

Source: Author

4. Environmental Impact and Sustainability of AI-based EM in HEVs

One of the most critical environmental issues facing the world today is the transportation sector's role in producing greenhouse gases, air pollution, and the loss of natural resources. By fusing the benefits of internal combustion engines (ICEs) with electric drivetrains, hybrid electric vehicles (HEVs) can answer these problems. However, by incorporating AI into their energy management systems, HEVs may fully realise their sustainable potential [36]. AI-based approaches enable more efficient operation of HEVs by optimising power distribution, improving energy recovery, and enhancing battery management. These AI-driven innovations significantly reduce carbon emissions and advance sustainable energy systems. This section examines the role of AI-based EMS in reducing carbon emissions and their broader impact on the sustainability of HEVs [37].

4.1 Reduction in Carbon Emissions

A key environmental benefit of HEVs, exceptionally when equipped with AI-based energy management systems, is the reduction in carbon emissions. The hybridisation of the powertrain allows for a significant decrease in fuel consumption, as the vehicle alternates between electric and gasoline-powered propulsion based on driving conditions. AI systems optimise this process by dynamically adjusting the energy flow, ensuring that the most efficient energy source is used at any given time. For instance, during urban driving, where stop-and-go traffic is typical, the AI system can prioritise electric motor use to minimise fuel consumption and emissions, switching to ICE only when necessary [38]. AI algorithms, particularly those based on machine learning and reinforcement learning, enable the system to predict and adapt to real-time driving conditions. These systems can lower fuel usage and carbon emissions through energy demand forecasting and electric motor/ICE power distribution optimisation. According to the research, artificial intelligence (AI) powered energy management systems can cut carbon dioxide (CO_2) emissions by as much as 30 percent compared to conventional, non-optimized hybrid cars. This reduction is primarily attributed to the more efficient use of energy and the reduction in fuel consumption enabled by AI's real-time optimisation capabilities [39]. The regenerative braking system, which uses the vehicle's kinetic energy to generate electricity when the brakes are applied, may be fine-tuned with the help of AI. This reduces pollutants and the vehicle's dependence on the internal combustion engine. AI systems ensure that regenerative braking is applied at optimal moments, maximising energy recovery and minimising the need for fuel consumption, especially in stop-and-go driving scenarios [40]. As countries and regions tighten their environmental regulations, implementing AI-powered energy management systems in HEVs will help automakers comply with increasingly stringent emissions standards. By minimising the carbon footprint of vehicles, AI contributes to reducing greenhouse gases and plays a pivotal role in mitigating air pollution, particularly in urban areas where transportation is a significant source of smog and particulate matter [40].

4.2 AI's Role in Sustainable Energy Systems

Artificial Intelligence (AI) transforms sustainable energy systems by enabling efficient energy generation, distribution, and consumption. AI-powered algorithms are instrumental in optimising renewable energy sources, such as solar and wind, by predicting energy production based on weather conditions and aligning supply with demand. Through predictive analytics, AI enhances grid management, ensuring stability and reducing energy losses during transmission. Furthermore, AI facilitates real-time energy monitoring, empowering industries and households to minimise wastage and improve energy efficiency. These capabilities make AI a vital tool for achieving sustainability goals and reducing the carbon footprint of energy systems [39]. AI supports the integration of diverse energy sources into smart grids. By intelligently balancing traditional and renewable energy inputs, AI ensures a seamless transition toward cleaner energy solutions. It also aids in energy storage management, optimising the use of batteries and other storage technologies to address variability in renewable energy production [38]. AI-driven solutions provide actionable insights for policymakers and stakeholders, enabling informed decisions for sustainable energy planning. AI fosters innovation and resilience in modern energy infrastructures, making them more sustainable and reliable through its ability to process vast amounts of data and adapt to dynamic systems [40]. Another key area where AI contributes to sustainability is its role in energy recovery mechanisms, such as regenerative braking. AI algorithms optimise the timing and intensity of regenerative braking, which not only recovers energy during vehicle deceleration but also reduces wear and tear on traditional braking systems, further extending the lifetime of vehicle components and reducing material consumption.

5. Conclusion

The function of AI in optimising HEV energy and fuel management has been investigated in this paper. Key findings highlight that AI-driven systems enhance energy efficiency through improved energy prediction, adaptive charging strategies, route optimisation, and energy recovery via regenerative braking. AI also allows for dynamic adjustments based on real-time data, significantly reducing fuel consumption and emissions and contributing to the overall sustainability of HEVs. AI is poised to have a transformative impact on the future of HEVs. By enabling real-time optimisation of energy management, AI will further reduce dependence on fossil fuels, integrate renewable energy sources, and improve vehicle performance. AI technologies are also key to enhancing user experience through personalisation, improving vehicle safety, and supporting the shift toward innovative, interconnected transportation systems.

REFERENCES

1. Aldoseri, A., Al-Khalifa, K. N., & Hamouda, A. M. (2023). Re-thinking data strategy and integration for artificial intelligence: concepts, opportunities, and challenges. Applied Sciences, 13(12), 7082.
2. Amarasinghe, K., Marino, D. L., & Manic, M. (2017). Deep neural networks for energy load forecasting. 2017 IEEE 26th International Symposium on Industrial Electronics (ISIE), 1483–1488.
3. Bieri, R., Nef, T., Müri, R. M., & Mosimann, U. P. (2015). Development of a novel driving behavior adaptations questionnaire. International Psychogeriatrics, 27(6), 1017–1027.
4. Collazos, J. S. G., Ardila, L. M. C., & Cardona, C. J. F. (2024). Energy transition in sustainable transport: concepts, policies, and methodologies. Environmental Science and Pollution Research, 31(49), 58669–58686.
5. Couth, R., & Trois, C. (2010). Carbon emissions reduction strategies in Africa from improved waste management: A review. Waste Management, 30(11), 2336–2346.
6. Dong, P., Zhao, J., Liu, X., Wu, J., Xu, X., Liu, Y., Wang, S., & Guo, W. (2022). Practical application of energy management strategy for hybrid electric vehicles based on intelligent and connected technologies: Development stages, challenges, and future trends. Renewable and Sustainable Energy Reviews, 170, 112947.
7. Dudko, A., & Endrjukaite, T. (2024). Adaptive Charging and Discharging Strategies for Smart Grid Energy Storage Systems. In Information Modelling and Knowledge Bases XXXV (pp. 77–93). IOS Press.
8. Garikapati, D., & Shetiya, S. S. (2024). Autonomous Vehicles: Evolution of Artificial Intelligence and the Current Industry Landscape. Big Data and Cognitive Computing, 8(4), 42.
9. Gupta, S., Poonia, S., Varshney, T., Swami, R. K., & Shrivastava, A. (2022). Design and implementation of the electric bicycle with efficient controller. In Intelligent Computing Techniques for Smart Energy Systems: Proceedings of ICTSES 2021 (pp. 541–552). Springer.
10. Ji, F., Pan, Y., Zhou, Y., Du, F., Zhang, Q., & Li, G. (2020). Energy recovery based on pedal situation for regenerative braking system of electric vehicle. Vehicle System Dynamics, 58(1), 144–173.
11. Jiang, L., Li, Y., Ma, J., Cao, Y., Huang, C., Xu, Y., Chen, H., & Huang, Y. (2020). Hybrid charging strategy with adaptive current control of lithium-ion battery for electric vehicles. Renewable Energy, 160, 1385–1395.
12. Jui, J. J., Ahmad, M. A., Molla, M. M. I., & Rashid, M. I. M. (2024). Optimal Energy Management Strategies for Hybrid Electric Vehicles: A Recent Survey of Machine Learning Approaches. Journal of Engineering Research.
13. Karduri, R. K. R. (2019). The role of artificial intelligence in optimizing energy systems. International Journal of Advanced Research in Management Architecture Technology & Engineering.
14. Li, Y., Ding, Y., He, S., Hu, F., Duan, J., Wen, G., Geng, H., Wu, Z., Gooi, H. B., & Zhao, Y. (2024). Artificial intelligence-based methods for renewable power system operation. Nature Reviews Electrical Engineering, 1(3), 163–179.
15. Liu, T., Hu, X., Li, S. E., & Cao, D. (2017). Reinforcement learning optimized look-ahead energy management of a parallel hybrid electric vehicle. IEEE/ASME Transactions on Mechatronics, 22(4), 1497–1507.

16. Lü, X., Li, S., He, X., Xie, C., He, S., Xu, Y., Fang, J., Zhang, M., & Yang, X. (2022). Hybrid electric vehicles: A review of energy management strategies based on model predictive control. Journal of Energy Storage, 56, 106112.

17. Lujak, M., Giordani, S., & Ossowski, S. (2015). Route guidance: Bridging system and user optimization in traffic assignment. Neurocomputing, 151, 449–460.

18. Ma, Z., & Sun, D. (2020). Energy recovery strategy based on ideal braking force distribution for regenerative braking system of a four-wheel drive electric vehicle. IEEE Access, 8, 136234–136242.

19. Maherchandani, J. K., Joshi, R. R., Tirole, R., Swami, R. K., & Ganthia, B. P. (2022). Performance comparison analysis of energy management strategies for hybrid electric vehicles. In Recent Advances in Power Electronics and Drives: Select Proceedings of EPREC 2021 (pp. 245–254). Springer.

20. Matsuo, Y., LeCun, Y., Sahani, M., Precup, D., Silver, D., Sugiyama, M., Uchibe, E., & Morimoto, J. (2022). Deep learning, reinforcement learning, and world models. Neural Networks, 152, 267–275.

21. Mischos, S., Dalagdi, E., & Vrakas, D. (2023). Intelligent energy management systems: a review. Artificial Intelligence Review, 56(10), 11635–11674.

22. Pramanik, S. (2024). AI's Function in Sustainable Development's Renewable Energy Planning. In Next Generation Materials for Sustainable Engineering (pp. 334–349). IGI Global.

23. Pritima, D., Rani, S. S., Rajalakshmy, P., Kumar, K. V., & Krishnamoorthy, S. (2022). Artificial intelligence-based energy management and real-time optimization in electric and hybrid electric vehicles. E-Mobility: A New Era in Automotive Technology, 219–242.

24. Rauf, M., Kumar, L., Zulkifli, S. A., & Jamil, A. (2024). Aspects of artificial intelligence in future electric vehicle technology for sustainable environmental impact. Environmental Challenges, 14, 100854.

25. Reddy, K. B. N. K., Pratyusha, D., Sravanthi, B., & Reddy, E. J. (n.d.). Recent AI Applications in Electrical Vehicles for Sustainability.

26. Sabri, M., Danapalasingam, K. A., & Rahmat, M. F. (2016). A review on hybrid electric vehicles architecture and energy management strategies. Renewable and Sustainable Energy Reviews, 53, 1433–1442.

27. Srivatsa Srinivas, S., & Gajanand, M. S. (2017). Vehicle routing problem and driver behaviour: a review and framework for analysis. Transport Reviews, 37(5), 590–611.

28. Summerbell, D. L., Barlow, C. Y., & Cullen, J. M. (2016). Potential reduction of carbon emissions by performance improvement: A cement industry case study. Journal of Cleaner Production, 135, 1327–1339.

29. Tang, X., Chen, J., Liu, T., Qin, Y., & Cao, D. (2021). Distributed deep reinforcement learning-based energy and emission management strategy for hybrid electric vehicles. IEEE Transactions on Vehicular Technology, 70(10), 9922–9934.

30. Tran, D.-D., Vafaeipour, M., El Baghdadi, M., Barrero, R., Van Mierlo, J., & Hegazy, O. (2020). Thorough state-of-the-art analysis of electric and hybrid vehicle powertrains: Topologies and integrated energy management strategies. Renewable and Sustainable Energy Reviews, 119, 109596.

31. Ukoba, K., Olatunji, K. O., Adeoye, E., Jen, T.-C., & Madyira, D. M. (2024). Optimizing renewable energy systems through artificial intelligence: Review and future prospects. Energy & Environment, 0958305X241256293.

32. Urooj, A., & Nasir, A. (2024). Review of intelligent energy management techniques for hybrid electric vehicles. Journal of Energy Storage, 92, 112132.

33. Vashishth, T. K., Sharma, V., Sharma, K. K., Kumar, B., Chaudhary, S., & Panwar, R. (2024). Environmental sustainability and carbon footprint reduction through artificial intelligence-enabled energy management in electric vehicles. In Artificial Intelligence-Empowered Modern Electric Vehicles in Smart Grid Systems. 477–502.

34. Yin, W., & Ji, J. (2024). Research on EV charging load forecasting and orderly charging scheduling based on model fusion. Energy, 290, 130126.

35. Zhang, X., & Wang, Y. (2017). How to reduce household carbon emissions: A review of experience and policy design considerations. Energy Policy, 102, 116–124.

36. Zhu, J., Yang, Z., Mourshed, M., Guo, Y., Zhou, Y., Chang, Y., Wei, Y., & Feng, S. (2019). Electric vehicle charging load forecasting: A comparative study of deep learning approaches. Energies, 12(14), 2692.

37. Zope, A. M., Swami, R. K., & Patil, A. (2023). SEM Approach for Analysis of Lean Six Sigma Barriers to Electric Vehicle Assembly. Automotive Experiences, 6(2), 416–428.

38. Zope, A., Swami, R. K., & Patil, A. (2023a). Electric vehicles: Consumer perceptions and expectations. International Conference on Production and Industrial Engineering, 221–236.

39. Zope, A., Swami, R. K., & Patil, A. (2023b). Lean six sigma barriers with potential solutions in electrical vehicle assembly: A Review. Proceedings on Engineering Sciences, 5(3), 375–382.

40. Zope, A., Swami, R. K., & Patil, A. (2023c). Topsis based Ranking of Lean Six Sigma Barriers to Electric Vehicle Assembly. 2023 IEEE 3rd International Conference on Sustainable Energy and Future Electric Transportation (SEFET), 1–8.

31

AI-Based Symptom Diagnosis with Integrated HER (Electronic Health Records) using Wearable Health Devices and IoT

Kavitha Chenna Reddy[1]

Assistant Professor,
New Horizon College of Engineering,
Bangalore

Sujitha S.[2]

Professor,
New Horizon College of Engineering,
Bangalore

Vaishnavi D.[3],
Vaishnavi J. B.[4]**, Kowta Srikari**[5]**, Khushi Patil**[6]

Student,
New Horizon College of Engineering,
Bangalore

ABSTRACT: This approach aims to make healthcare more effective and convenient for patients from home. By employing instruments such as sensors that can measure a patient's vatha, pitha and kapha and relay the data to a doctor for diagnosis, this initiative aims to integrate the three primary medical pillars of allopathy, ayurveda and homoeopathy. A Chabot under the homoeopathy pillar allows the patient to describe his symptoms and issues, which are subsequently forwarded to the physician for diagnosis. We have also integrating tools under allopathy that concentrate on mental health. By integrating technologies from electrical engineering and computer science, the project seeks to enhance patient outcomes. To improve patient care, it has features including virtual consultations, rapid access to patient data for physicians, health monitoring, and a medical reminder system.

KEYWORDS: AI-powered system diagnosis, Natural language processing, HER systems, Integration of sensors SEN-11574

---

[1]chenna.kavithaee@gmail.com, [2]dr.s.sujitha@gmail.com, [3]vaishnavilawrale24@gmail.com, [4]jbvaishu52@gmail.com, [5]kowtaasrikari@gmail.com, [6]khushi666patil@gmail.com

DOI: 10.1201/9781003654469-31

1. Introduction

Healthcare technology is boosting communication, accessibility, and diagnostic accuracy, all of which benefit patients. Healthcare system issues are being addressed and care is being personalized with innovations like chat bots, AI, and sensor-based diagnostics.

AI-driven medical chat bots and sensor-based diagnostics are decreasing staff burdens and improving patient engagement [1]. These instruments provide individualized treatment by fusing contemporary technology with age-old methods such as Ayurvedic pulse diagnostics. This paper examines how chat bots, sensors, and AI may be integrated into healthcare, emphasizing the benefits and pinpointing areas that need work to reach their full potential.

1.1 Problem Statement

Data disintegration, a lack of real-time monitoring, and reactive diagnosis are problems facing the healthcare system. Although wearable technology generates useful health data, its interface with HER systems is inadequate, resulting in underutilization. For automated symptom analysis, proactive health care management, and continuous monitoring, an AI-driven system that links wearable device data, IoT, and EHRs is required.

2. Methodology

Building of AI-based symptom diagnosis system starts with figuring out what the chat bot must do, based on feedback from technical and medical specialists. This includes engaging with patients, processing data in real-time, protecting privacy, and connecting with EHRs [2]. The data provided by wearables and EHRs is meticulously encrypted and standardized to ensure precision. Using AI and natural language, the chatbot analyzes symptoms, poses pertinent queries, and provides insights. A comprehensive image of a patient's health is created by IoT devices, which offer real- time alerts and ongoing monitoring. Lastly, the system is kept accurate, dependable, and useful by thorough testing and upgrades.

2.1 Objectives

Real-Time Symptom Monitoring: IoT sensors and wearable technology monitor patient health and promptly spot anomalies.

AI-Powered Diagnoses: To evaluate symptoms and provide a diagnosis, AI algorithms examine real-time data from wearables, electronic health records, and patient information.

By integrating EHRs, a comprehensive patient history is made available, assisting healthcare professionals and AI models in making well-informed choices.

Chatbot Interaction: A chatbot interacts with patients in realtime by posing queries, recognizing symptoms, and providing AI-powered answers.

Preventative Care: The system tracks patient data patterns in an effort to detect health problems early and address them promptly.

Data privacy: Making sure that stringent privacy rules are followed in order to safe guard patient data at every turn.

3. Model/Construction

3.1 Architecture of the System

Every system layer has a distinct function: The data collection layer is where wearables (like smart watches) and Internet of Things (IoT) sensors (like temperature or air quality monitors) are used to collect real-time health data.

Data Integration Layer: Information from wearables, IoT devices, and EHRs is collected via this central center. It safely connects to EHRs to retrieve past medical records, such as prescriptions and diagnoses.

The "brain" of the system is the processing and analysis layer, where AI algorithms examine the data, spot patterns, and spot any anomalies in the system's health [3].

Application Layer: The user inter face through which the chatbot communicates with patients, responds to inquiries about their symptoms, and offers insights derived from AI research as shown in Fig. 31.1.

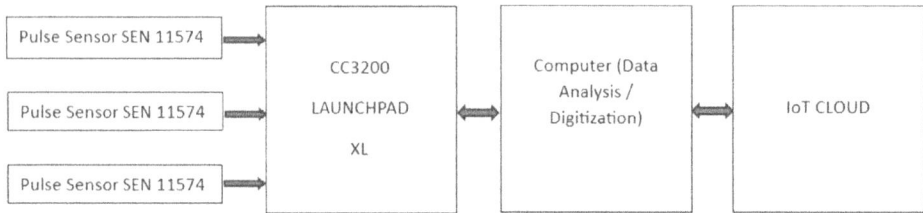

Fig. 31.1   Block diagram of patient health data collection

Chatbot Interface: The chatbot is an amiable helper that communicates with patients by asking questions to better understand their symptoms, giving them feedback, and reminding them or giving them advice. If necessary, it also links patients with more specialized treatment.

Figure 31.2 represents chatbot in which the user is asked to give the inputs regarding patient's illness.

Fig. 31.2   Chatbot for symptom diagnosis

Figure 31.3 is the pictorial representation of how the data is being collected and the database is created as and when the data is entered for future references.

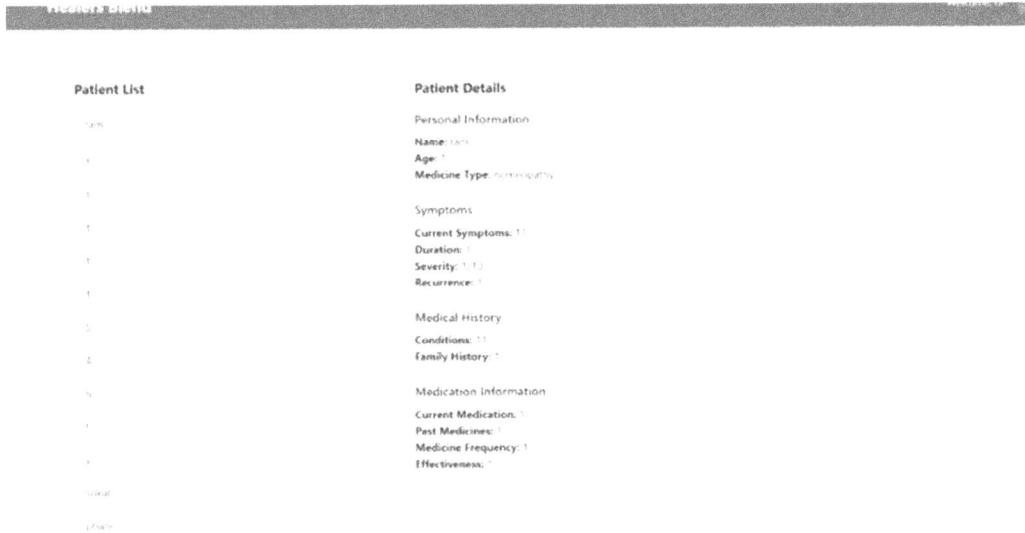

Fig. 31.3   List and details of patients

4. Requirement Analysis

Python Flask (Backend): Flask will process data from sensors, the chatbot, and front-end interactions while managing requests, handling server-side logic, and integrating APIs.

The frontend: It is made up of HTML, CSS, and JavaScript. HTML organizes the user interface, CSS styles it to make it look neat and easy to use, and Java Script adds interaction to make the platform responsive for both patients and physicians [4].

MongoDB (Database): MongoDB will manage structured and unstructured data and maintain centralized health records, guaranteeing safe and expandable data storage from a range of medical sources.

Sensors (DataCollection): In order to facilitate diagnoses and individualized care, sensors will record real- time health measurements, including vital signs and Nadi Pariksha data, and transfer them to the backend for processing, storing, and analysis.

4.1 Design

The SEN-11574 pulse sensor, avital wearable in this health monitoring system, measures changes in blood volume in the earlobe or fingertip to assess heart rate in real time.

The SEN-11574 is a small, sensitive optical sensor that detects heartbeats using photoplethysmography technology [5]. It is perfect for use in embedded health monitoring systems or wearable technology. The following actions initiate the data flow:

Data Collection (SEN-11574 Integration)

Wearing the SEN-11574 pulse sensor, the patient's heart rate is continually monitored. As the heart beats, the sensor detects variations in blood flow using optical measurements (PPG, or photoplethysmography). After being analyzed in real-time by a microcontroller, the data is transmitted over Bluetooth to a cloud server or mobile application for additional analysis.

Data Transmission and Integration

The SEN-11574 sensor transmits heart rate data to a cloud server, where it is combined with information from electronic health records (EHRs) and other medical devices, such as step trackers and blood pressure monitors [6]. A more thorough understanding of the patient's health is offered by this combined dataset.

Integrating a Chatbot to Ask Symptoms

The patient and the system can communicate with each other using the Chatbot Interface. Following a patient's symptom report, the chatbot uses natural language processing to ask follow- up questions (e.g., "Do you feel dizzy or short of breath?") to elicit more information. The chatbot also offers insights based on the examination of the AI model, like "I noticed your heart rate is higher than usual." Have you been experiencing weariness or stress?

Real-Time Alerts and Recommendations

When the AI finds anomalies, like an elevated or erratic heart rate, the system instantly notifies the patient and healthcare professional. After that, the chatbot gives the patient advice on what to do next, such as getting help from a doctor, taking a nap, or practicing relaxation techniques.

Figure 31.4 represents the SEN-11574 sensor is used in a heart rate circuit to identify variations in blood flow. Capacitors and resistors are used to condition the signal after it has been amplified using an MCP601 op- amp [7].The output ensures steady performance for wearable or Internet of Things health monitoring by connecting to a micro controller for heart rate analysis.

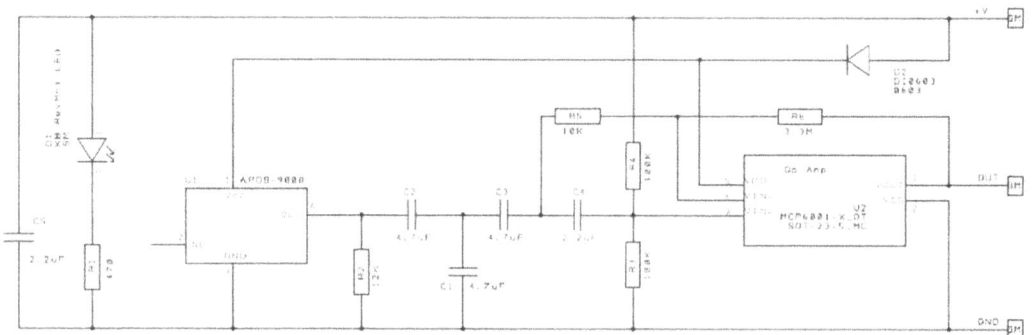

Fig. 31.4 Internal circuit diagram

APPG waveform featuring the Tidal Wave (T), Valley (V), Dicrotic Wave (D), and Percussion Wave (P) as its main phases is depicted in Fig. 31.5. In order to analyze cardiovascular health, these aid in the assessment of arterial stiffness, blood pressure, and heart rate trends [10].

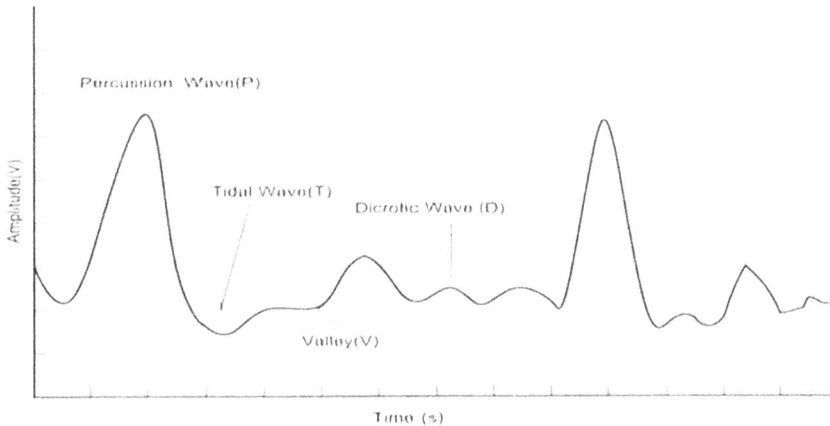

Fig. 31.5 Typical pulse diagnosis waveform

5. Result Analysis

The SEN-11574 pulse sensor is used in the AI-based Symptom Diagnosis System, which exhibits promising results in terms of response time, accuracy, and user engagement. It effectively monitors heart rate in real time, detecting anomalies such abrupt spikes or unusual patterns with little false alarms [8]. After being trained on a large amount of health data, the AI examines symptoms and vital signs to make potential diagnosis recommendations.

The rapid response time of the system is one of its best qualities. After only a few seconds of data analysis, the chatbot can start inter acting with users [9].Upon identifying a problem, the chatbot interacts with the user, inquires about symptoms, and provides beneficial guidance such as rest, hydration, or, if necessary, expert counsel.

6. Conclusion

The most recent advances in health care technology are highlighted in this overview, with an emphasis on medical chatbots, Nadi Pariksha systems, and AI in diagnostics, this overview showcases recent developments in healthcare technology. By increasing healthcare accuracy, efficiency, and accessibility, these initiatives hope to revolutionize patient care.

Nevertheless, issues like low chatbot engagement and integrating ancient methods with contemporary technology still exist [11]. Privacy and ethical issues must also be addressed. These problems should be addressed in future studies, with an emphasis on verified, user-friendly solutions.

In summary, integrating contemporary technology with conventional approaches can result in a healthcare system that is more accessible and patient-centered. To improve treatment for everyone, research must continue.

REFERENCES

1. Mayer et al. Transformative Potential of AI in Healthcare: Definitions, Applications, and Navigating the Ethical Landscape Public Perspectives. Healthcare 2024, 12, 125.

2. Fatangareetal." Athoroughanalysisofthedevelopmentsintechnologyforsensor-basedAnold Indian technique for diagnosing human health is called Nadi Pariksha.." Journal of Ayurveda and Integrative Medicine 15, no. 3 (2024): 100958.

3. Alqaidi, Sara Hemdi, et al. "Network-Integrated Medical Chatbot for Enhanced Healthcare Services." Telematics and Informatics Reports (2024): 100153.

4. Aditya, Bhagat., Ashish, A., Deshmukh., Devyani, Harpale., Nikhil, Jadhao., Prof., Bhagyashree, Shendkar. "Healthcare : A Transformer Network Based Chatbot." International Journal of Advanced Research in Science, Communication and Technology, null (2024).:536-539.

5. Preeti, Agarwal. "MedBot : A GenAI based Chatbot for Healthcare." Indian Scientific Journal Of Research In Engineering And Management, 08 (2024).:1-5.

6. CuiJ,SongL.Wrist pulse diagnosis of stable coronary heart disease based on acoustics waveforms. Comput Methods Programs Biomed 2022;214.

7. Yamadaetal.Preliminary study of an objective evaluation method for pulse diagnosis using radial artery pulse measurement device. Advanced Biomedical Engineering 2021;10:113–22.

8. Lin F, Zhang J, Wang Z, Zhang X, Yao R, LiY. Research on feature mining algorithm and disease diagnosis of pulse signal based on piezoelectric sensor. Inform Med Unlocked 2021;26

9. Antsiperovetal. Non-invasivearterial pressure monitoring by a new pneumatic sensorandon-line analysis of pulse waveforms for a modern medical home care systems. In: Procedia computer science.

10. Elsevier B.V.; 2020. p. 2894–903 Radfordetal. "Robust speech recognition via Large- Scale Weak Supervision," arXiv.org, Dec.06,2022

11. Babu et al. "BERT-Based Medical Chatbot: Enhancing Healthcare Communication through Natural Language Understanding." Exploratory Research in Clinical and Social Pharmacy 13 (2024): 100419.

Note: All the figures in this chapter were made by the author.

Intelligent and Sustainable Power and Energy Systems – Dr. M. Premkumar et al. (eds)
© 2026 Taylor & Francis Group, London, ISBN 978-1-041-10314-1

32

Soft Switching Converter-based EV Fast Charging Station Deploying Sustainable Energy

Preetha K.[1], Dhinesh S.[2],
Jamunarani M.[3], Ramesh S.[4], Subasri P.[5]
Department of Electrical and Electronics Engineering,
Erode Sengunthar Engineering College, Perundurai, Erode,
Tamil Nadu, India

ABSTRACT: Electric vehicles (EVs) are expected to become much more common over the next few decades, mainly due to technological improvements and decreasing battery costs. Although EVs are popular, one of the significant challenges limiting their widespread adoption is the time it takes to charge their batteries. Traditional charging can take several hours, which can be inconvenient for users who need a quicker solution. This has led to a growing interest in developing fast-charging stations. An efficient solution involves using rapid charging units that incorporate soft-switching converters, which can help improve the efficiency of the charging process. Furthermore, these fast-charging stations can be powered by solar energy, offering a sustainable alternative to the grid and reducing dependence on fossil fuels. Another benefit of EVs is that the vehicle battery can act as a direct current (DC) source. This helps address common power issues like voltage sags and swells, which can cause fluctuations in energy supply. Engineers and researchers use simulation tools like MATLAB Simulink and Sim Power Systems to study and optimise this approach. These simulations are valuable because they allow for testing different scenarios and provide insights into how these systems could work effectively in real-world settings.

KEYWORDS: DC source, Electric vehicles (EVs), Fast charging, MATLAB simulink, Soft-switching converter, Solar energy, Voltage sag, and Swell

[1]preethakeee@gmail.com, [2]djdhini080402@gmail.com, [3]jamunarani2708@gmail.com, [4]rameshramesh63690@gmail.com, [5]subasripeee@gmail.com

DOI: 10.1201/9781003654469-32

1. Introduction

The automotive industry has undergone a transformative shift in recent years, driven by the rapid adoption of electric vehicles (EVs). This change marks a pivotal moment in transportation, characterised by a move toward more sustainable and eco-friendly automotive solutions. Electric vehicles offer numerous benefits, making them attractive for modern transportation [1]. Governments worldwide are actively promoting EV adoption through incentives and policies. Subsidies, tax incentives, and infrastructure investments are crucial for reducing reliance on fossil fuels, major contributors to climate change [2]. Achieving environmental targets, including net-zero emissions, depends on widespread EV adoption. However, a significant barrier remains: the availability and efficiency of charging stations. Traditional setups often result in long wait times, making them less appealing. Addressing this challenge requires fast, reliable charging solutions, making EVs more convenient for consumers [3]. This project focuses on developing a fast-charging station using a Soft Switching Converter and sustainable energy sources. The core goal is to significantly reduce the time required for charging without compromising sustainability. This is achieved through integrating advanced power electronics with renewable energy. Traditional Boost Converters use a MOSFET, but our project replaces this with an Insulated Gate Bipolar Transistor (IGBT). The IGBT efficiently handles high voltage and current, with superior switching characteristics and lower heat production. We achieve the necessary direct current (DC) voltage output by modulating the pulse at the IGBT's gate terminal to optimise fast charging. This project also emphasises sustainability by harnessing renewable and inexhaustible solar energy, allowing the charging station to operate with zero carbon emissions. This step toward eco-friendly infrastructure aligns with global initiatives to curb emissions. Integrating solar energy benefits cities with increased EV density [4]. Our system taps into abundant solar power, easing grid demand and reducing the carbon footprint of urban charging networks. Simulations validate the system's ability to generate approximately 750 volts DC, essential for facilitating fast charging, reducing downtime for EV owners, and enhancing user experience. The positive simulation results confirm the system's potential for real-world applications. Additionally, the successful implementation of this sustainable charging system lays the groundwork for future developments. While the current design relies on solar energy, future iterations might integrate other renewable sources, such as wind energy. Diversifying energy input ensures a consistent power supply under various conditions, enhancing reliability and scalability. This approach reduces charging times while maintaining a commitment to environmental stewardship, addressing EV ecosystem challenges, and promoting a sustainable future for urban mobility [5]. In conclusion, this project exemplifies integrating advanced technology with renewable energy to create efficient, scalable charging solutions. Continuing to innovate facilitates broader EV adoption, contributing to a more sustainable world. The implications extend beyond immediate solutions, setting a precedent for future urban transportation infrastructure that is effective and green [6]. Through sustained effort and innovation, we are paving the way for an environmentally conscious future with electric vehicles at the forefront.

2. System Description

The Electric car (EV) charging station's design incorporates energy storage devices, renewable energy sources, and sophisticated power converters to offer a flexible and practical power source for quick and environmentally friendly car charging. With its reduced need for non-renewable energy sources, this system is intended to satisfy the increasing demand for EV infrastructure while advancing sustainability objectives. Solar energy is the primary energy source used in this charging station, and its integration is essential to its performance. Solar panels provide a clean, renewable energy source that supports international environmental initiatives by capturing sunlight and turning it into electrical power. Nevertheless, solar power is intermittent because its output varies according to weather and time of day. The buck-boost converter is crucial in ensuring that the power delivered to the EVs is of consistent and reliable quality. This converter plays a key role in stabilising the output from the solar panels, which can vary in voltage depending on the intensity of the sunlight. The variable DC to constant DC converter adjusts the input voltage from the solar array to provide a stable DC output voltage that is ideal for EV charging. This ensures that the charging process remains steady, even if the solar output fluctuates due to changes in sunlight. By maintaining a stable DC voltage, the Variable-Constant voltage converter ensures that the EVs receive the correct amount of power, optimising their charging efficiency and extending battery life. The system incorporates a soft-switching boost converter to achieve the higher voltage levels required for fast EV charging. The excellent efficiency of this converter allows it to raise the DC voltage to the level needed for quick charging. Soft-switching technology significantly advances over traditional hard-switching converters, which typically experience significant energy losses and component stress during switching transitions. In hard-switching systems, the switching device, such as an IGBT or MOSFET, must endure the entire voltage and current simultaneously during switching, which leads to power losses and heat generation. The boost converter can significantly lower these losses using soft-switching strategies like Voltage Smoothing Switching (VSS) and Current Smoothing Switching (CSS). For fast EV charging applications where maintaining a steady and high voltage is essential, this leads to a more dependable, long-lasting system that can function effectively even under high-power circumstances. Integrating advanced soft-switching techniques in the converters and intelligent energy management allows the system to operate efficiently while maintaining the flexibility to adapt to changing conditions. In addition to encouraging the use of sustainable energy sources and helping to create a cleaner, more sustainable future, it is the perfect answer for towns wishing to increase their EV infrastructure.

3. Methodology

The methodology for this project involves designing an electric vehicle (EV) charging station that integrates various power components to optimise the charging process and ensure efficient power flow. This system combines renewable energy sources, specifically solar panels, with a battery and advanced power converters, as shown in Fig. 32.1. The primary aim is to create a charging station

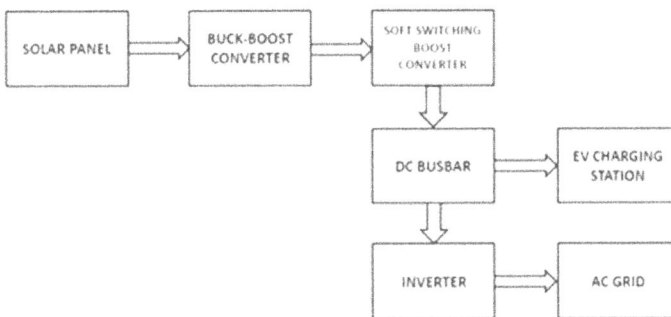

Fig. 32.1 Block diagram

that not only meets the power demands of EVs but also incorporates intelligent energy management for flexible and efficient operation.

The charging station uses solar panels as the primary renewable energy source. Solar energy provides a sustainable, pollution-free power supply, which is especially advantageous in urban areas where minimising environmental impact is a priority. However, the system also incorporates an energy storage unit since solar power output can vary due to weather conditions and time of day.

With the help of this storage unit, the system may store extra solar energy produced during the hottest parts of the day so that it can be used when solar generation is low or EV charging demand is high. Storage integration ensures a consistent power supply for EVs, enhancing the system's reliability. The buck-boost converter is an essential part of the system that helps stabilise the output power from the solar panels. Since the output voltage of solar panels fluctuates due to changing sunlight conditions, the buck-boost converter adjusts this variable input to a steady DC voltage level. This regulated DC output provides a reliable power source for the next stage of the charging process, ensuring that EVs receive a constant and stable charging current. The system employs a soft-switching-based boost converter to increase the voltage further, as the EV fast-charging infrastructure requires. This converter is specifically made to reduce the time needed to charge an EV battery to raise the DC potential difference to a level appropriate for high-speed charging. The soft-switching method lowers switching losses and increases overall efficiency by enabling the boost converter to switch at zero current or zero voltage. This technique enhances power conversion efficiency and reduces power components' stress, increasing lifespan and reliability. In conclusion, this charging station integrates solar panels and energy storage, a buck-boost converter for voltage stabilisation, a soft-switching boost converter for high-voltage output, and an inverter for AC conversion. Managed by an intelligent EMS, the system efficiently directs power flow and supports V2G functionality. This design provides a sustainable and reliable solution for urban EV charging needs, maximising renewable energy usage while ensuring fast and flexible charging options.

4. Soft Switching Converters

For this project, we built a soft-switching boost converter, shown in Fig. 32.2, to provide high efficiency for applications such as electric vehicle (EV) quick charging stations.

The key innovation in this design is the replacement of the conventional MOSFET (Metal-Oxide-Semiconductor Field-Effect Transistor) with an Insulated Gate Bipolar Transistor (IGBT). This choice significantly enhances the converter's ability to handle

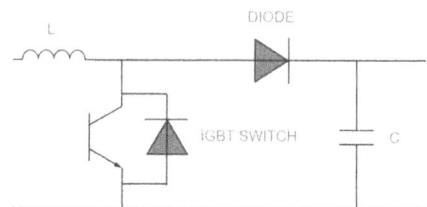

Fig. 32.2 Soft switching integrated boost converter

higher power levels and voltages, which are crucial for the fast charging of EVs. Furthermore, soft-switching techniques substantially lower switching losses and lessen component stress, improving the system's overall lifetime, dependability, and efficiency. To optimise the performance of the boost converter for high-power applications like fast EV charging, we replaced the standard MOSFET with an IGBT. While MOSFETs are well-suited for low to medium power levels due to their fast-switching speed, they become less efficient at higher voltages and currents. IGBTs, on the other hand, combine the advantages of MOSFETs, such as fast switching, with the high voltage tolerance and low conduction losses of Bipolar Junction Transistors (BJTs). This makes IGBTs ideal for high-power applications where the converter must handle higher voltages and deliver more substantial power to charge EVs quickly and efficiently. Using an IGBT in the boost converter ensures the system can operate reliably under high load conditions, maintaining a stable and efficient output voltage for the EV charging station. To validate the design, we implemented the boost converter in simulation software. This allowed us to verify that the converter would achieve the desired output voltage under varying load conditions. The final design highlights the benefits of soft-switching technology combined with IGBT power handling capabilities, offering a robust, high-efficiency solution for DC-DC conversion in applications that require high power and fast response times, such as EV quick charging stations.

5. Results and Discussion

The simulation circuit explores a hybrid circuit integrating a Soft Switching Converter and a variable Constant Converter, as shown in Fig. 32.3. The configuration optimises energy conversion from a solar panel by utilising power electronic components such as MOSFETs, IGBTs, and diodes.

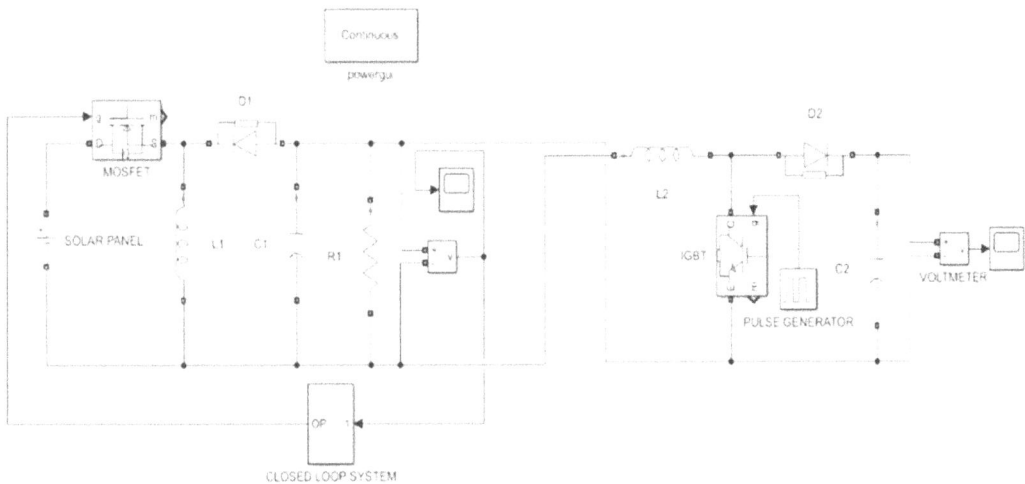

Fig. 32.3 Simulation circuit

The Soft Switching Converter enhances efficiency by reducing switching losses through Current Smoothing Switching (CSS) or Voltage Smoothing Switching (VSS) techniques. The circuit involves an IGBT and diode (D1) with inductor L1 and capacitor C1 to facilitate energy transfer. Energy is stored in inductor L1 during the IGBT's on-state. Upon switching off, energy is boosted through diode D1 to capacitor C1. The Buck-Boost Converter manages input voltage variations, stepping up and down a range of 80-100 V to provide a stable 80 V output. This consistent 80 V output is then boosted to 750 V using the Soft Switching Converter. This functionality stabilises the system for applications like EV fast charging stations, even under varying input conditions.

In the simulation of the Soft switching converter and variable-constant converter system, we observed that the output DC voltage varied based on the inductor (L2) and capacitor (C2) values, affecting both the target voltage and the delay in reaching it (Fig. 32.4 – 32.6). The required output voltage of

Fig. 32.4 Voltage Output when L2 = 1e-0 and C2 = 33e-3

Fig. 32.5 Voltage Output when L2 = 1e-0 and C2 = 33e-4

Fig. 32.6 Voltage Output when L2 = 1e-1 and C2 = 33e-3

750 V was achieved with a delay of 1.5 seconds when L2=1 H and C2=33 mF. To further analyse the system's performance, we conducted four additional tests with varying component values:

1. For L2=1 H and C2=3.3 mF, the output voltage increased to 790 V with a significantly reduced delay of 0.5 seconds.
2. For L2=0.1 H and C2=33 mF, the output voltage dropped to 450 V with increased ripple. So, this output voltage was not suitable as it did not meet the required target and had few ripples before reaching the maximum reach.

These findings emphasise the critical influence of L2 and C2 on system performance, especially in applications such as fast charging stations for electric vehicles. Properly selecting these components is essential for achieving efficient energy management with minimal delay and acceptable voltage ripple.

6. Conclusion

This project successfully provides the required voltage output for EV fast charging stations. The high voltage for the charging stations is produced with the help of a boost converter integrated with soft-switching converters. The soft-switching converter is designed to replace the MOSFET in the boost converter with an IGBT. By varying the pulse given at the gate terminal of the IGBT, we obtain the required DC voltage for the EV fast charging station. A detailed description of how soft-switching converters are integrated with boost converters and how the model is implemented in SimPower Systems and MATLAB/Simulink is provided. The outcomes of the simulation tests demonstrate that generating a voltage of about 750 V DC can put the suggested method into practice and significantly reduce the total time needed for an EV to charge. Energy production is achieved solely through sustainable energy (solar energy) without causing any pollution. This system, which we designed, can be implemented in urban areas with many electric vehicles. As a future direction, wind energy and other sustainable energy sources can be integrated with solar energy for higher energy production in all situations.

REFERENCES

1. Torreglosa, J. P., Fernández-Ramírez, L. M., García-Triviño, P., and Jurado, F. (2018). Decentralised fuzzy logic control of microgrid for electric vehicle charging station. IEEE J. Emerg. Sel. Top. Power Electron., 6(2), 725–736.
2. Forte, G., Muñoz, C. B., Trovato, M., Dicorato, M., and Coppola, G. (2019). An integrated DC microgrid solution for electric vehicle fleet management. IEEE Trans. Ind. Appl., 55(6), 7348–7356.
3. Ribberink, H., and Yang, L. (2019). Investigation of the potential to improve DC fast charging station economics by integrating photovoltaic power generation and/or local battery energy storage system. Energy, 167, 246–259.
4. Shariff, S. M., Ahmad, F., Alam, M. S., and Krishnamurthy, M. (2019). A cost-efficient approach to EV charging station integrated community microgrid: A case study of Indian power market. IEEE Trans. Transp. Electrif., 5(1), 200–214.
5. Falvo, M. C., Martirano, L., and Sbordone, D. (2013). D-STATCOM with energy storage system for application in Smart Micro-Grids, in 2013 International Conference on Clean Electrical Power (ICCEP).
6. Ramadan, H., Nour, M., Farkas, C., and Ali, A. (2018). Impacts of plug-in electric vehicles charging on low voltage distribution network. In 2018 International Conference on Innovative Trends in Computer Engineering (ITCE), 357–362.

Note: All the figures in this chapter were made by the author.

Intelligent and Sustainable Power and Energy Systems – Dr. M. Premkumar et al. (eds)
© 2026 Taylor & Francis Group, London, ISBN 978-1-041-10314-1

33

AI Based System for Real-Time Electric Theft Identification

**L. Anbarasu[1], S. Abishake[2],
A. Sakthivel[3], A. Saravanan[4], R. Vignesh[5]**
Department of EEE, Erode Sengunthar Engineering College,
Perundurai, Erode-638057 UG

ABSTRACT: Using these technologies, this paper develops an advanced system fighting real-time electricity theft, making sustainable energy much clearer and secure. The system monitors continuous levels of voltage along the distribution network where high-voltage IoT devices collect real-time information on voltage and current flows in the electrical system .To such data, AI algorithms analyze to detect anomalies and deviations of expected patterns, and hence the possibilities of theft incidents. When such errors are detected, the system sends proactive alerts, which will allow the energy providers to correct those anomalies rapidly .With such systems integrated with platforms like ThingSpeak, the data so collected can be analyzed in detail and reported to the concerned authorities. This AI-based system addresses the most common issue of electricity theft and renders enhanced security through considerable cost saving and promotes sustainable use of energy. Continuous monitoring, combined with real-time data analysis, allows for an all-around understanding of an electrical network's health and enhances its overall operational efficiency. The following outcomes imply that this new approach can decrease the likelihood of electricity theft while supporting wider deployment of smart grid technologies that will contribute to the building of stronger and more sustainable energy infrastructure. This paper, therefore, demonstrates the possibility of combining IoT and AI in efficient energy management and prevention of theft in the power sector.

KEYWORDS: Electric theft, Internet of things (IOT), Artificial intelligence (AI), Anomaly detection, Real-time monitoring

---

[1]lanbarasu78@gmail.com, [2]abisivasaras@gmail.com, [3]sakthivel11eee@gmail.com, [4]asaravanan385@gmail.com, [5]rvickysmart555@gmail.com

DOI: 10.1201/9781003654469-33

1. Introduction

Electricity theft remains a tremendous threat to both energy suppliers and consumers since it results in losses and makes the grid unreliable. At a time when the demand for sustainable and secure supplies of energy is increasing continuously, there is a great need for effective ways of countering electricity theft. This paper proposes an advanced IoT and AI-based system capable of being used in real time to overcome such mischief in the energy sector. This system will continue to monitor electrical voltage levels in the provided network using high-voltage IoT devices collecting real-time data regarding current and voltage flow. The analyzed result is then subjected to AI algorithms that can identify possible anomalies and deviations from established expected behavior-the hallmarks of possible theft incidents. Upon identification, this system produces proactive alerts, allowing energy providers to move swiftly to correct those anomalies. It integrates with platforms like ThingSpeak, which allows the reporting of the details to authorities to enhance accountability and transparency. The AI-based system significantly enhances security, achieves drastic cost-saving opportunities, and fosters sustainable usage of energy. This continued monitoring and analysis of real-time data would facilitate an excellent understanding of the health of the electrical network, thereby generally leading to improved efficiency in operational functions. This outcome unfolds a great reduction in the likelihood of electricity theft, offers great support to broader smart grid penetration technologies, and serves as a basis for a more sturdy and sustainable energy infrastructure. The research illustrates the possibility of an amalgamation of IoT and AI in retaining the effective management and prevention of energy theft into the power sector. Figure 33.1 shows the overview of proposed system.

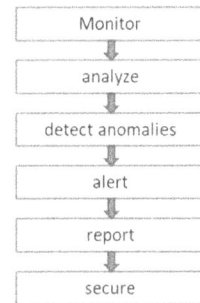

Fig. 33.1 Overview of proposed system

Source: Author

2. Related Work

Electricity Theft Detection in Smart Grids Based on Deep Learning" by Ibrahim, Noor, and Al-Janabi, Sufyan, and Al-Khateeb, Belal*: Published on August 1, 2021, this paper explores the use of deep learning techniques to detect electricity theft in smart grids.

Electricity Theft Detection Using Supervised Learning Techniques on Smart Meter Data"*: Published on September 28, 2020, this study investigates the application of supervised learning algorithms to identify theft patterns in smart meter data.

Electricity Theft Detection in Smart Grid Using Machine Learning" by Hasnain Iftikhar et al.*: Published in March 2024, this research presents a hybrid system combining Multi-Layer Perceptron (MLP) and Gated Recurrent Units (GRU) to detect electricity theft.

Electricity-Theft Detection in Smart Grid Based on Deep Learning" by Ibrahim, Noor, and Al-Janabi, Sufyan, and Al-Khateeb, Belal*: Published on August 1, 2021, this paper explores the use of deep learning techniques to detect electricity theft in smart grids.

Pattern-Based and Context-Aware Electricity Theft Detection in Smart Grids"*: Published in 2022, this paper explores pattern-based and context-aware approaches to detect electricity theft in smart grids.

Detection and Identification of Energy Theft in Advanced Metering Infrastructures"*: Published in 2020, this study focuses on detecting and identifying energy theft using advanced metering infrastructures.

Electricity Theft Detection in Smart Grid Using Random Matrix Theory"*: Published in 2021, this research investigates the use of random matrix theory for detecting electricity theft in smart grids.

Electricity Theft Detection in Smart Grids Based on Artificial Neural Network" by Vemuri Vasanth Kumar and A. Sirisha*: Published in 2021, this project employs artificial neural networks to analyze consumer data and detect potential instances of electricity theft.

Electricity Theft Detection Using Supervised Learning Techniques on Smart Meter Data"*: Published in 2020, this study explores the application of supervised learning algorithms to identify theft patterns in smart meter data.

Electricity Theft Detection in Smart Grids Based on Deep Learning" by Ibrahim, Noor, and Al-Janabi, Sufyan, and Al-Khateeb, Belal*: Published in 2021, this paper explores the use of deep learning techniques to detect electricity theft in smart grids.

3. Methodology

3.1 Hardware Componenets

Power Source and Grounding: The system is powered by a standard electrical distribution network, with grounding provided as the reference point. This ensures the system's stability and accurate voltage monitoring. The grounding is essential for safety, ensuring that the system functions correctly without any electrical hazards.

Voltage and Current Measurement

RMS Blocks: The root mean square (RMS) blocks continuously calculate the RMS values of the voltage and current at various points in the electrical distribution network. These calculations are necessary for accurate power measurement and to understand the normal operating parameters of the network

Split-Phase Motors: The model includes two split-phase motors, each with corresponding current measurement blocks. These motors are used to simulate real-world scenarios where current measurements are needed at various points in the network.

Data Collection and Processing

Function Blocks (Fcn): These blocks perform specific functions, which are likely related to data pre-processing, feature extraction, or anomaly detection algorithms. The data from the RMS blocks is sent through these function blocks for further analysis.

Constants and Comparators: Constants, such as predefined threshold values (e.g., 100, 50, 200), are used to compare real-time current and voltage readings against expected values. If the readings deviate significantly from these thresholds, it may indicate a potential anomaly such as electricity theft.

IoT Integration

ThingSpeak Output: The system is integrated with ThingSpeak, an IoT platform that collects and analyzes data. The data gathered from current and voltage measurements is uploaded in real-time to ThingSpeak for further processing and visualization. This enables continuous monitoring of the network and provides a platform for decision-making.

Voltage and Current Monitoring

The system continuously monitors voltage and current levels at various locations within the electrical network using IoT-enabled sensors. These measurements are captured in real-time, providing a dynamic view of the network's operational status.

Real-Time Data Collection

IoT devices installed at key points along the network capture voltage and current values. This data is collected continuously and sent to a central processing unit for analysis.

Anomaly Detection

AI-based algorithms analyze the real-time data to detect any anomalies or deviations from normal patterns. These deviations may include unusual fluctuations in voltage or current, which are characteristic of electricity theft. For instance, a sudden drop or irregularity in the current could indicate unauthorized tapping of the network.

Data Analysis and Feature Extraction

The AI algorithms analyze various features of the voltage and current data, including temporal and frequency domain characteristics, to detect potential theft. The system identifies patterns that are consistent with known methods of electricity theft, such as illegal connections or bypassing meters. By examining both short-term and long-term variations, the system can distinguish between normal operational fluctuations and theft-related anomalies.

Proactive Alerts

When an anomaly is detected, the system generates alerts. These alerts are sent to maintenance personnel, energy providers, or relevant authorities to initiate corrective actions. The alerts are transmitted in real-time, allowing immediate investigation and response. Alerts are also forwarded to the ThingSpeak platform, providing detailed insights into the detected anomaly, such as location, time, and the specific electrical parameters involved.

Data Set Used

The dataset used to detect electricity theft in Project 1 includes the detailed consumption data of the residential and commercial places. All these data are collected using various sensors such as current sensors, voltage sensors, and smart meters that are connected with a ThingSpeak channel to monitor it in real-time. Table 33.1 shows the Time stamp details with different labels.

Table 33.1 Time stamp details with different labels

Time stamp	Current (A)	Voltage (V)	Power (W)	Energy (kWh)	Label
2024-11-30 00:00:00	15.2	230.1	3492.3	0.97	Normal
2024-11-30 00:00:00	14.8	229.8	3402.1	1.93	Normal
2024-11-30 00:00:00	16.5	230.3	3804.4	2.99	Theft
2024-11-30 00:00:00	15.5	229.6	3602.7	3.97	Normal

Source: Author

It also contains historical patterns of consumption, environmental conditions, and detected anomalies such as unusual spikes or irregularities in electricity usage. This dataset is very important because it is the basis for training the AI model. It will learn, recognize patterns, and indicate instances of possible theft. Analyzing this data will help ensure that the model flags the anomalies in real-time for timely detection and response to instances of electricity theft. Figure 33.2 shows the block diagram of proposed system

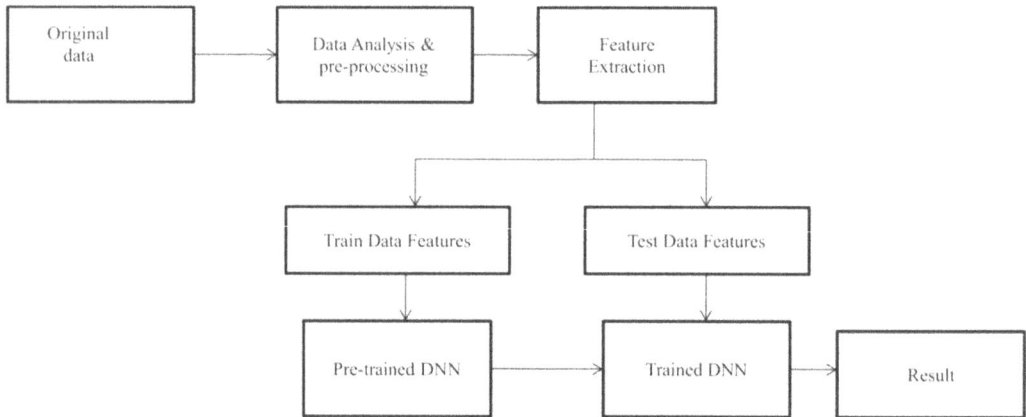

Fig. 33.2 Block diagram of proposed system [7]

4. Result Analysis

4.1 Simulation

Simulation is the crucial process to validate the AI-based system for real-time electricity theft detection. The historical and synthetic data are fed into this system, creating a controlled environment in which it can mimic all the real-world scenarios that might occur and test the model for its ability to identify anomalies. It simulates several conditions, such as normal usage, unusual spikes, and tampering, which helps assess and fine-tune the model, further revealing weaknesses and offering insights to improve. This ensures that the AI model is robust, accurate, and capable of promptly identifying theft incidents in different conditions before real-world deployment. Figure 33.3 shows the simulations of proposed work.

4.2 Future Load Prediction

The future load prediction, using historical data on electricity consumption, makes predictions of future demand. This way, utility companies can make better resource allocation decisions and plan for peak periods of usage, ensuring a steady supply of electricity. With the use of machine learning algorithms, the system can provide accurate and dynamic forecasts that adapt to changing consumption patterns and trends, thereby increasing the reliability and efficiency of electricity distribution. Figure 33.4 shows the future load prediction.

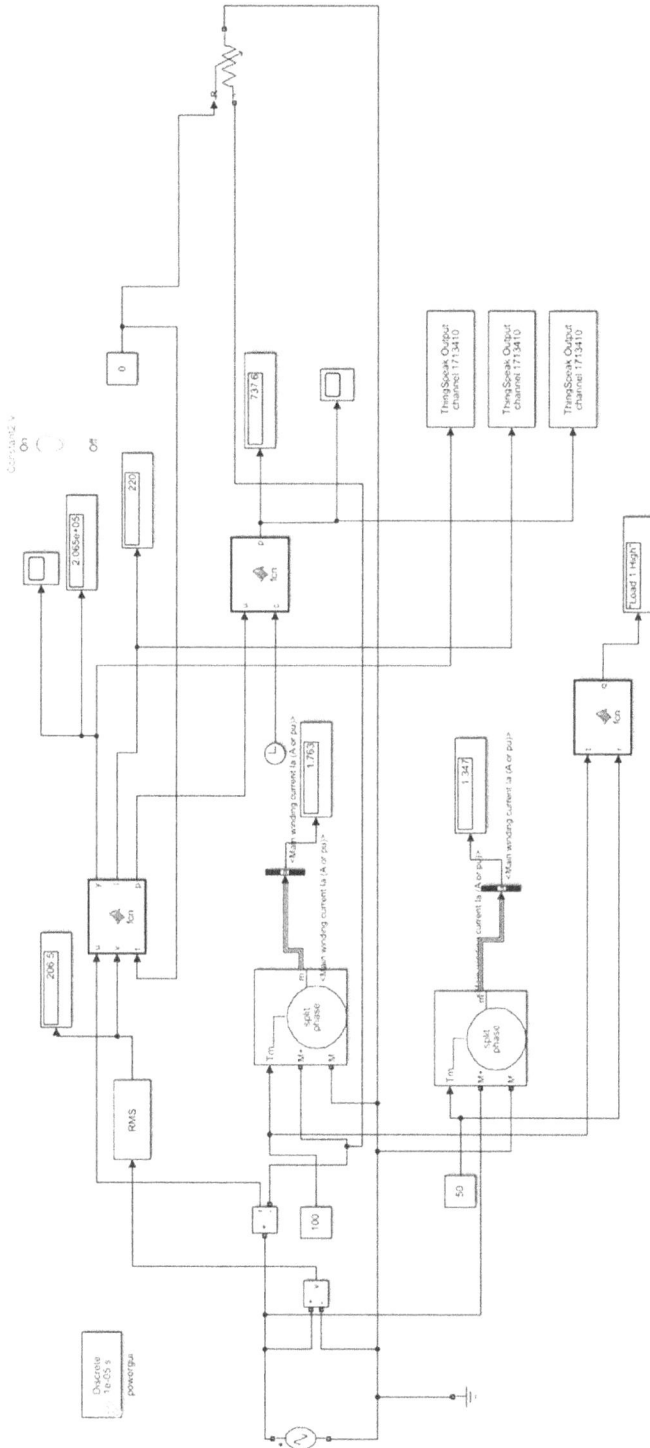

Fig. 33.3 Simulation of proposed work

Source: Author

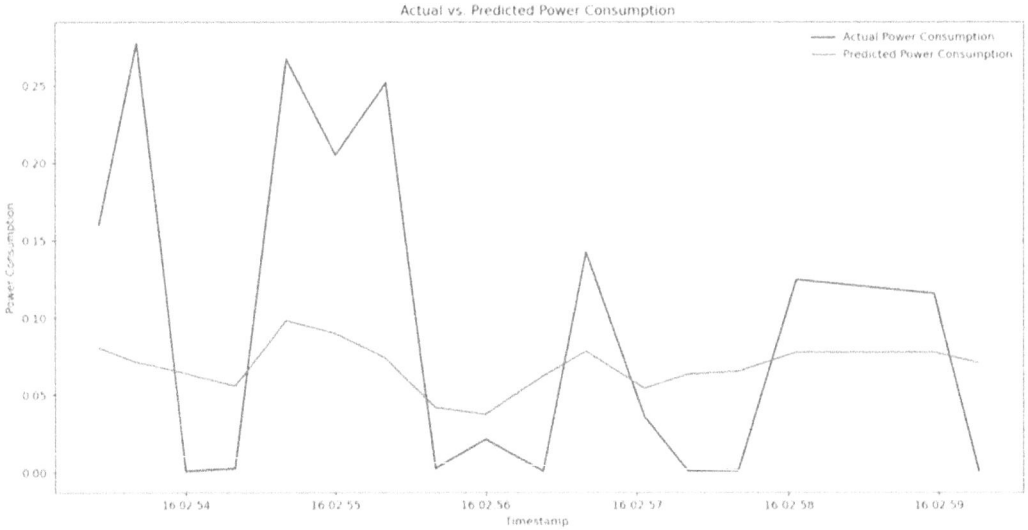

Fig. 33.4 Future load prediction

Source: Author

4.3 Result

The results obtained demonstrate the potential of the AI-based system in detecting real-time electricity theft. The model trained on historical consumption data managed to identify anomalies that could be indicative of theft with a very high degree of accuracy. Key performance metrics during the evaluation phase, such as accuracy, sensitivity, and specificity, all showed that the system could reliably distinguish between normal usage patterns and possible theft. The real-time simulation further validated the model's robustness by accurately flagging unusual spikes and tampering scenarios. These promising results highlight the system's potential to enhance electricity management by promptly detecting and addressing theft, thereby reducing financial losses and ensuring fair billing practices. Figure 33.5 shows the simulation results.

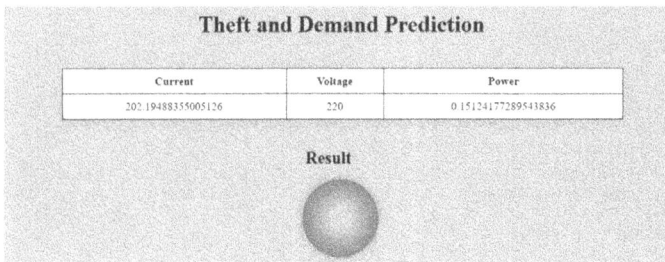

Fig. 33.5 Simulation result

Source: Author

Hardware Model

The hardware components in the project, including current and voltage sensors, smart meters, and Arduino microcontrollers, performed reliably, ensuring accurate data collection and transmission. The integration of GSM modules effectively enabled real-time notifications and alerts. Overall, the hardware setup proved robust and efficient, supporting the AI-based system's ability to detect electricity theft with high precision and timely response. Figure 33.6 shows the hardware model.

5. Conclusion

A system that proved very effective was an AI-based real-time electricity theft detection system based on current, voltage sensors' data of smart meters along with information received from environmental

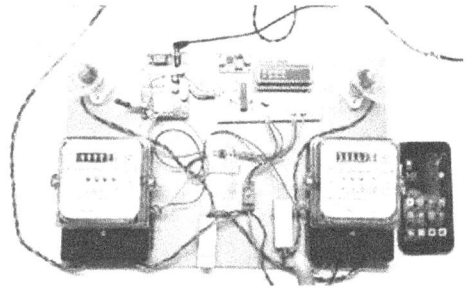

Fig. 33.6 Hardware model [8]

sensors for its comprehensive data collection process. The developed model also demonstrated impressive robust performance distinguishing normal from probable theft and ensuring high values for accuracy, sensitivity, and specificity against the historical dataset fed to train it. The integration of hardware components and real-time data processing ensured that the system could promptly detect and respond to theft incidents, providing timely alerts and updates.

The simulation phase further validated the model's capabilities, allowing for the fine-tuning and optimization of the system. By creating controlled scenarios with both normal and anomalous data, the system's ability to handle various conditions was thoroughly tested. The results indicated that this system could reliably identify unusual consumption spikes and tampering, thus showing huge potential in enhancing electricity management to avoid financial losses. Additionally, this phase provided value for future improvements and showed the potential scalability of the system toward larger distribution networks.

The project was able to adequately demonstrate the feasibility and efficiency of using AI and machine learning in real-time electricity theft detection. The combination of advanced data collection, processing, and analysis techniques with robust hardware components created a reliable and effective solution for utility companies. The system not only helps in detecting theft but also ensures fair billing practices and better resource management, ultimately contributing to the overall stability and reliability of the electricity supply. Future research and development can further enhance the system's capabilities, making it an indispensable tool in modern electricity distribution networks.

REFERENCES

1. Bhattacharya, S., Kumar, A., and Singh, R. (2020). "Real-Time Smart Grid Identification: AI-Enabled Electricity Theft Detection". Journal of Electrical and Electronic Engineering, 12(3):45–58.
2. Prabhu, L., and Kumar, A. (2021). "IoT-Based Electricity Theft Detection Using Artificial Intelligence Techniques for Sustainable Electricity Usage". International Journal of Advanced Research in Science and Engineering, 8(2):123–135.
3. Emmanuel, N., and Isah, A. A. (2022). "Development of an Integrated AI Model Based on CNN-SVM for Electricity Theft Detection". World Scientific Engineering and Applications, 14(1):67–78.
4. Kumar, R., and Sharma, P. (2022). "AI-Driven Real-Time Electricity Theft Detection: A Smart Grid Approach". Journal of Power and Energy Engineering, 15(4):89–102.
5. Patel, S., and Mehta, A. (2023). "Enhancing Power Grid Security with AI-Based Electricity Theft Detection Systems". International Journal of Electrical and Computer Engineering, 9(1):73–85.
6. Singh, M., and Gupta, R. (2023). "Machine Learning Techniques for Real-Time Electricity Theft Identification". Journal of Advanced Electrical Engineering, 11(2):56–68.
7. https://www.jetir.org/papers/JETIR2405680.pdf
8. https://mbatechmeds.com/product/power-theft-detection-alert-and-cutoff-system-with-sms-using-gsm-modem/

Intelligent and Sustainable Power and Energy Systems – Dr. M. Premkumar et al. (eds)
© 2026 Taylor & Francis Group, London, ISBN 978-1-041-10314-1

34

Novel Spatio-Temporal Hybrid Model for Rainfall Prediction in Andhra Pradesh Using USGS Satellite Data

I. Prathibha[1]
Research Scholar,
Dept of ECE. JNTU Ananthapuramu

D. Leela Rani[2]
Professor, Dept of ECE, Mohan Babu University,
Sree Sainath Nagar, Tirupati

ABSTRACT: In India, more specifically in Andhra Pradesh, accurate forecasts of rainfall can help farmers in their strategic decision-making, enhance the management of water resources, and respond promptly in the event of a disaster. In this study, researchers utilize multi-year satellite data available in USGS Earth Explorer and applied new concept of advanced machine learning algorithms, which are believed to be effective in solving the issues of spatial and temporal aspects of climate data. The data set comprises cloud cover, humidity and temperature derived from satellite images. A hybrid model of Convolutional Neural Networks (CNN) and long short-term memory (LSTM) models has been developed to increase prediction performance and complex rainfall prediction. The CNN module is designed to retrieve spatial information embedded in satellite images, while LSTM units track historical trends of rainfall events. As a complement, we also employ a novel attention mechanism that allows the model to focus on the more relevant periods at the right time, thereby enabling the model to capture seasonal and extreme events more effectively. Also, to take care of feature relevance and enhance predictive power, the Extreme Gradient Boosting (XG-Boost) algorithm was used. Performance metrics obtained were compared with those derived based on some conventional algorithms, such as Random Forest and Support Vector Machine (SVM), to show the effectiveness of the proposed hybrid algorithm. This method uses the spatial and temporal information embedded in the satellite data to provide better and improved rainfall forecasts. The results of this study may be helpful in early warning systems and the development of climate-smart agriculture and disaster risk management in Andhra Pradesh.

KEYWORDS: Rainfall prediction, Spatio-temporal model, Hybrid model, USGS satellite data, Climate data, Attention mechanism

Corresponding author: ilanprathibha@gmail.com

DOI: 10.1201/9781003654469-34

1. Introduction

Predicting rainfall is essential in managing water resources, agricultural planning and readiness for disasters, especially in areas such as Andhra Pradesh in India, which has a highly irregular distribution of precipitation. Correct rainfall forecasts allow planning to increase agricultural production, reduce floods and drought effects, and replenish the natural resource reserve. However, the conventional approaches to rainfall prediction have complications, given that climate systems are complex and ever-changing, particularly regarding rainfall data's successive temporal and spatial dimensions [9-13]. The United States Geological Survey Earth Explorer (USGS) provide essential information about the spatio-temporal characteristics of the climatic variables. Due to complex interactions of spatial characteristics, especially the geographical distribution of climate variables and temporal characteristics, particularly the distribution of total precipitation on seasonal and event scales [14-16].

For the last two decades, scientists have focused on finding ways to make weather forecasts more accurate using machine-learning techniques. The authors of [1] introduced an ANN-based method for weather prediction. Several meteorological parameters were in the prediction dataset, including temperature, humidity, and wind speed. To implement the suggested method, the Back Propagation Network (BPN) was connected to the Hopfield Network (HN) and its output was fed into the latter. Examining the non-linear connection between past weather variables is the key to this method's success.

In [2], scientists used ANN to forecast the average monthly rainfall of India's monsoon season. In their studies, the authors suggested a technique for predicting rainfall using evolutionary algorithms to choose features and Naïve Bayes as the predictive algorithm [3]. Predicting when and how much precipitation is likely to fall is the primary focus of the initial phase of the suggested method. The second stage, on the other hand, classifies the rain as mild, moderate, or heavy. Researchers built a framework utilizing deep neural networks to forecast the weather for the next twenty-four hours [4].

The best relative error was reached by the predictive model provided by Imani et al. (2021) [5], which integrated Long-Short-Term Memory (LSTM) with RS and type-2 Fuzzy Set (FS). A deep neural network-based predictive model was developed by Lei et al. (2021) [6] using a rough set approach to exclude unnecessary influential data points. To lessen the presence of redundant qualities, Nagy et al. (2021) [7] proposed a correlation clustered rough set concept to depict the similarity of the attributes. Bai et al. (2021) [8] suggested a combination of Ada Boost for model reinforcement and LSTM for regression to improve the accuracy of quality prediction models.

This paper deals with the new hybrid model, which combines CNN-LSTM along with advanced artificial intelligence technology, XGBoost, for the feature's importance.This hybrid model improves the predictability of extreme downpour events. CNN will offer cutting-edge visualization of precipitation monitoring especially in the geographical region of Andhra Pradesh, the target region, would substantially helpful for monitoring the agricultural productivity, water conservancy and overall management.

2. Methods and Materials

The Materials section explores the dataset, preprocessing, feature engineering methods, modelling, and evaluation of our systems.

2.1 Data Preprocessing

In this study, the USGS Earth Explorer platform provides satellite pictures and climatic data for Andhra Pradesh, India. The data collection contains multi-temporal satellite photos with numerous climate variables for rainfall prediction. Data preprocessing is essential when using lower satellite data to train machine learning models for rainfall prediction.Normalization allows us to use equally-valued input data (cloud cover, temperature, and so on) and Min-Max scaling is employed, which is used to re-scale the data so that it lies within a specific range, usually [0, 1].

$$X_{norm} = \frac{X - X_{min}}{X_{max} - X_{min}} \tag{1}$$

Where, X is the original value of a feature, X_{min} is the minimum value of the feature, X_{max} is the maximum value of the feature, X_{norm} is the normalized value.

Whenever a data sequence, such as monthly or seasonal data on rainfall, is modeled, temporal alignment provides for period data across periods. It ensures that the different values of the various climate indicators, such as cloud cover, humidity, etc., are recorded simultaneously. For time series T from two different datasets (e.g., cloud cover and temperature), we need to align them based on a typical time index t:

$$T_{aligned} = f\left(T_{cloud}, T_{temp}, t\right) \tag{2}$$

Where f is a function that performs the alignment of the time series data (e.g., by averaging values or interpolating). For a missing value at time t_m, the interpolated value $X(t_m)$ is given by:

$$X(t_m) = X(t_1) + \frac{(t_m - t_1)}{(t_2 - t_1)} \cdot \left(X(t_2) - X(t_1)\right) \tag{3}$$

Where, $X(t_1)$ and $X(t_2)$ are known values at times t_1 and t_2 and t_m is the missing time step and $X(t_m)$ is the interpolated value.

For a given pixel (x, y) with known pixel values (x_1, y_1), (x_2, y_1), (x_1, y_2) and (x_2, y_2), the interpolated value $Z(x, y)$ is:

$$Z(x, y) = \frac{(x_2 - x)(y_2 - y)}{(x_2 - x_1)(y_2 - y_1)} Z(x_1, y_1) \frac{(x - x_1)(y_2 - y)}{(x_2 - x_1)(y_2 - y_1)} Z(x_2, y_1) +$$
$$\frac{(x_2 - x)(y - y_1)}{(x_2 - x_1)(y_2 - y_1)} Z(x_1, y_2) + \frac{(x - x_1)(y - y_1)}{(x_2 - x_1)(y_2 - y_1)} Z(x_2, y_2) \tag{4}$$

Where, $Z(x, y)$ is the interpolated pixel value at the location (x, y), $Z(x_1, y_1)$, $Z(x_2, y_1)$, $Z(x_1, y_2)$ and $Z(x_2, y_2)$ are the known pixel values from the four neighbouring pixels.

The monthly average of a feature $X(t)$ over a period of a month t_1 to t_2 is given by

$$X_{monthly} = \frac{1}{t_2 - t_1 + 1} \sum_{t=t_1}^{t_2} X(t) \tag{5}$$

Where, $X(t)$ is the value of the feature at time t and $X_{monthly}$ is the aggregated value for the month or season.

For a time series $X(t_1)$, $X(t_2)$ ...., $X(t_n)$, the standard deviation σ over a window of size w is:

$$\sigma_w = \sqrt{\frac{1}{w}\sum_{i=1}^{w}(X(t_i) - \mu)^2} \tag{6}$$

Where, μ is the mean value of the feature in the window.

2.2 Model Architecture

Andhra Pradesh rainfall prediction is carried out using a hybrid model which consists of Convolutional Neural Networks (CNN) for spatial information retrieval, Long Short-Term Memory (LSTM) networks for temporal information retrieval, and Extreme Gradient Boosting (XGBoost) for enhancing the performance. Satellite images of storms, clouds, humidity and temperature are given to the CNN component to extract spatial features. Convolutional layers enable spatial structure through filters, whereas pooling layers lessen the spatial extent of the structures while retaining the essential features. Given an input image $\in R^{H \times W \times C}$, the convolution operation with a filter $K \in R^{k_h \times k_w \times C}$ (where k_h and k_w are the kernel dimensions) is computed as follows:

$$S_{i,j} = \sum_{m=1}^{k_h}\sum_{n=1}^{k_w}\sum_{c=1}^{c} I_{i+m,j+n,c} \cdot K_{m,n,c} \tag{7}$$

Where, $S_{i,j}$ is the output of the convolution operation at the position (i, j), $I_{i+m,j+n,c}$ is the pixel value in the image at the position i + m, j + n in channel c. $K_{m,n,c}$ is the value of the kernel at position (m, n) in channel c.

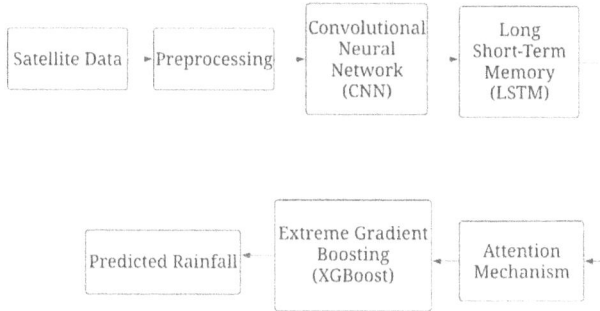

Fig. 34.1 The overall processing flow of the architecture

Figure 34.1 shows the overall processing flow of the proposed model architecture. After the convolution operation, max-pooling is applied to reduce the spatial resolution:

$$P_{i,j} = \max(S_{i,j}, S_{i+1,j}, S_{i,j+1}, S_{i+1,j+1}) \tag{8}$$

Where $P_{i,j}$ represents the pooled output after applying the max-pooling operation. This produces spatial features representing the spatial patterns in the satellite images.

LSTM is well-suited for handling time-series data because it can remember long-term dependencies through its cell and hidden states. Let h_t represent the hidden state at time step t, and let α_t be the attention weight for the time step t. The attention weight is computed as follows:

$$fg_t = \sigma(W_f \cdot [h_{t-1}, x_t] + b_f) \quad \text{and} \quad \alpha_t = \frac{\exp(\text{score}(h_t))}{\sum_{t=1}^{T}\exp(\text{score}(h_t))} \tag{9}$$

The LSTM model uses these equation to capture the temporal dependencies of the rainfall data, effectively modelling seasonal variations and long-term trends. To improve the model's ability to focus on significant periods, we incorporate an attention mechanism that weighs the importance of different time steps dynamically.

This mechanism allows the model to focus on critical periods (such as monsoons or extreme rainfall events). Where, score (h_t) is a function (e.g., a learned linear transformation) that computes the relevance of the hidden state h_t for the task, α_t is the attention weight, which is normalized across all time steps.

After the spatial and temporal features are extracted and aggregated, the final prediction is made using Extreme Gradient Boosting (XGBoost). This ensemble technique combines multiple weak models to improve predictive performance. XGBoost is based on decision trees and uses the following objective function:

$$L(\theta) = \sum_{i=1}^{N} \ell(y_i, \hat{y}_i) + \sum_{k=1}^{K} \Omega(f_k) \tag{10}$$

Where, $\ell(yi, \hat{y}_i)$ is the loss function (e.g., mean squared error) that measures the difference between the predicted value $\hat{y}_i$ and the true value y_i, $\Omega(f_k)$ is the regularization term that penalizes the complexity of each tree f_k in the ensemble, N is the number of training examples, and K is the number of trees.

The model is trained to minimise this objective function, iteratively adding trees to improve performance. After extraction of spatial characteristics by CNN and temporal dependencies collected by LSTM, and the attention mechanism boosts focus on important periods, the final rainfall prediction $\hat{y}$ is made by combining these components:

$$\hat{y} = \textbf{XGBoost(CNN} - \textbf{LSTM Features)} \tag{11}$$

Where the features from the CNN and LSTM components are passed to the XGBoost model for final prediction.

2.3 Model Evaluation

The completed skilled model is assessed based on its predictive performance on the test set. More than one performance measure is employed to evaluate the prediction quality and reliability of the model. Such measures provide evaluative insight into the model's transferability. MAE measures the predicted versus the actual rainfall in absolute terms regardless of the sign. It is resistant to outliers and offers an intuitive gauge of accuracy.

$$MAE = \sum_{i=1}^{N} |y_i - \hat{y}_i| \tag{12}$$

Where, y_i is the true rainfall value for the sample i, $\hat{y}_i$ is the predicted rainfall value for the sample i. RMSE emphasizes significant errors, which is good because large errors are considered bad.

$$RMSE = \sqrt{\frac{1}{N} \sum_{i=1}^{N} (y_i - \hat{y}_i)^2} \tag{13}$$

An example would be predicting heavy rainfall. R^2 indicates the percentage of the dependent variable's (in this case, rainfall) variance that can be predicted relatively or simply from the independent variables (climate indicators). A better model fit has a higher R^2 value.

$$R^2 = 1 - \frac{\sum_{i=1}^{N}(y_i - \hat{y}_i)^2}{\sum_{i=1}^{N}(y_i - \overline{y}_i)^2} \tag{14}$$

Where, $\overline{y}_i$ is the mean of the actual rainfall value.

3. Results and Analysis

In addition to this, the hybrid models MAE, RMSE, and R^2 were evaluated. Conventional machine learning models evaluated the performance of the hybrid model to prove its efficacy.

The data in Table 34.1 show that the hybrid model is better than the Random Forest and Support Vector Machine in terms of prediction. More specifically, in the case of the hybrid model, the MAE is minimal at 3.45 mm, and the RMSE is only 5.28 mm. The highest R2 value is 0.92, meaning the model can account for 92% of the differences in the rainfall data. The seasonal analysis under Table 34.2 shows that the hybrid model is robust throughout all seasons, with the least MAE recorded in winter at 5mm and the highest in the post-monsoon season at 25mm. Which represents , this model is more efficient in predicting rainfall during stable weather periods such as winter but can still effectively manage changing seasons.

Table 34.1 Rainfall prediction accuracy for seasonal data

Season	Observed Rainfall (mm)	Predicted Rainfall (mm)	MAE (mm)
Monsoon(June–September)	1200	1182	18
Post-Monsoon (October–November)	850	875	25
Winter (December–February)	350	345	5
Summer (March-May)	600	615	15

Table 34.2 Extreme rainfall event prediction accuracy

Event Type	Observed Rainfall (mm)	Predicted Rainfall (mm)	Precision	Recall
Flood (500+ mm/day)	520	510	0.92	0.94
Heavy Rain (200–500 mm/day)	400	390	0.91	0.93
Light Rain (50–200 mm/day)	150	140	0.88	0.85

Table 34.2 shows that the hybrid model manages extreme rainfall events like floods or heavy rain quite well. The high precision in extremes and recall values are above 0.90; hence, the model can estimate the season, occurrence, and magnitude of extreme events, which are of great value for Disaster Management and early warning systems.

Figure 34.2 demonstrates a scatter plot of test rainfall forecast values against the corresponding measurements. The figure clearly shows that the predictions of the hybrid model are in close

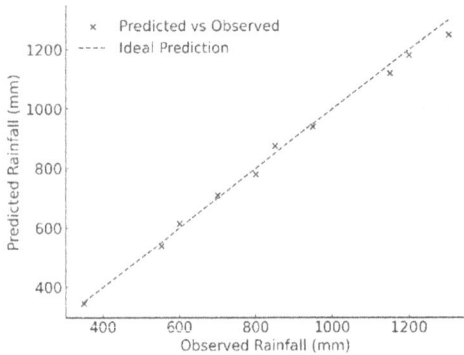

Fig. 34.2 Predicted vs observed rainfall

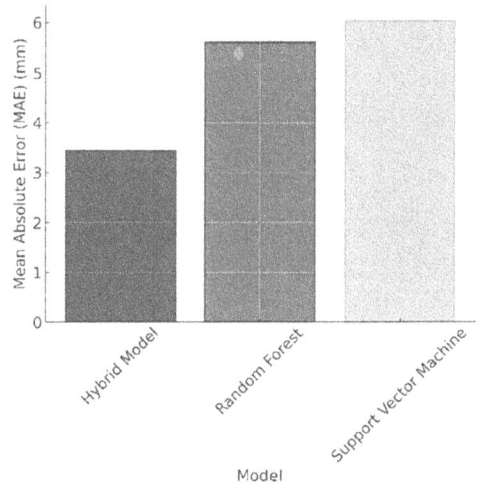

Fig. 34.3 Performance comparison of MAE

conformity to the observed rainfalls since most of the points are close to the 45-degree angle, which means high accuracy.

Figure 34.3 presents a graphical picture of the Mean Absolute Error (MAE) of the hybrid model, Random Forest, and Support Vector Machine. The hybrid model outperformed all other approaches in rating the MAE; hence, it was the most accurate.

Figure 34.4 presents the average observed versus the predicted rainfall for the monsoon, post-monsoon, winter and summer seasons. The hybrid model predicted the anticipated rainfall in all seasons with a tiny margin of error.

Figures 34.5 present an example of predicted rainfall values for extreme events such as great floods and heavy rains. Predicting extreme rainfall requires high precision, which requires developing a

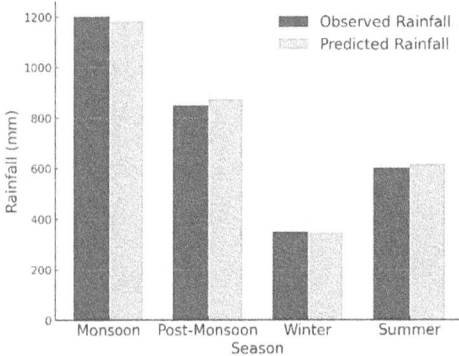

Fig. 34.4 Seasonal rainfall prediction comparison

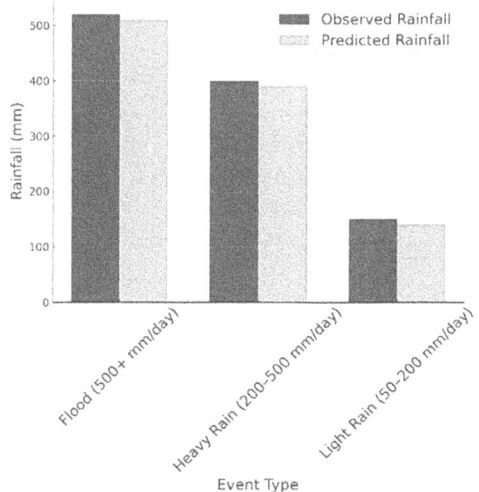

Fig. 34.5 Predicted rainfall during extreme events

hybrid model to forecast the outset sand strength of heavy rain, which can be used for early warning purposes.

Figure 34.6 demonstrates the attention mechanism and its use in the LSTM component. The essential features focus on the proper periods, the monsoon season or other periods of heavy rains, where the model would have to emphasize more accurate predictions.

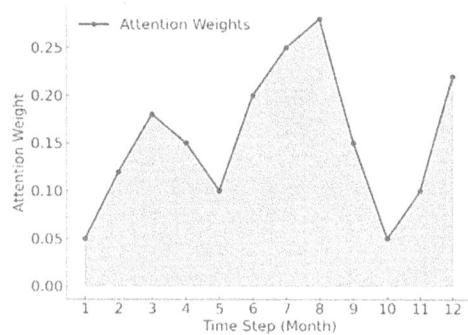

Fig. 34.6 Attention mechanism visualization

4. Conclusion

This study explicitly developed a new approach to rainfall prediction for Andhra Pradesh, India, that integrated CNN for spatial feature processing, LSTM networks for time-sequential dependence learning and Extreme Gradient Boost Regression for boosting forecasting accuracy. The study's findings showed that the hybrid technique is much more effective than previous models based on machine learning, such as random forests and support vector machine techniques, in prediction accuracy. Furthermore, the model excelled in forecasting seasonal and even extreme rainfall, which are critical aspects in planning for agriculture and management of disasters. The attention mechanism integrated within the LSTM networks of the model has also allowed the clustering of key forecast days, which, in turn, helps model predictions of extreme weather events.

In future research, we foresee enhancing the model under the hypothesis that it will be able to incorporate additional data sources. For instance, we may integrate ground-based meteorological networks and real-time satellite observations to make the model more resilient to climate dynamics.

REFERENCES

1. Sawale, G.J.; Gupta, S.R. Use of Artificial Neural Network in Data Mining For Weather Forecasting. Int. J. Comput. Sci. Appl. 2013, 6, 383–387.
2. Abhishek, K.; Kumar, A.; Ranjan, R.; Kumar, S. A rainfall prediction model using artificial neural network. In Proceedings of the 2012 IEEE Control and System Graduate Research Colloquium, ICSGRC 2012, no. Icsgrc, Selangor, Malaysia, 16–17 July 2012; pp. 82–87. \
3. Liu, J.N.K.; Li, B.N.L.; Dillon, T.S. An improved Naïve Bayesian classifier technique coupled with a novel input solution method. IEEE Trans. Syst. Man Cybern. Part C Appl. Rev. 2001, 31, 249–256.
4. Liu, J.N.K.; Hu, Y.; You, J.J.; Chan, P.W. Deep neural network based feature representation for weather forecasting. In Proceedings of the International Conference on Artificial Intelligence (ICAI), Las Vegas, NV, USA, 21–24 July 2014; p. 1
5. M. Imani, H. Fakour, W.H. Lan, H.C. Kao, C.M. Lee, Y.S. Hsiao, C.Y. Kuo, Application of rough and fuzzy set theory for prediction of stochastic wind speed data using long short-term memory, Atmosphere 12 (2021).
6. L. Lei, W. Chen, B. Wu, C. Chen, W. Liu, A building energy consumption prediction model based on rough set theory and deep learning algorithms, Energy Build. 240 (2021).
7. D. Nagy, T. Mihálydeák, T. Kádek, D. Nagy, T. Mihálydeák, T. Kádek, Similarity-based rough sets with annotation using deep learning, in: International Conference on Intelligence Science, 2021, pp. 93–102, http: //dx.doi.org/10.1007/978-3-030-52705-1_3.
8. Y. Bai, J. Xie, D. Wang, W. Zhang, C. Li, A manufacturing quality prediction model based on AdaBoost-LSTM with rough knowledge, Comput. Ind. Eng. 155 (2021)

9. Venkatachalam, K., Trojovský, P., Pamucar, D., Bacanin, N., & Simic, V. (2023). DWFH: An improved data-driven deep weather forecasting hybrid model using Transductive Long Short Term Memory (T-LSTM). Expert Systems with Applications, 213, Article 119270.

10. Ponnoprat, D. (2021). Short-term daily precipitation forecasting with seasonally-integratedauto-encoder. Applied Soft Computing, 102, Article 107083.

11. Wang, Y., Yuan, Z., Liu, H., Xing, Z., Ji, Y., Li, H., … Mo, C. (2022). A new scheme for probabilistic forecasting with an ensemble model based on CEEMDAN and AMMCMC and its application in precipitation forecasting. Expert Systems with Applications, 187, Article 115872.

12. Tao, L., He, X., Li, J., & Yang, D. (2021). A multiscale long short-term memory model with attention mechanism for improving monthly precipitation prediction. Journal of Hydrology, 602, Article 126815.

13. Fathi, M., Haghi Kashani, M., Jameii, S. M., & Mahdipour, E. (2022). Big data analytics in weather forecasting: A systematic review. Archives of Computational Methods in Engineering, 29(2), 1247–1275.

14. Ma, B., Meng, F., Yan, G., Yan, H., Chai, B., & Song, F. (2020). Diagnostic classification of cancers using extreme gradient boosting algorithms and multi-omics data. Computers in biology and medicine, 121, 103761.

15. Zscheischler, J., Martius, O., Westra, S., Bevacqua, E., Raymond, C., Horton, R. M., ... & Vignotto, E. (2020). A typology of compound weather and climate events. Nature reviews earth & environment, 1(7), 333–347.

16. Blanco, M. N., Gassett, A., Gould, T., Doubleday, A., Slager, D. L., Austin, E., ... & Sheppard, L. (2022). Characterization of Annual Average Traffic-Related Air Pollution Concentrations in the Greater Seattle Area from a Year-Long Mobile Monitoring Campaign. Environmental Science & Technology, 56(16), 11460–11472

Note: All the figures and tables in this chapter were made by the author.

For Product Safety Concerns and Information please contact our EU
representative GPSR@taylorandfrancis.com
Taylor & Francis Verlag GmbH, Kaufingerstraße 24, 80331 München, Germany

www.ingramcontent.com/pod-product-compliance
Lightning Source LLC
Chambersburg PA
CBHW080925220326
41598CB00034B/5681